CAMBRIDGE LIBRARY COLLECTION

Books of enduring scholarly value

Life Sciences

Until the nineteenth century, the various subjects now known as the life sciences were regarded either as arcane studies which had little impact on ordinary daily life, or as a genteel hobby for the leisured classes. The increasing academic rigour and systematisation brought to the study of botany, zoology and other disciplines, and their adoption in university curricula, are reflected in the books reissued in this series.

Life of Alexander von Humboldt

Alexander von Humboldt (1769–1859) was a naturalist, explorer and philanthropist now well known for his multidisciplinary approach to science. First published in English in 1873, this two-volume biography was translated from the German edition of 1872, edited by Karl Bruhns, which had been compiled in commemoration of the centenary of Humboldt's birth. Incorporating numerous extracts from Humboldt's own enthusiastically written letters and anecdotes from his many acquaintances, it charts his travels in South America, Asia and Europe. Volume 1 covers his early exploratory trips and some of his lesser-known exploits, including becoming Superintendent of Mines in Northern Bavaria, his research on the sixteenth-century eruption of Nevado del Altar in Ecuador and his friendship with the young Prince William of Prussia. Of interest to students and researchers in the history of science, this is a minutely detailed and compelling insight into the life of the man behind the scientist.

Cambridge University Press has long been a pioneer in the reissuing of out-of-print titles from its own backlist, producing digital reprints of books that are still sought after by scholars and students but could not be reprinted economically using traditional technology. The Cambridge Library Collection extends this activity to a wider range of books which are still of importance to researchers and professionals, either for the source material they contain, or as landmarks in the history of their academic discipline.

Drawing from the world-renowned collections in the Cambridge University Library and other partner libraries, and guided by the advice of experts in each subject area, Cambridge University Press is using state-of-the-art scanning machines in its own Printing House to capture the content of each book selected for inclusion. The files are processed to give a consistently clear, crisp image, and the books finished to the high quality standard for which the Press is recognised around the world. The latest print-on-demand technology ensures that the books will remain available indefinitely, and that orders for single or multiple copies can quickly be supplied.

The Cambridge Library Collection brings back to life books of enduring scholarly value (including out-of-copyright works originally issued by other publishers) across a wide range of disciplines in the humanities and social sciences and in science and technology.

Life of Alexander von Humboldt

VOLUME 1

JULIUS LÖWENBERG
ROBERT AVÉ-LALLEMANT
ALFRED DOVE
EDITED BY KARL BRUHNS
TRANSLATED BY
JANE AND CAROLINE LASSELL

CAMBRIDGE
UNIVERSITY PRESS

CAMBRIDGE UNIVERSITY PRESS

Cambridge, New York, Melbourne, Madrid, Cape Town,
Singapore, São Paolo, Delhi, Mexico City

Published in the United States of America by Cambridge University Press, New York

www.cambridge.org
Information on this title: www.cambridge.org/9781108041751

© in this compilation Cambridge University Press 2012

This edition first published 1873
This digitally printed version 2012

ISBN 978-1-108-04175-1 Paperback

ALEXANDER von HUMBOLDT.

FIRST VOLUME.

LONDON: PRINTED BY
SPOTTISWOODE AND CO., NEW-STREET SQUARE
AND PARLIAMENT STREET

A. Krausse sc.

Alexander von Humboldt
aged 26.
(1796.)

LIFE

OF

ALEXANDER von HUMBOLDT.

COMPILED

IN COMMEMORATION OF THE CENTENARY OF HIS BIRTH

BY

J. LÖWENBERG, ROBERT AVE-LALLEMANT, and ALFRED DOVE.

EDITED BY

PROFESSOR KARL BRUHNS,

DIRECTOR OF THE OBSERVATORY AT LEIPZIG.

IN TWO VOLUMES.

TRANSLATED FROM THE GERMAN BY

JANE AND CAROLINE LASSELL,

TRANSLATORS OF SCHELLEN'S 'SPECTRUM ANALYSIS.'

VOLUME I.

WITH PORTRAIT OF HUMBOLDT AT THE AGE OF TWENTY-SEVEN.

LONDON:

LONGMANS, GREEN, AND CO.

1873.

TRANSLATORS' PREFACE.

THE NEED experienced in Germany of an adequate biography of
Alexander von Humboldt has perhaps been felt with scarcely
less urgency in this country: and there is therefore every
reason to hope that this attempt to place before the world in a
true light the life and labours of the Author of 'Cosmos,' will
meet in England with a welcome no less cordial than that which
has been granted to the work in its original form.

In preparing this biography for English readers, it has been
deemed advisable to omit the third volume, devoted to a
critical investigation of Humboldt's scientific labours, since
these are given with sufficient minuteness for the general
reader in the biographical portion; nor has it been thought
desirable to include the last section of the second volume,
consisting of an elaborate catalogue of his voluminous works.
With these exceptions, the omissions consist only of official
documents and some passages of purely local interest.

In rendering quotations from 'Faust,' the translators have
followed the metrical version of Mr. Bayard Taylor. The
thermometric readings are given according to Fahrenheit's
scale. The notes within brackets have been added by the
translators.

RAY LODGE, MAIDENHEAD: *March* 1873.

PREFACE.

———•◦•———

WHEN, on May 6, 1859, Alexander von Humboldt closed his
eyes in death, in the ninetieth year of his age, the numerous
eulogies, biographical notices, and sketches of character that
appeared in periodicals of every description, both in his own
country and in foreign lands, afforded striking evidence of the
universal recognition of his genius, and of the unanimous
acknowledgment of the influence he had exerted on the
advancement of science during the present century.

Many attempts have already been made to record the life of
this illustrious man, and trace the process of development of
his master mind. A skeleton biography was furnished by
Humboldt, in the year 1850, at the request of the editor, for
Brockhaus' 'Conversations-Lexikon.' An abridgment of this
biography appeared in the tenth edition of that work, while
the original article was published entire in the eighth volume
of the periodical entitled 'Die Gegenwart' (1853). A biographical memoir, compiled, however, with no direct references
to authorities, nor even with much acquaintance with scientific
subjects, was brought out by H. Klencke in 1851, and in 1859
W. F. A. Zimmermann published his 'Humboldtbuch.' In
addition, fragmentary portions of his biography have appeared
in various forms. But nowhere has the indefatigable Student
of Nature been depicted in the daily routine of his investiga-

tions, nowhere has the man of science been represented as he *lived* and *laboured* ; for even the biography published in 1860 by W. C. Wittwer: ' Alexander von Humboldt. His Life and Labours as a Man of Science,' furnishes in reality little more than that which had already appeared elsewhere.

The fact is that the compilation of a biography of Humboldt which should in every sense be worthy of the subject and embrace the whole range of his scientific labours, was an undertaking of no small difficulty. Even the men who were most closely associated with him are compelled to confess that they are not in a position to do justice to the activity of his universal genius. The necessary sources of information for the accomplishment of such a purpose were unattainable, were even hermetically sealed. Humboldt himself was ' painfully shy of communicating anything relating to family affairs ; ' and, though admitting their necessity for the history of science, he detested all biographies, especially eulogies. Thus, in writing to Dr. Spiecker, he remarks :—' I have so often refused to take part in any biographical notice of either myself or my brother, whether by furnishing materials or by revising the compilations of others, that I feel compelled to preserve the same line of conduct in Germany that I have hitherto maintained both in France and England. My horror of biographies is almost as great as that which I feel for the portraits of old men painted by the sun, although both may be viewed in the history of science as a disagreeable necessity. Life is daily losing for me its charm, and I need your kindest indulgence, my worthy friend, for this idiosyncracy of an old man. I have even inserted a clause in my will forbidding that any eloge should be pronounced over me at the Institute.' His will, bearing date May 10, 1841, is preserved in the royal palace at Potsdam, and the clause above referred to runs as follows :—' I request that my dear relatives and friends will endeavour to prevent the appearance of any biographical notice of me or laudatory article in either the " Staatszeitung " or other public journal over which they can exercise any control. I have also drawn up a letter for transmission to the Institute at Paris requesting that the eloge usually delivered upon the death of a foreign associate may be omitted in my case.' As in consequence of these prohibitions his nearest relatives and friends published immediately after his death a ' protest against the publication of any confidential letters,' it was simply impossible to attempt the compilation of a life of Humboldt that should be based upon the investigation of authentic sources. For the inner life of a man is to be seen only in those confidential communications which are hastily committed to paper, and are addressed only to his most intimate friends. A biography compiled without the aid of correspondence or other manuscript records could no more give a life-like representation of Humboldt than a cold marble bust can approach the glow of living flesh.

Meanwhile this prohibition could not be long maintained. Notwithstanding its frequent sad infringement, it is impossible with any justice to deny the right of posterity to the letters of men whose public career is connected inseparably with the history of their country.

The first series of Humboldt's letters, published scarcely a year after his death, failed to represent him in a favourable aspect, inasmuch as they revealed some of those weaknesses of character from which even great men are not always exempt. By these letters, and still more by the censorious remarks contained in the annotations by Varnhagen, whose undoubted genius was unfortunately marred by a spirit embittered by disappointment, many secrets were brought to light which were quite unsuitable for general publication, or even for being privately circulated amongst his personal friends, although it cannot be denied that the ' Letters from Alexander von Humboldt to Varnhagen von Ense ' have made us acquainted with many facts and exhibited various traits of character, without which the portrait of Humboldt would seem now to be incomplete. The letters had been penned by Humboldt for the gratification of a friend who delighted in gossip, and they

should in any case, after perusal by him, have been committed
to the flames. In comparison with this rich and piquant collection, the small work which came out the same year (1860)
entitled 'Letters from Alexander von Humboldt to a Young
Friend, with Notes of Conversations between the Years 1848
and 1856,' appeared insignificant, and excited but little attention. Many years passed before the scientific aspect of Humboldt's life was again brought into prominence, through the
publication of other letters—those, for instance, which were
edited by H. Berghaus and De la Roquette. Meanwhile, the
history of the years 1864 and 1866 drew the attention of
Germany to that political stage where Prussia, the once insignificant country of Humboldt, was now to take the lead,
and it was only on the approach of the centenary of his birth
that the interest once felt towards him was reawakened.

At the meeting of the scientific association at Dresden on
September 18, 1868, it devolved upon me to make a few in-
troductory remarks upon the progress of science, especially
with reference to astronomy, in the course of which I made
some allusion to the approaching centenary of Humboldt's
birth. I availed myself of the opportunity to point out the
need of an adequate biography of our distinguished country-
man, the want of which was the more to be regretted as the
number of those who enjoyed personal intercourse with him
and retained the most vivid recollection of his peculiar characteristics was every year diminishing.

Since that period several fresh sources of information have
been thrown open to the compiler of such a work. In the
course of the year 1868 a valuable collection of letters from
Humboldt to Marie Auguste Pictet made their appearance in
the geographical organ 'Le Globe,' constituting the seventh
volume of that periodical, which were followed in the succeeding year, 1869, by the publication of the 'Correspondence
between Alexander von Humboldt and Count George von
Cancrin,' and of the exceedingly interesting 'Letters of Hum-

boldt to Christian Carl Josias Baron von Bunsen.' Various
eloges and speeches commemorative of the centenary of
Humboldt's birth were published by different academies, geographical societies, and other associations, not only in Europe,
but also in America, among which we can refer only by name
to those of Agassiz, Bastian, Dove, and Ehrenberg; nor did
there fail to appear on the occasion of the celebration many
other smaller essays and popular biographies.

During the previous year I had already received a communication from the firm of F. A. Brockhaus of Leipzig, offering to
undertake the publication of the work in the event of my
preparing a life of Humboldt. This intelligence helped to
mature my scheme, and I resolved to attempt to enlist the
co-operation of a number of scientific men in the work of
composing a comprehensive biography of Humboldt, in which
justice should be done to his labours in the various branches of
science. The centenary of his birth appeared to me to offer a
most appropriate opportunity for the commencement of such a
literary memorial—the more so, as at that time there had been
no proposition to erect in his native city a monument in
bronze.

My scheme met with ready sympathy from my friends in
Berlin. Professor W. Förster placed in my hands the manuscripts deposited at Humboldt's death in the Berlin Observatory, and promised to afford me all the assistance in his power ;
Frau von Bülow, a niece of Humboldt, granted me the use of
the Journals kept by him during his travels in America and
Asia, and allowed me to have access to all the letters in her
possession ; while Professors Dove and Du Bois-Reymond expressed their willingness to join in the undertaking.

The work is divided into two Parts. The first consists of the
story of his life, the second, of a discussion in detail of his
labours in the various departments of science. The first part
is subdivided into three sections, treating of his Youth, his
Manhood, and his Old Age ; while the second part, consisting

of eight distinct treatises, is devoted to the consideration of
each department of science into which his investigations extended. Various portraits, published for the first time in this
work, afford interesting records of the personal appearance of
Humboldt during the three principal stages of his career.

The prospectus of the work was issued on the centenary
anniversary of Humboldt's birth. I was able to present it in
person before the Congress of Astronomers at Vienna, convened
on that day in honour of Humboldt, upon which occasion I
was entrusted with the pleasing duty of presenting a summary
of his valuable services in the department of astronomy.

Upon my return from Vienna I proceeded without delay to
the execution of my scheme. The task of narrating the events
of Humboldt's youth and of the years he spent in travel was
undertaken by Herr Julius Löwenberg, who had been occupied
with the subject for many years, and was in possession of numerous documents, letters, and other material indispensable for a
biography. The history of Humboldt's sojourn in Paris, where
he spent many years in scientific labour and in intercourse
with the distinguished men there assembled, was confided to
Dr. R. Avé-Lallemant, who has himself spent many years amid
the scientific circles of the French capital. Finally, in Dr.
Alfred Dove I secured the assistance of one who, from his intimate acquaintance with the society of Berlin, is peculiarly
fitted to depict in an authentic manner, free from every trace
of traditionary false colouring, the closing period of Humboldt's
life, dating from the time of his settlement at Berlin. I was
thus able satisfactorily to complete the arrangements for the
compilation of the purely biographical portion of the work.
With regard to the scientific part of my undertaking, I have
been fortunate enough to obtain the co-operation of Professors
H. W. Dove and J. W. Ewald of Berlin, of Professor A. H. R.
Grisebach of Göttingen, and of Professors J. W. Carus, O.
Peschel, and G. H. Wiedemann of Leipzig. Professor du Bois-
Reymond, to whom had been entrusted the analysis of Hum-
boldt's achievements in the province of physiology, has, to

my great regret, been compelled to withdraw his valued co-operation from the work; but I have been fortunately able to secure in his stead the help of Professor W. Wundt of Heidelberg.

Her Majesty the Empress-Queen Augusta has in the most gracious manner placed at the disposal of the Editor many of Humboldt's unpublished letters ; unfortunately, a valuable collection of letters perished in the flames during the year 1848. Besides the Journals, Frau von Bülow placed in my hands twenty-two valuable letters from Humboldt, nineteen of which were written to his brother during the expedition to Russia, and three are addressed to his sister-in-law.

A series of more than 80 letters, addressed by Humboldt to the friend of his youth, Freiesleben, subsequently Director of Mines at Freiberg, embracing a period from the time of Humboldt's departure from Freiberg to the year 1845, have been placed in my hands by his son, Herr Freiesleben, of Dresden. From Professor Borchardt of Berlin I have received 13 letters addressed to Jacobi, the mathematician ; from Professor Kronecker 80 written to Lejeune-Dirichlet ; 30 addressed to Gauss, through the kindness of Professor Schering of Göttingen ; 54 to Schumacher, lent by his sons at Valparaiso and Altona ; 30 to Karsten, Privy Counsellor of Mines, through his son, Professor Karsten, of Kiel ; 11 to W. Struve and Fuss, through Herr Otto von Struve, Director of the Imperial Observatory at Pulkowa ; and 330 to Encke, which have been lent to me by his heirs. From Dr. G. von Boguslawski I have received 28 letters addressed by Humboldt to his father, the late Director of the Observatory at Breslau ; from Herr Carus of Dresden 12, addressed to his father, late surgeon to the king and President of the Academy of Leopold-Charles ; and from Dr. H. Vogel 9, addressed to the late Herr Vogel, Inspector of Schools, the father of the unfortunate African explorer. For a valuable collection of 50 letters I am indebted to Madame Richards-Gagiotti, of Florence, and to Von Mädler, formerly Director of the Observatory at Dorpat, I am indebted for 19.

My thanks also are due to Professor Galle of Breslau, Dr.
Luther of Bilk, Herr Reich of Freiberg, W. Weber of Göttingen,
Herr Wild of St. Petersburg, and Dr.Focke of Bremen, from
whom I have received various letters addressed either to them
or to their scientific friends.

From General Count von Helmersen of St. Petersburg, from
Herr Paschen of Schwerin, and others, I have received various
communications concerning Humboldt's scientific labours. I
am indebted to Herr Hermann of Cologne for 16 letters
addressed by Humboldt while a student at Göttingen to his
fellow-collegian Wegener ; while I have been furnished with a
series of 61 letters addressed to Eisenstein, the mathematician,
by the committee organised for the erection of a memorial to
Humboldt at Berlin.

I have also been so fortunate as to obtain possession of all
the papers belonging to Humboldt which at the time of his
death fell into the hands of his confidential attendant, Herr
Seifert. These consisted of 500 letters of recent date, most of
them written within ten years of his death, from royal personages, statesmen, men of science, and artists, besides various
sketches of maps and many original treatises, poems, &c., that
were presentations to Humboldt. I am also indebted to Herr
Seifert for the communication of several orders in Council, the
loan of some private letters and an insight into some of the
details of his domestic history.

In addition to the foregoing I may also mention that I have
received various letters and other documents from Herr Fried-
länder of Berlin, from Herr von Hermann of Munich, from
Herr von Carnall, from Professor Rudolph Wagner of Göttingen,
from Dr. Henry Lange of Berlin, from Römer of Löthein, near
Meissen, from Herr von Locher, from Herr Lehfeldt, from
Frau Goldschmidt née Kunth, from Fraülein Schwenken of
Langendeinbach, and others.

From the Superintendents of the Royal Archives of Berlin I
received permission to inspect the public records containing any
reference to Humboldt, and I am indebted to them for much

valuable information in reply to my numerous inquiries. Through the kindness of Herr von Weber, Director of the Royal Archives at Dresden, I was allowed to have access to the State documents, and make abstracts from the various papers having reference to Humboldt. By Professor Gneist of Berlin I was offered the use of 295 letters to Böckh, which were found among his papers. From Herr G. Rose I received a valuable addition to the Cancrin correspondence, consisting of a number of unpublished letters and several important orders in Council; through the kindness of Professors Bellermann, Curtius, Dove, and others, I have been granted a perusal of the letters addressed to them by Humboldt.

To all those who have thus kindly assisted in the compilation of this biography I would here express my most grateful thanks.

The words of William von Humboldt, which have been selected as the motto for the first volume: ' My only conception of biography is that of historical truth,' have served as a guiding principle throughout the work.

In portraying the early life of Humboldt, in particular, the Author has regarded it as a duty, not so much to controvert the false and unfounded representations that have prevailed hitherto regarding this period, but rather to support the statements of a very different character, by the introduction of proofs never before published.

For the compilation of the second section, containing the narrative of the expeditions to America and Asia, the principal source of information has been the works published by Humboldt himself, which, however, were never wholly completed. These have therefore been supplemented by a careful investigation of his manuscript journals, and of many published and unpublished letters relating to that period. The expedition to Asia, though separated by half a lifetime from the travels in America, presented so much similarity in subject and mode of treatment that it was deemed advisable to include them in the same section.

The third section, which contains an account of his residence in Paris, where for eighteen years he was closely occupied with the preparation of his works and the arrangements necessary for their publication, is also founded partly on information derived from printed matter already before the public and partly from manuscript letters and documents.

In the preparation of the fourth section, which treats of Humboldt's life at Berlin, from the year 1827, when he took up his residence in his native city, till his death, a vast amount of material has been at the disposal of the Author, enabling him to introduce much new matter relating to this period, as well as to correct many errors that have largely prevailed.

The elaborate catalogue of all the works, treatises, and miscellaneous writings of Humboldt, constituting the fifth section, will, as a first attempt to reduce this literary chaos into anything like system and order, be welcomed by all those who value accuracy even in the most trivial facts in literature.

It was found almost impossible in arranging for the compilation of the sixth section, so to classify the various scientific subjects that in the eight treatises, each of which was entrusted to a different author, there should be on the one hand no omissions of importance and on the other hand no unnecessary repetitions. The names of the several authors are a sufficient guarantee for the complete and accurate treatment of the subjects they have taken in hand.

Of the three portraits illustrating the work, that in the first volume is engraved from a chalk drawing in the possession of Frau von Bülow ; it was taken in the year 1796, and has never before been published. The portrait in the second volume, which also appears through the kind permission of Frau von Bülow, was taken at Paris in 1814, and is interesting from the fact that it is copied from a drawing made by Humboldt himself from the looking-glass. The third portrait is from an oil-painting by Eduard Hildebrandt, with whom Humboldt was on terms of intimate friendship ; and this picture, in the possession of Herr Seifert, is one of the few portraits painted by this highly-gifted artist.

According to the scheme originally planned by the Editor, the work was to have appeared in the Easter of 1871; but no more elaborate excuse need be furnished for the delay than is to be found in the exciting events affecting the greater part of Europe during the summer of 1870 and the spring of 1871.

May the united efforts of the various Authors to afford to the present generation a complete and faithful picture of Alexander von Humboldt, both in his life and labours, meet with the kind reception due to their exertions.

It is in this hope that, upon the recurring anniversary of his death, I present this work, not only to every votary of science, but to every friend of intellectual progress, as an intellectual memorial of Humboldt in commemoration of the centenary of his birth.

KARL BRUHNS.

LEIPZIG : *May* 6, 1872.

CONTENTS

OF

THE FIRST VOLUME.

———◆◇◆———

CHAPTER IV.

WEIMAR AND JENA.

CHAPTER V.

THWARTED PLANS AND THEIR ULTIMATE ACCOMPLISHMENT.

II.

ALEXANDER VON HUMBOLDT. TRAVELS IN AMERICA AND ASIA. BY JULIUS LÖWENBERG.

TRAVELS IN AMERICA.

CHAPTER I.

PRELIMINARY REMARKS.

APPENDIX.

Errata.

Page 220, line 3 from below, *for* genial *read* gifted.

 „   222 „ 10 „   above, *for* most genial *read* full of genius.

I.

ALEXANDER VON HUMBOLDT:

YOUTH AND EARLY MANHOOD.

BY

JULIUS LÖWENBERG.

'Mais surtout, mon digne ami, faites une biographie et non un eloge; en voulant m'honorer vous me feriez du tort. Je n'ai été déjà que trop loué dans le public, et cela irrite toujours.'—ALEXANDER VON HUMBOLDT *to* M. AUG. PICTET. (*Le Globe*, Journal géogr., vii. 177.)

'Wenn von Biographie die Rede ist, habe ich nun einmal den Begriff nur von historischer Wahrheit.'—WILHELM VON HUMBOLDT. (*Briefe an eine Freundin*, Dec. 16, 1828.)

CHAPTER I.

EARLY HOME.

Signification of Name —Traditional History—Ancestry and Parentage—
Tegel and Childhood — Tutors and Bent of Mind — Universal Enthusiasm for Geographical Discovery—Overwork and Illness—Preparations for the University—State of Society at Berlin.

THE NAME of HUMBOLDT shines in twofold splendour, illustrious alike in the world of science and in the history of modern
development ; for intimately associated with Alexander von
Humboldt must ever stand his elder brother William, who, distinguished as a statesman for the nobility of mind of a Pericles,
was even further renowned as a philologist and critic. By the
name of Humboldt a chord is struck, the vibrations of which
extend in ever-widening circles, awakening as they spread
thoughts of the deepest import, and arousing visions of the
loftiest efforts in philosophic enquiry and the most varied
achievements in every branch of scientific investigation.

The biographies hitherto published of the two distinguished
men who bore the name of Humboldt trace their descent
from an ancient noble family of Eastern Pomerania, resident
at Zamenz or Zemmenz, in the principality of Cammin, who
had been for generations in possession of property in the circle
of Neustettin. It is stated by Berghaus that their ancestors ' in
ancient times fought under Flammberg, in the wars between the
Germans and Sclavonians.' Pott interprets the etymological
meaning of the name ' as a fabulous Hun of gigantic propor-
tions.' French biographies represent the father of William
and Alexander von Humboldt to have been so wealthy, that by
lending the whole of his property to the King, he had actually
borne half the expenses of the Seven Years' War ; eventually it

became customary to prefix the title of Baron to the name of Humboldt.

Whether it was intended by such fabulous tales to confer additional lustre upon the Dioscurean brothers, William and Alexander von Humboldt, or whether any actual honour was thus conferred upon them, need not be discussed here. In the meantime we shall offer no injustice to historic truth, nor be guilty of any want of due reverence, if we wait for additional evidence before accepting myths of this kind, which, without further proof, have been copied and recopied even to the typographical errors.

The earliest and best source of information concerning the genealogy of the Humboldt family is Krone's 'Allgemeines teutsches Adelslexikon,' published in the year 1774, and the details contained in this work have supplied material for the various biographies hitherto published. It is to be regretted that the ancient classic authorities for the history of the Pomeranian nobles have not given even brief notices of the family of Humboldt and the property attributed to them, as the correctness of Krone's statements might thus have been tested, and additional facts probably obtained. Gundling, indeed, enumerates 'the Humboldts of Zemmenz' among the nobility in the circle of Neustettin, but accompanies the statement with no further remark. Brüggemann, in quoting from Gundling and some books of heraldry, includes the Humboldts among the noble families of Pomerania, but mentions them as no longer resident in that province. It is true that Zamenz or Zemmenz is alluded to in his work, and is described as an estate—a small farm—annexed to the manor of Juchow, in the circle of Neustettin, but it is nowhere stated that this property, an old feoff of Kleist's, was ever in the possession of the Humboldts; on the contrary, it is again referred to as belonging, in 1744, to two brothers of the name of Kleist. The estate at Zamenz has been described also by Klempin and Kratz as appertaining with Juchow and Zeblin to Kleist's freehold, without any allusion to the Humboldts in reference to any portion of the property.

Meanwhile, though the ancestral property and the early ennoblement of the Humboldt family have not been satisfac-

torily proved, it is yet evident from authentic and circumstantial accounts that several of their ancestors were among the
bravest and best men of their time.

Whether Heinrich Humboldt, who in 1442 was in possession
of a small farm at Grunow, in the circle of Angermund, is to
be reckoned as one of the ancestors of the illustrious brothers,
is uncertain. There is, however, no doubt that one of their
immediate progenitors was Johann Humboldt,[1] who lived during
the most disastrous period of the Thirty Years' War, and died
as Burgomaster of Königsberg, in the New Mark, on February
11, 1638, in the sixty-third year of his age. Of his son
Clemens—great-grandfather of William and Alexander von
Humboldt—the following particulars are gathered from the
unpublished Chronicles of the years 1400–1750 of George
Christ. Gutknecht, pastor of Hermsdorf and Wulkow :—

'Clemens Humpolt, Bailiff of Neuhoff, in the Electorate of
Brandenburg, died on the 2nd of January, 1650, and was,
together with his daughter, who died a few days afterwards,
conveyed in a hearse to Vircho, and interred in the church
there. In fulfilment of his wish, and at the request of his
widow, the funeral sermon, from 2 Timothy iv. 6, was printed
at Stettin. Being ambitious of a university education, he had
been sent to Frankfort, but at the end of a year was obliged,
owing to the troublous times consequent upon the war, to bid
farewell, *nolens volens,* to the University, since his father,
Consul at Königsberg in the New Mark, had become so greatly
impoverished by the frequent quartering of soldiers, the intolerable contributions laid upon him, and the robbery and
plunder consequent upon the passage of countless troops, that he
was unable to supply the young student any longer with the
sumptus necessarios for the prosecution of his studies. Mean-

---

[1] The names of various members of the Humboldt family occur in numerous mortgage and other deeds belonging to the seventeenth century, proving
them to have been persons of consideration in the districts where they
resided. A descendant of a distant branch of the family is still living at
Berlin, so that Alexander von Humboldt was scarcely correct in speaking
of himself and his brother as 'long the last of their name,' 'My brother
and myself are the sole representatives of the name of Humboldt.' ('Briefe
von A. v. Humboldt an Varnhagen,' p. 113 ; 'Lettres d'Alexandre de Humboldt à M. Aug. Pictet,' p. 181.)

while he was appointed to the office of Clerk in the town of Syndico in Crossen, which he retained for two years, and was afterwards promoted to the rank of Bailiff. He was pious and charitable, and listened to sermons with peculiar devotion, writing them down most indefatigably in a neat little book, and on his return home not consenting to dine until he had carefully transcribed them, so that on his death there were found five neatly-stitched separate volumes, in which the sermons for five years had been written out with surprising industry. He always accounted it a pleasure to do a service to any servant of the Church of God, and he bequeathed a handsome legacy to three churches for the purposes of restoration, and for reinstating the bells of which they had been plundered during the war : he was gentle, obliging, and modest, and possessed a true German heart, often labouring hard, and travelling night and day, in the service of his retainers, to secure them against oppression and injustice ; yet his unenlightened neighbours frequently attempted his life, and by their persecution, to escape which he had often to flee during the night, were ultimately the cause of his death at the age of forty-five years.'

His gravestone is still to be seen in the church of Virchow, in the circle of Drammburg, and bears an inscription in Latin, on either side of which stands in German the text of the funeral sermon, with the remark appended : ' Preached from and expounded by Pastor Christian Grützmachern, on the 30th of January, 1650.'

Of his son, Conrad Humboldt, the following notice occurs in a note in König's manuscript 'Collect. geneal.' vol. xxxix.:—
' Cyriacus Günther von Rehebergk, Captain of Neuhoff, urges the appointment of his step-son Humbold to the office *formerly held by his father*, since he has travelled in foreign lands, has studied at universities, has twice served on a mission to Moscow with Schultetz (?), Counsellor of Legation, has contracted a marriage with the daughter of Herr Beeorks (?), Electoral Resident of this neighbourhood, and, on the advice of President von Schweder, has completed his qualifications by a visit to France during the year 1676.' A second entry runs as follows :—
' 1682, 11 March. Appointment of Conrad v. Humbolt as Counsellor ;' to which is appended the remark, ' The original draft

of this appointment is deserving of notice, since it is evident that the prefix " v." (von) has been recently added.' He himself subscribed his name in official documents as simply Conrad Humboldt. Abundant evidence of the ability he displayed in his official duties is to be found in various enactments preserved in the Royal ministerial archives respecting the Starost of Draheim, in which he is represented as opposing, with characteristic energy and untiring perseverance, the arbitrary assumptions of the neighbours, especially of the Manteuffel family.

This Conrad, Bailiff of Draheim, had an only son, Hans Paul Humboldt, who entered the army in 1703, and upon his retirement as captain with a pension of eight thalers a month, took up his residence in the neighbourhood of Cöslin. His application for the grant of a patent of nobility is still extant, whence it appears that the prefix 'von' does not date farther back than the year 1738; so that where it now and then occurs earlier, it must have been only used conventionally, on account of the high position borne by the individual to whom the prefix is given. Even as late as 1830 it was still doubtful whether the title of Baron was due to William and Alexander von Humboldt. Alexander himself cared neither for prefixes nor titles; they are mostly wanting in the signatures of his letters, particularly in his communications to his intimate friends and in his scientific correspondence. He submitted with reluctance to the title of ' Excellency' in the dedication to Berghaus's ' Charts of the Coast of Peru,' making it a condition that all the honours represented by his ' decoration hieroglyphics,' as he termed his orders, as well as the honourable distinction of ' Actual Privy Councillor,' should be omitted.[1] In giving to Pictet at Geneva, in 1805, his ' Confessions,' for the preparation of a biographical sketch to be affixed to an English translation of his American works, he writes:—' In mentioning me, I should much prefer that you named me simply M. Humboldt, or at least M. Alexander Humboldt. It sounds more English, since the constant repetition of the *de* is very unpleasant to the ear. For the sake, however, of pre-

[1] 'Briefwechsel A. von Humboldt's mit Heinrich Berghaus,' vol. ii. pp. 163, 285.

serving the honours of our family—you see I am considering
your work in a diplomatic light—please mention me on *one
single occasion* as Frederick Alexander Baron von Humboldt,
but *only on one*. This is a matter connected with certain *principles* with which you do not altogether sympathise, but which
have been maintained by my brother and myself throughout
life, leading us never to make use of any *title*, except in the
most extraordinary cases, therefore *never* on the title-page of a
book.' [1]

Of the children of Hans Paul Humboldt, several died young ;
only four sons and a daughter survived him.

One of his sons, Alexander George, born at Zamenz, in
Pomerania, in 1720, is the father of the brothers William and
Alexander von Humboldt. The following account of him is
given by Büsching, the geographer [2] :—

'After receiving an excellent education in his father's house,
he entered the Prussian military service in 1736, and served in
a regiment of dragoons under Lieut-General von Platen. Although he immediately distinguished himself in three campaigns, he does not appear to have met with any favourable
opportunity for the display of his talents whereby he could
obtain promotion ; he therefore left the army with the rank of
Major in 1762. He was appointed by the King to the office
of Chamberlain in 1764, and was attached to the household of
the Prince of Prussia. Attracted by the eminent qualities of
Maria Elizabeth von Colomb, widow of Baron von Hollwede,
he was united to her in marriage in 1766; two sons were the
issue of this union. In 1769 he resigned his appointment in
the household of the Crown Prince, and lived henceforth without official employment, but not without useful occupation.
He let his property in the New Mark upon a lease while he
devoted himself to the improvement of his residence at Tegel,
and it is manifest from the result of his labours that he was a
man of intelligence and taste. He was exceedingly benevolent,
affable, and charitable, and won the respect and esteem of all

---

[1] 'Lettres d'A. de Humboldt à M. Aug. Pictet,' p. 189.
[2] Description of his journey from Berlin to Kyritz (Leipzig, 1780),
p. 28.

classes. His death, which occurred on the 6th of January, 1779, at the age of 59, was an event universally deplored.'

The 'Vossische Zeitung,' of January 9, laments his loss in these terms :—' Not only the highest in the State, but the people also mourn in him a friend, and the country a patriot.'

To these admirable qualities is to be ascribed the confidential relationship in which Major von Humboldt stood towards the Great King, with whom, as Adjutant to the Duke of Brunswick, during the worst times of the Seven Years' War, he frequently held personal intercourse. In a letter concerning Wedel's disaster, the king writes, ' I have told Humboldt all that there is to be told at this distance.'

Even his retirement from the prince's court at Potsdam, in consequence of the domestic troubles of the heir-apparent, did not impair this confidence that was so honourable to Humboldt. In a letter from the English ambassador, in the year 1776, Major von Humboldt is described as ' a man of good understanding and estimable character,' and pointed out as one of the foremost in the list of capable men who might be expected to occupy the post of Minister in the future reign of Frederick William II.[1]

His various connections with the courts of the other princes, especially with that of Prince Ferdinand, procured him, among other undertakings, an interest in the farming of Lotteries, which subsequently became very lucrative, both to him and his heirs; he was also concerned in a tobacco-magazine enterprise (for leasing the sale of tobacco), which since November 1, 1766, had been undertaken by the ministers Count Reuss and Count Eickstädt and the chamberlain Baron von Geuder.

Major von Humboldt entered upon domestic life, as already stated, in 1766, by his marriage with the widow of Captain Ernst von Hollwede, who was but recently deceased; she was the daughter of Johann Heinrich von Colomb, Director of the East Friesland Chamber, and was cousin to the lady who subsequently became Princess of Blücher. To her mainly did the family of Humboldt owe the possession of their considerable landed property: from her mother she inherited, in 1764, the

[1] 'Briefe von Alexander von Humboldt an Varnhagen,' p. 113. 'Briefe von Chamisso, Gneisenau, Haugwitz, W. v. Humboldt, &c.,' vol. i. p. 5.

house in Jägerstrasse, No. 22, where Alexander was born; through her first husband, Captain von Hollwede, she came into possession of the estate at Ringenwalde; through her was transmitted to her second husband, Major von Humboldt, the residence at Tegel which had been held by Von Hollwede on an inheritable lease; lastly, the property at Falkenberg was purchased by her in 1791 from Lieut.-Colonel von Lochow.

The issue of the marriage of Major von Humboldt with the widow of Von Hollwede was a daughter, who died in infancy, and two sons.

Frederick *William* Christian Charles Ferdinand, born at Potsdam, on June 22, 1767.

Frederick William Henry *Alexander*, born at 22, Jägerstrasse, Berlin, on September 14, 1769.

The year marked by the birth of Alexander—1769—was also that in which the following illustrious men first saw the light:—Napoleon, Cuvier, Chateaubriand, Canning, and Wellington. At the time of his birth, the great King of Prussia was at the height of his victorious career, the genius of Lessing had begun to illuminate the intellectual world of Germany, the philosophic mind of Kant was establishing the laws of pure reason, while in Goethe, as a youth of twenty years, already raged the storm and passion of the classic literature of Germany. These were the brilliant stars of his horoscope.

At his baptism, which took place on October 19, 1769, when Sack, the court chaplain officiated, several distinguished persons were named as sponsors, including the following royal personages : [1]—the Prince of Prussia, afterwards King Frederick William II., Prince Henry of Prussia, the Hereditary Prince of Brunswick, and Duke Ferdinand of Brunswick.

Born with the advantages connected with a high social position, Alexander von Humboldt was a noble in the most elevated and extensive sense of the word, ennobled more by his qualities of mind and heart than by the pedigree of his ancestors.

It is a remarkable circumstance that the mother of the ' scientific discoverer of America,' the Columbus of the nineteenth

---

[1] ' Baptismal Register of the Cathedral at Berlin,' vol. vii. p. 252.

century, bore the same name as the geographical discoverer of the fifteenth century. She was descended from an ancient noble family, Von Colomb, who fled from Burgundy on the revocation of the Edict of Nantes, and settled in the Margraviate.[1] This lady, however, was able to confer upon her sons advantages of a more substantial character than the high sound of her distinguished name. Naturally gifted with remarkable administrative talent, 'she had received an education befitting women of her rank, and united to these advantages an extensive knowledge of the world, and the possession of a considerable fortune. Her endeavours were latterly directed towards the reformation of her son by her first marriage, who had frequently caused her great anxiety, while her desire for her two younger sons was to see them distinguished by everything that was attainable in intellectual and moral culture.'[2] Upon her, therefore, in consequence of the premature death of her husband, devolved the education of her sons, in the prosecution of which she became involved in heavy expenses that had to be met by mortgages on her property and real estate. It may be remarked here, that the mortgage on one of these estates was only cancelled officially in 1845, though this engagement had long since been released, as it is termed in official language, 'by act of notary.' Truly no act of notary has ever become so notorious as in this private transaction.

To the biographer there is a peculiar charm in tracing the earliest impressions received upon the mind, and showing that of all the influences which help to form the character, none are more direct or more powerful than those exercised by a mother. The mother of the two Humboldts was not, however, one of those gifted women capable of transmitting genius and force of character to their children. It had been originally her intention to introduce her sons early into the great world, where their connections and interest at court promised them a brilliant career; yet she yielded her own wish in compliance with the judicious counsel offered by Kunth: indeed the liberal and

---

[1] A. von Humboldt, 'Kritische Untersuchungen über die historische Entwickelung der geographischen und nautischen Kenntnisse im 14. und 15. Jahrhundert,' translated by Ideler, vol. ii. p. 277, note.

[2] Kunth's 'Manuscript Autobiography.'

unprejudiced spirit that she manifested in the choice of tutors for her sons is worthy of all commendation, for she herself belonged to a station in life where at that time riding, fencing, and dancing were considered the chief requisites in the education of youths of distinction. She spared no expense in securing the services of the best masters, and in maintaining an intercourse with the most gifted and intellectual men of the time.

Of all the possessions of Major von Humboldt none was more closely associated with the earliest recollections of his two sons, nor more intimately connected with the future history of William and Alexander von Humboldt, than the country-house at Tegel.

On the banks of the Havel, about eight miles from Berlin, from which it is separated by extensive pine woods, is situated the village and mansion of Tegel. The river at this point expands into a beautiful lake, studded with numerous islands and surrounded by a richly wooded shore. On the high bank forming one side of the shore stands the house, looking towards the south upon the town and fortress of Spandau. The house had been originally a hunting-box of the Great Elector, and was afterwards incorporated into the district of Schönhausen.

The crown lands at Tegel, while under the rangership of Bürgsdorf, had been employed as an extensive nursery of foreign trees,[1] whence in 1786, 500 varieties, mostly of North American species, were supplied for the ornamental plantations of the Royal Gardens at Potsdam, Charlottenberg, and Schönhausen: previous to this arrangement the crown had devoted the estate to the culture of silkworms, and had let the house and land on the nominal rent of 138 thalers (28l.), on condition of this plan being carried out, and of 100,000 mulberry trees[2] being planted. Since the year 1738, the various tenants,

[1] Leonardi, 'Beschreibung der preussischen Monarchie,' vol. iii. part i. p. 746.

[2] [The rearing of silkworms excited so much attention throughout Europe at this time, that Government bounties were given in almost every country for the encouragement of schemes for this purpose; but in nearly every instance the attempt failed. It may be interesting to English readers to be reminded that many of the old mulberry trees in our own country were planted in consequence of a circular letter issued by James I. to persons of influence throughout the kingdom, recommending their cultivation.]

Thielow, Moering, Imbert, Struwe, and Von Hollwede, had made many unsuccessful attempts to fulfil these conditions. Even in the first year of his lease Major von Humboldt expended about 1,200 thalers (250*l.*) on mulberry trees, and laid out more than any of his predecessors in improving the buildings necessary for this enterprise. But the cultivation of mulberry trees and the rearing of silkworms were alike unsuccessful, and at length, in 1770, the tenants were released from this obligation. To Major von Humboldt is especially due the credit of having rendered the house at Tegel a delightful residence ; ' the beautiful pleasure-grounds were partly laid out in the English taste and partly left in the wild beauties of nature, interspersed with numerous plantations of American shrubs.'[1]

Humboldt's house in town, as well as at Tegel, was constantly open for the reception of distinguished visitors. Not unfrequently even the heir to the throne honoured Major von Humboldt with a visit at his country seat.

In May 1778, Goethe was received as a welcome guest at Tegel on the occasion of his only visit to Berlin. By his good genius the poet was led away from his discomforts in the Margravian Athens to Potsdam, whither he went on foot, passing through Schönhausen and Tegel. There he took his mid-day rest, attracted as it were by the intellectual charms of the spot where the two youths, William and Alexander, of a genius akin to himself, played at his feet as boys of nine and eleven years of age. He has honoured the place, as is well known, by an allusion in the first part of ' Faust,' in the verses spoken by the proktophantasmist Nikolai :—

> Vanish, at once ! We've said the enlightening word.
> The pack of devils by no rules is daunted :
> We are so wise, and yet is Tegel haunted.[2]

The ghost alluded to, however, was not in the house at Tegel, but in the forester's lodge in the village, where it had been raised by a waggish gamekeeper.

---

[1] Büsching.

[2]   ' Verschwindet doch; wir haben ja aufgeklärt!
      Das Teufelspack, es fragt nach keiner Regel :
      Wir sind so klug, und dennoch spukt's in Tegel ! '

It is unnecessary to give a description here of Tegel as it afterwards became when converted by William von Humboldt into his Tusculum, adorned with treasures of modern and ancient art; and to him this residence was rendered doubly attractive by the charm of early association.[1] 'Here I passed my childhood and a portion of my youth . . . the place is preeminently adapted for the exhibition of the manifold charms which fine and well-grown trees of every variety continuously display through the changing seasons of the year.'

On his return to Berlin from Freiberg, Alexander von Humboldt describes[2] this delightful residence to his friend Freiesleben in the following melancholy strain :—

'Vine-clad hills which here we call mountains, extensive plantations of foreign trees, the meadows surrounding the house, and lovely views of the lake with its picturesque banks awaiting the beholder at every turn, render this place undoubtedly one of the most attractive residences in the neighbourhood. If, in addition, you picture to yourself the high degree of luxury and taste that reigns in our home, you will indeed be surprised when I tell you that I never visit this place without a certain feeling of melancholy. You remember, no doubt, the conversation we held in returning to Töplitz from Milischauer, when you listened with so much interest to the description of my youthful days. I passed most of that unhappy time here at Tegel, among people who loved me and showed me kindness, but with whom I had not the least sympathy, where I was subjected to a thousand restraints and much self-imposed solitude, and where I was often placed in circumstances that obliged me to maintain a close reserve and to make continual self-sacrifices. Now that I am my own master, and living here without restraint, I am unable to yield myself to the charms of which nature is here so prodigal, because I am met at every turn by painful recollections of my childhood which even the inanimate objects around me are continually awakening. Sad as such recollections are, however, they are interesting from the thought that it was just my residence here which exercised so powerful an influence in the formation of my character and the direction of

[1] 'Briefe an eine Freundin,' pp. 123, 156.
[2] This letter is dated June 5, 1792.

my tastes to the study of nature, &c. But enough of this. I shall weary you with so much about myself.'

Complaints of this kind often escaped Alexander when in a melancholy mood, and even William occasionally gave expression to similar feelings. They had their origin mainly in the depressing illness of their mother, to whom the solitude and retirement of Tegel were often indispensable. It would perhaps be scarcely justifiable to withhold every allusion to these outbursts of discontent; but in giving this one instance, we may remark that even the rarest gifts of fortune are often accompanied by much that is sad and distressing,—a painful truth early experienced by the youths, but which their magnanimous natures taught them patiently to endure and almost to ignore.

Alexander von Humboldt passed his childhood and early youth in inseparable companionship with his elder brother, William. These years flew by, to all appearance, as pleasantly as the favourable circumstances in which they were placed by the position and wealth of their parents would indicate. The winter was spent in their own house at Berlin, while in summer they lived occasionally at Ringenwalde, but more frequently at Tegel, on account of its vicinity to the capital. It may be remarked here, that Campe, the writer of books for the young, was tutor in the household of Major von Humboldt before he joined Basedow in the Educational Institution at Dessau.

On this subject William von Humboldt writes as follows to his friend and correspondent Charlotte, in a letter dated Tegel, December, 1822 :—' Campe was private tutor in my father's house, and there is still standing a row of great trees here which he planted. From him I learned reading and writing, and some amount of history and geography, according to the fashion of those times, which consisted in a knowledge of the capital cities, the seven wonders of the world, &c. Even at that time he had a very happy knack of stimulating the mind of a child.'

On another occasion he writes :—' I am not mistaken about Campe. He was at one time tutor, or, as it was then termed, governor to an elder step-brother of mine, Hollwede, a son of my mother's by a former marriage. From him I learned, when I was three years old, reading and writing. He must have left our house somewhere about 1770 or 1771. On quitting us he entered the Church, but soon left his charge to engage with Basedow in the Philanthropin at Dessau. His journey to Paris, in which I accompanied him, did not take place till the year 1789.'

From these statements it is clear that Alexander von Humboldt was scarcely more than an infant at the time of Campe's residence with the family. It is therefore more than doubtful if *he* ever received any instruction from Campe, and quite certain, notwithstanding the frequent assertions to the contrary, that Campe could not have exercised a 'lasting influence' upon *both* brothers, nor have 'first aroused in Alexander the unconquerable passion for the exploration of foreign lands.'

Alexander von Humboldt would assuredly have become the great traveller had he never even read Campe's 'Robinson,' the first edition of which was published in the year 1780.

It may perhaps seem somewhat remarkable that the two leading characteristics of Campe's mind—the love of language, which led him, after Klopstock, to be one of the first in Germany to engage in the study of philology, though principally in regard to the German language, and the love of adventure which made him delight in presenting before the minds of children the histories of bold adventurers and great explorers—should have been so strikingly manifested in William and Alexander von Humboldt. It may nevertheless be worthy of notice that in the journey to Paris in 1789 above alluded to, Campe assumed the post of governor to the elder Humboldt. But it is equally certain that the bent of mind relatively distinguishing the two Humboldts was *original* in the truest and most characteristic sense of the word, and was in both cases developed quite independently at an early age. At a time when it was the fashion to educate youths of distinction according to the method described by Rousseau in 'Émile,' the educational powers evinced by Campe were far too highly estimated, and consequently the influence he exerted over the two Humboldts exaggerated beyond measure. Alexander, who always mentions his tutors with kindness and gratitude, never refers to Campe in these terms, and did not scruple even to allude to him in a tone of ridicule.[1]

[1] 'Campe has a project of going to America,' writes Humboldt to

Should further evidence be needed to disprove the assertion that Campe exercised a decided influence over the intellectual development of the two Humboldts, surely the exclamation uttered by George Forster[1] upon receiving a visit from William von Humboldt and Campe, on their return from Paris in 1789, will be deemed conclusive:—'Good heavens! Is it not astonishing that there are any men left in Germany, when their tutors are all like Campe!'

Alexander von Humboldt's first tutor was Johann Heinrich Sigismund Koblanck—a fact which has hitherto escaped the notice of biographers; he died as first preacher of the Louisenkirche in Berlin. In his manuscript *curriculum vitæ*, still preserved by his grandson, Dr. Koblanck, Privy Councillor of the Board of Health, there occurs the following passage:— 'Within a month of quitting the University of Halle, in 1773, Koblanck became a resident in the household of Major von Humboldt, as successor to Campe, the noted pedagogue, in the capacity of governor and tutor to the young Baron von Hollwede and the two sons of Herr von Humboldt, William and Alexander, who have since earned for themselves so high a reputation, the one by his attainments in literature, the other by his travels in every quarter of the globe. In the year 1775, he was appointed military chaplain to the Von Arnim Royal Regiment of Infantry, and was ordained at Potsdam on October 20, in the same year.'

According to to family tradition, Alexander von Humboldt

Sömmering from Hamburg, January 28, 1791; 'whether he carries it out or not is quite uncertain. Conceive, my dear friend, the motives that he specifies; not that he may enchant the intelligent youth of that country by the introduction of his children's books, his Robinsoniads, &c.; not that he may disseminate among the savages his new proof of the immortality of the soul; not that he may regulate dancing in Philadelphia according to the laws of chastity—nay, but that he may enter upon a close study of the constitution of the United States, so as to be able in the course of a year (for Europe must even be deprived of him for that length of time) to publish the result of his observations to the Old World, in order that truth and freedom may be extended to all mankind. Can you fancy anything more truly absurd? I am daily expecting to hear from Campé, inviting me to accompany him.'

[1] George Forster's 'Sämmtliche Schriften,' vol. viii. p. 89. See also K. von Raumer, 'Geschichte der Pädagogik,' vol. ii. p. 308; Schlosser, 'Geschichte des 19. Jahrhunderts,' vol. iii. Part ii. p. 163.

received from Koblanck his first instructions in reading and writing.

The second tutor to whom the care of Alexander's education was committed was Johann Clüsener, now mentioned for the first time in this connection; he afterwards became private secretary to the Princess Ferdinand, and Counsellor in the local Government of Sonnenburg, and in the year 1828 was still living at Berlin. An autograph letter written to him by Major von Humboldt, dated Ringenwalde, November 25, 1776, is still preserved, and bears the address : 'À Monsieur, Monsieur Clüsener, Gouverneur des Messieurs de Humboldt à Schloss Tegel.' Kunth in his 'Manuscript Autobiography' expressly mentions him as his predecessor in the household of Major von Humboldt.

Though not possessed of any special powers of instruction, Kunth, who in later years attained the dignity of Actual Privy Councillor, was gifted with a character of remarkable excellence, and is deserving of lasting fame for the care he bestowed upon the education and culture of the two Humboldts. In the year 1777, at the age of twenty, he entered the family of Major von Humboldt in the capacity of tutor, when William was ten and Alexander eight years of age.

Kunth, son of the Protestant pastor of Baruth, had received a liberal education, and early acquired an excellent knowledge of Latin, French, and Italian, with a considerable amount of external culture; by intercourse with the distinguished society he was privileged to meet at the musical and dramatic entertainments given at the manorial residence in his father's neighbourhood, his manners acquired that polish and assurance for which he was afterwards distinguished. During his university career he had devoted more attention to modern languages and elegant literature than to the study of theology, and was in quest of an appointment as Secretary of Legation when he accepted the engagement of tutor to the sons of Major von Humboldt, as successor to Herr Clüsener. He made himself so acceptable in his new position, that he was almost immediately requested to undertake some important duties in connection with the household, as well as to conduct much of the necessary correspondence. In Major von Humboldt's absence it not unfrequently devolved upon him to entertain persons of distinction, and on one occasion he had the honour to receive the Duke of Brunswick—a circumstance which furnishes sufficient proof of the versatility of his powers and the confidence which was reposed in him.[1] ' It is rare,' remarks the celebrated political economist, State-Councillor Hoffmann, in his obituary of Kunth,[2] 'to find such brilliant hopes receiving so happy a fulfilment. The bond existing between Kunth and his pupils was of a more indissoluble nature than that formed merely through the solicitude of a faithful teacher to communicate to highly gifted pupils the knowledge he himself possessed.'

It would seem that Kunth but rarely gave instruction personally to his pupils, and his unassuming nature never allowed him to claim more than a limited influence in producing the intellectual distinction which they subsequently attained. It is related by Henriette Herz,[3] that on one occasion during the winter of 1827–8, as Alexander von Humboldt was delivering an admirable lecture, excellent both in purport and arrangement, before a mixed audience at Berlin, when every eye was beaming with excited admiration, Kunth whispered in her ear, ' He is not indebted to me for this!' and once, when allusion was made in the presence of William von Humboldt to the extensive knowledge of history possessed by Kunth, accompanied by a remark upon his almost painful amount of prolixity, he observed :—' That is true ; his lectures upon history make one almost wish to have been Adam, when history had only just begun.'[4] There can be no doubt, however, that the views held by Kunth upon citizenship, political life, and philanthropy, and the sympathy he invariably manifested with all endeavours after freedom, must have exercised a very considerable influence upon the minds of his pupils, and it is evident from the expressions he subsequently made use of in controverting the views of the minister Stein in the retrospect of his latter years, that he formed, as it were, the prototype of their characters. Kunth

---

[1] Kunth's ' Manuscript Autobiography.'
[2] ' Staatszeitung ' of November 3, 1829.
[3] Fürst, ' Henriette Herz,' p. 148.
[4] ' Aus dem Nachlasse Varnhagen's. Briefe von Chamisso, Gneisenau, Haugwitz, Wilhelm von Humboldt, &c.,' vol. i. p. 11.

was possessed of qualities which made him a diligent worker and an excellent official; and his views on the claims of industry and other branches of political economy were marked by sense and clearness. He devoted himself with the most indefatigable energy to the establishment of a free and enlightened legislation, to the elevation of the industrial classes, and the furtherance of trade and commerce, and he deserves especial credit for the unremitting industry with which he laboured, not only towards the suppression of the corrupt practices of many of the trade corporations, but also towards the introduction of a system of free trade, and the formation of the Zollverein or tariff union throughout Germany.[1]

The chief service that Kunth rendered to his pupils in his character of tutor consisted in the judicious efforts he constantly displayed to procure the most valuable educational advantages afforded by Berlin, such as private instruction, social intercourse, and suitable companionship, in this way supplying the advantages they missed from never having been at a public school. These methods were peculiarly calculated to encourage the development of the individual gifts and mental tastes which so soon began to manifest themselves in the youths.

No record has been preserved of the progressive order of their studies; but Alexander very early evinced a taste for natural history. Flowers and plants, butterflies and beetles, shells and stones, were his favourite playthings; and the collecting, arranging, and labelling of these treasures was carried on with so much zeal, that as a child he acquired, in jest, the name of 'Little Apothecary.'

The principal instruction they received in these early years was derived from Ernst Gottfried Fischer, Professor in the Gymnasium of the Grey Friars, whose mathematical schoolbooks continued to be in use long after his death. In his manuscript journal, still preserved, he remarks :—' In addition to my arduous official duties, I was obliged, for the sake of increasing my income, to engage in some extraneous occupation, consisting principally in private tuition. Such labours, undertaken as a means of livelihood, are not generally favour-

---

[1] Pertz, 'Leben des Ministers Freiherrn von Stein,' vol. vi. p. 789.

able for private study, yet in the first few years I was fortunate enough not to have to expend my strength on dull and stupid pupils. The instruction of such youths as William and Alexander von Humboldt, and Joseph Mendelssohn, cannot be classed among those uninteresting labours to which duty and necessity so often reduce a man of learning. I recall with extreme pleasure the hours passed almost daily for several successive years with the family of Major von Humboldt, engaged in giving instruction in Latin, Greek, and mathematics, and cheered by the bright hopes which enchanted me at that time as much by anticipation as now they do by their happy fulfilment.' It may here be remarked that Alexander could not have been included in these instructions in Greek, since it was not till June 1788, therefore only after his first academical term, that he began, ' in his nineteenth year, under the tuition of Bartholdi, to decline εχιδνα.' [1]

Löffler, who became afterwards Professor in the University of Frankfort-on-the-Oder and later Councillor of the Upper Consistory at Gotha, a free-thinker and author of a pamphlet on the Neo-Platonism of the Fathers, gave instruction at one time to the family of Major von Humboldt in Latin and Greek.

Among the lectures attended by the youths may be mentioned those of Engel, Professor at the Joachimsthal Gymnasium and Æsthetic Director on the Board of Management of the Royal Theatres—the author of 'Popular Philosophy,' who, with a delivery almost rivalling that of Garve and Mendelssohn, familiarised the minds of his youthful hearers with that modest, practical philosophy which he so ably advocated as adapted equally to the guidance of the conscience and the control of the reason. In his biography of William von Humboldt, Haym remarks: 'The power of communicating knowledge in an attractive manner existed in Engel in an eminent degree ; a transparent clearness of thought characterised his understanding, a correct and elegant taste regulated his feelings, while he gave expression to both thought and feeling in chaste and appropriate language. In his ' Popular Philosophy,'

---

[1] Letter from Alexander von Humboldt to Wegener, his fellow-student at Frankfort, dated Berlin, June 9, 1788.

Knowledge appeared in her most engaging form. Engel was himself most truly a philosopher for the world, and without doubt he was an admirable instructor of youth.'

In the year 1785, as we learn from his biographer Gronau,[1] Dohm, then officially engaged in the Department of Trade and Commerce, gave a series of lectures upon political economy, at the request of the minister Von Schulenburg, for the instruction of a young Count von Arnim.

'At the wish of their excellent mother,' writes Gronau, 'the two brothers William and Alexander von Humboldt attended these lectures, which were in all points similar to an ordinary university course, and were continued from the autumn of 1785 to June 1786.'—'Dohm retained through life the happiest recollections of this early connection with his youthful auditors; and it is evident that Alexander von Humboldt also maintained a strong interest in his former tutor, since after an interval of twenty years, in a beautiful spirit of grateful affection, he sought him out towards the end of the year 1806, and, for the sake of doing him a trifling pleasure, devoted several spare hours of a morning and evening to "his good tutor," as he always called Dohm, in giving him a narrative of his travels in America, illustrating his descriptions by an exhibition of some of the treasures he had brought back with him.'

It would appear that Alexander did not join his brother William in attending the lectures on law and jurisprudence delivered by Klein, Counsellor of the Supreme Court of Judicature, and one of the compilers of the New Code of Law in Prussia. Yet it is probable, from the interest that Moses Mendelssohn is known to have taken [2] in the studies of William von Humboldt, that there may be some truth in the tradition

[1] Gronau, 'Chr. Wilh. v. Dohm, nach seinem Wollen und Handeln,' p. 127.

[2] In the private library of the King of Saxony there is still preserved a very elaborate treatise of nearly 750 quarto pages, together with some letters of Klein and remarks of Moses Mendelssohn upon 'extorted treaties' ('über erzwungene Verträge'). A letter of Alexander von Humboldt, appended to the book, runs as follows:—

'The book, entirely in my brother's handwriting, contains the essays which he wrote out after every private lesson on the laws of nature from Counsel-

that he used to spend the 'morning hours' in walking about
the garden in company with the two brothers, and, while engaged in friendly intercourse, discoursing to them valuable
lessons in philanthropy and philosophy. Ample proof, however, exists that intercourse of a similarly instructive character
was maintained between the Humboldts and David Friedländer,
since several letters are still extant that were addressed to
him by the two brothers. Upon his death on December 25,
1834, his eldest son, Benoni Friedländer, received the following
expressions of condolence from Alexander and William von
Humboldt :—

'Berlin: December 27, 1834.

'. . . . The attractive image of your noble and intellectual
father is distinguishable among the earliest and most grateful
reminiscences of my youth. The marked kindness which he
invariably showed me enhances the pleasure of these recollections : for to your lamented father I am indebted for much
valuable influence in my education and in the direction of my
thoughts and feelings. He was, with Engel, the constant friend
of our house. He possessed an extensive knowledge of antiquity, a love of speculative philosophy, a delicate and correct
taste for the beauties of poetry, and considerable ability in
solving, by the wonderful flexibility of our language, the difficult
problem of translating from the sacred tongues ; to these intellectual gifts of a high order were united the most liberal
opinions upon the political events of the day, and the most
sincere and devoted affection towards his oppressed people.
His long, happy, and successful life has closed within the
circle of his family, who could well appreciate the value of his
intellectual worth, since through him they have been fashioned
into a similar mould.

lor Klein. The marginal notes commenting upon the work, for the most
part in praise, but occasionally in refutation, are in Klein's handwriting.
I cannot but wish that this manuscript, not meant for publication, may fall
into the hands of those who will know how to appreciate the philosophical
jurisprudence contained in this youthful production of William von Humboldt.

'Berlin: February 1854.'

'In the midst of your sorrow pray receive, with every good wish, the assurance of the most ardent affection and grateful friendship of

'Yours, &c.,

'ALEXANDER VON HUMBOLDT.'

'Tegel: January 2, 1835.

'In my retirement here, I heard only recently of the death of your revered father, and I was on the point of writing to you, to express my sincere and heartfelt grief, when your letter reached me. I was truly glad to receive the assurance that a gentle and painless release had been granted to him. If his life, as you remark, has been a happy one, it has also been one of usefulness and honour, for by his worth and talents he had won for himself a position in which he will never cease to be remembered. It is a great gratification to me to learn that during the illness of our deceased friend my brother and I had often been in his thoughts, and it can never be forgotten by either of us how diligently he laboured for our improvement. From the kind interest which he manifested towards us from the first, he kept us always in a pleased and eager mood, while his quick intelligence, the almost uninterrupted cheerfulness of his disposition, and his constant incitement to some useful occupation, either mental or physical, rendered our studies both interesting and attractive. He early directed our minds to correct views on several important points connected with life and society—views, however, which at that time were by no means universally received. I cannot sufficiently thank you for your kindness in sending me some particulars of his last hours. I beg the favour of the continuance of your kind consideration, which, I assure you, will ever be most sincerely and affectionately reciprocated by me.

'I remain, with the highest esteem,

'Yours,

'WILLIAM HUMBOLDT.'

Another tutor to the two Humboldts is referred to by Kunth under the name of Meyer; probably this was the mathematician Meyer Hirsch, whose books on algebra and geometry are still in use as educational works, and who at that time was engaged in the tuition of some young princes of the court.

A hasty but not uninteresting glimpse of the youthful days of Alexander von Humboldt is given in Heim's ' Life of Kessler.' In the journal of this noted physician, the following entry occurs under date of July 30, 1781 :—' Rode over to Tegel, and took an early dinner with Frau von Humboldt; explained to the young Humboldts the twenty-four classes of the Linnæan system of botany, which the elder one readily comprehended, retaining the names without difficulty.' Again, on May 19, 1783, he writes :—' Rode over to Spandau to witness the review with my friends from Tegel, Kunth and his distinguished pupils.'

Since William von Humboldt at that time was scarcely sixteen years of age, and Alexander not yet fourteen, we must suppose that the term ' distinguished ' is a polite interpolation of a later date.

In modern languages their instructor, according to Alexander von Humboldt, was Professor Le Bauld de Nans, tutor to the royal family ; he was editor of the ' Gazette littéraire de Berlin,' a publication started by the actor Francheville, in which Humboldt's first literary effort appeared.

It would seem that the fine arts were also studied by Alexander with some degree of success ; for in the catalogue of the first exhibition of the Berlin Academy, in the year 1786, the following entry occurs under the division ' Amateurs : '—' No. 290. Herr von Humboldt, junr. Friendship weeping over the Ashes of the Dead. Drawn in black chalk after Angelica Kauffmann.'

Those who are acquainted with the scientific labours of Alexander von Humboldt are doubtless familiar with his later drawings, in every variety of style, illustrative of his studies in botany, zoology, anatomy, and natural science, besides maps and configurations of country. It may not, however, be so generally known that in later years he studied both drawing and painting in Paris, under Gérard, when he made severe studies from models as well as from life, and laboured with so much success that he was able to produce some admirable portraits. A half-length portrait, life-size, drawn in black chalk, and inscribed with his own hand: 'Alexander von Humboldt, of myself in a looking-glass. Paris, 1814,' is one of the best portraits of him existing. He also made an excellent portrait in pencil of Professor Kunth, his indefatigable assistant in the preparation of his botanical works.

In the arts of etching and engraving on copper he received instruction from the celebrated Chodowiecki, and several impressions from plates engraved by Humboldt are still preserved, to which we shall have occasion again to allude.

Neither of the brothers had the smallest appreciation for music: to William it was absolutely intolerable, while Alexander regarded it as a 'calamité sociale.'[1]

This is all that is certainly known concerning the tutors engaged in the tuition of Alexander von Humboldt; Willdenow, though often included among them. was certainly not one of the number, and it was only in later years that his influence operated so powerfully in the development of Humboldt's botanical tastes. Humboldt writes of his early education to Pictet in the following terms in the year 1806[2]:—

'Until I reached the age of sixteen, I showed little inclination for scientific pursuits. I was of a restless disposition, and wished to be a soldier. (!) This choice was displeasing to my family, who were desirous that I should devote myself to the study of finance, so that I had no opportunity of attending a course of botany or chemistry; I am self-taught in almost all the sciences with which I am now so much occupied, and I acquired them comparatively late in life. Of the science of botany I never so much as heard till I formed the acquaintance in 1788 of Herr Willdenow, a youth of my own age, who had just been publishing a Flora of Berlin. His gentle and amiable character stimulated the interest I felt in his pursuits. I never received from him any lessons professedly, but I used to bring him the specimens I collected, and he gave me their classifica-

---

[1] Anton Springer, 'Friedrich Christian Dahlmann' (Leipzig, 1870), p. 237.

[2] 'Lettres d'Alexandre de Humboldt à Marc-Auguste Pictet, 1795-1824,' in 'Le Globe, Journal géogr. de la Soc. de Géogr. de Genève (1868), vol. viii. p. 180. See also Brockhaus' 'Conversations-Lexikon,' art. 'Alexander von Humboldt.'

tions. I became passionately devoted to botany, and took especial interest in the study of cryptogamia. The sight of exotic plants, even when only as dried specimens in an herbarium, fired my imagination with the pleasure that would be derived from the view of a tropical vegetation in southern lands. Owing to his intimacy with Chevalier Thunberg, Herr Willdenow was often in receipt of plants from Japan, and I could never see them without indulging the hope that some time or other I might visit foreign countries.'

These last words furnish a glimpse of those early days when the desire for foreign travel was first awakened in the youth's heart. Humboldt always recalled with interest the rise of these early tastes which exercised so powerful an influence upon his later life. Thus he writes [1] :—

'From my earliest youth I had an intense desire to travel in those distant lands which have been but rarely visited by Europeans. This impulse is characteristic of a certain period in our existence, when life appears as a boundless horizon, when nothing so completely captivates the fancy as the representations of physical danger and the excitement of sensational emotion. Although educated in a country which held no direct communication with the colonial settlements of either the East or West Indies, and afterwards called to reside at a distance from the coast among mountains famous for their extensive mines, I yet felt the passion for the sea and for long voyages growing within me with ever-increasing strength.'

And further :—' The study of maps and the perusal of books of travel exercised a secret fascination over me which was at times almost irresistible, and seemed to bring me in close relationship with places and things in regions far remote. The thought that I might possibly have to renounce all hope of seeing the splendid constellations that shine in the southern hemisphere invariably sent a pang to my heart.'

In the 'Aspects of Nature'[2] he states :—'In the longing for a sight of the Great Pacific from the high peaks of the Andes, was mingled the interest with which, as a boy, I had listened to

[1] 'Reise in die Aequinoctialgegenden des neuen Continents,' vol. i. p. 47. Hauff's edition, vol. i. pp. 2, 3.
[2] 'Ansichten der Natur,' 3rd ed. vol. ii. p. 363.

the narrative of the bold expedition of Vasco Nuñez de Balboa. The reedy shores of the Caspian Sea, as I viewed them from the delta formed by the mouths of the Volga, are certainly not picturesque ; yet the first sight of this vast inland sea of Asia yielded me great delight from the fact that in my youthful days I had drawn its outline in a map. The tastes first awakened by the impressions of childhood, and moulded by the circumstances of after life, often become, when imbued with the deep earnestness of later years, the incentive to scientific labour or to undertakings of vast import.'

And in a passage in ' Cosmos '[1] he remarks :—' The pleasure I derived as a child from the contemplation of the form of continents and seas as delineated in maps, the yearning to behold those southern constellations which never appear above our horizon, the representations of palms and cedars of Lebanon occurring in the illustrations of a pictorial Bible, may all have contributed to excite within me the desire to travel in foreign lands. Were I to ask myself, while reviewing such early reminiscences, what first awoke in me the insatiable longing to view the glories of a tropical region, I should reply :—The perusal of George Forster's vivid descriptions of the islands in the Pacific; the sight of some paintings by Hodge, in the house of Warren Hastings in London, representing scenes on the Ganges ; and the admiring wonder excited by the contemplation of a gigantic dragon-tree in an old tower of the Botanic Garden at Berlin.'

Some influence may also be due to the fact that the youthful years of Alexander von Humboldt were passed during a time of general excitement for geographical discovery, when the passion for exploration had taken possession of the leading nations of Europe, causing them to emulate each other in acquiring a knowledge of distant lands and seas.

Even the disasters that befel the unfortunate enterprises of La Perouse and D'Entrecasteau, of Bligh and Malaspina, were powerless to damp the extraordinary zeal for travel and discovery which had been excited by Byron, Wallis, Carteret, Bougainville, and Cook. Owing to the persevering energy of

---

[1] 'Kosmos,' vol. ii. p. 5.

Cook, who between the years 1768 and 1779 had explored in three successive voyages the whole extent of the Southern Ocean, the veil that had concealed one half of our globe had been rent asunder and the whole civilised world fired with enthusiasm. The noble example of Cook and his companion Banks, of such men as Solander, Sparrmann, and the two Forsters, inflamed the zeal of other navigators, and led to the expeditions of Vancouver and Flinders to the coasts of Australia and New Zealand.

The zeal thus manifested in maritime discovery was equally displayed in the investigation of unknown continents. Through the policy of Catherine of Russia, various expeditions had been despatched for the exploration of different parts of the continent of Asia; those to the northern districts were under the care of Gmelin, Pallas, Georgi, and Güldenstädt, associates of the Academy of St. Petersburg, while Thunberg, also a distinguished academician, was entrusted with the exploration of the provinces of Eastern Asia: and while the Asiatic Society of London undertook the exploration of India, and the various embassies of Great Britain furnished geographical information concerning the countries of Thibet, China, and Java, Niebuhr, Volney, Choiseul-Gouffier, and Le Chevalier contributed, under the auspices of the French Government, most valuable data in reference to Palestine, Syria, and Asia Minor, not only as regarded geographical discovery, but also concerning the history of those countries and their natural characteristics. Since the formation of the African Association of London, in 1768, the continent of Africa had in like manner been subjected to extensive exploration, and the most important information obtained of the northern and eastern districts by the travels of Sonnini, Niebuhr, Forskal, Hoest, Poiret, Desfontaines, Volney, Bruce, Houghton, and Hornemann; of Western Africa by the expeditions of Norris, Isert, Golberry, and Grandpré; and of the southern regions by the explorations of Thunberg, Sparrmann, Paterson, and Le Vaillant. The spirit of the age led Hearne and Mackenzie, the one as discoverer, the other as explorer, to visit even the polar regions of North America. The time indeed was rapidly approaching when almost the entire globe should be thrown open to the spirit of exploration then animating Europe, and already events which the boldest minds scarcely ventured to anticipate were transpiring which should crowd into one generation more discoveries than had been made in the three preceding centuries.

This intense excitement for discovery in geography and natural science was accompanied by an enthusiasm which made itself felt even in the literature of the day, supplying language with the most glowing images, the most gorgeous colouring, and inspiring even the prose writings of that time with the most daring flights of imagination. We have but to call to mind Buffon's 'Époques de la Nature' (1778), Bernardin de St.-Pierre's 'Études de la Nature' (1784), and 'Paul et Virginie' (1788), as well as the writings of Playfair or the descriptions of George Forster.

Can we wonder, then, if under such exciting circumstances the youth's passion for travel, his thirst for knowledge and the investigation of nature, should inflame with ever increasing enthusiasm!

On January 6, 1779, occurred the unexpected death of Major von Humboldt, the result of a merely casual illness; this event was the more unlooked for, since his usual good health had given promise of a long life. It is possible that his removal may be regarded almost as a fortuitous circumstance, as, from his high military position, he might probably have viewed with impatience the tastes developed in his sons, and considered the gratification of such tastes derogatory to their position in society. The loss of their father produced no change in their mode of life. To their mother, as the natural guardian of her sons, was committed the charge of their property and the management of all their affairs, so that the youths continued to remain under her watchful care, and under the intellectual guidance of their tutor Kunth.

The physical constitution of Alexander was of such a nature as to prevent so early a development of intellectual power as occurred in the case of his brother William. In making a retrospect of his early education, William remarks concerning himself, 'that from his earliest years he could scarcely repress the desire to see and know as much as possible about every-thing that surrounded him, and was not content until every idea that presented itself before him had been worked out in his own mind and thoroughly appropriated.' Alexander, on the contrary, could only master his daily tasks by dint of extraordinary effort. As a boy, he was much less robust than William, and suffered from an amount of debility which not unfrequently produced great prostration. He himself confesses to Freiesleben, his fellow-student at Freiberg and subsequently Director of Mines, ' that in the first years of his childhood his tutors were doubtful whether even ordinary powers of intelligence would ever be developed in him, and that it was only in quite later boyhood that he began to show any evidence of mental vigour.'[1]

In reference to this subject, the following passage, occurring in a letter written by George Forster to Heyne, on July 14, 1790, is full of interest :—' Herr von Humboldt, who desires to be specially remembered to you, is still with me, and has been tolerably well throughout the journey, though not so well as I could wish. He says, indeed, that he has been constantly ailing for the last five years, and is never much better excepting immediately after a severe illness ; then he gradually becomes worse again until the outbreak of another attack of indisposition, when the system is for a time freed from the accumulation of unhealthy humours. I am, however, fully persuaded that in his case the body suffers from a too great activity of mind, and that his brain has been sadly overworked by the logical course of education adopted by the Berlin professors.'

And again, in a letter of later date, Forster writes to Jacobi on August 6, 1791, in a similar strain:—' Alexander von Humboldt is at Freiberg, and I am afraid I shall lose sight of him. William has been lost to me for long ; he is at Erfurt,

---

[1] As if for consolation, history furnishes many examples of the late development of some of the finest intellects, which then seemed only to expand the more suddenly and the more gloriously. Albertus Magnus, the learned philosopher of the middle ages, was as a child so dull of understanding that he appeared incapable of learning to read ; the genius of Newton lay for a long time so entirely concealed that his mother took him away from school and destined him to agricultural pursuits ; Linnæus, for the same reason, was intended by his father to be apprenticed to a shoemaker ; Molière learnt to read only in his fourteenth year. The case was scarcely so desperate, however, with Alexander von Humboldt.

getting married to a Fräulein von Dacheröden, and in his present humour declines all public employment, which, with his talents, is much to be regretted. Alexander, on the contrary, is most anxious to work, but he has not the physical power.'

Even in 1795 Humboldt is obliged thus to allude to his health in a letter to Willdenow :—' You have a right to be angry with me for writing so seldom. Yet if you knew the circumstances in which I am placed—that I am constantly travelling about, and that I was laid up for three months last winter with a serious illness, which obliges me to devote the little leisure I now have to study—you would at least accept my excuses, even if you could not justify me.'

At a still later period even, Kunth expresses some anxiety about his health in a letter to Von Moll of September 17, 1799 :—' If his health,' he writes, ' does not succumb to the climate and the hardships of travel, how much valuable information may not be expected in various departments of physical science from the observations of one who, possessed of such extensive knowledge, and burning with zeal for the investigation of nature, is intending to spend years together in the midst of scenes of such sublimity.'

Fortunately, however, delicate constitutions often suffer less from the change to a foreign climate and the hardships of travel than those natures which appear to be more robust. Thus Irwin survived all the privations of the desert surrounding Thebes, while Ledyard, though much more vigorous, fell a victim to the climate before leaving Cairo; and Seetzen, constitutionally delicate, successfully combated the dangers of travel in Syria, Egypt, and Arabia, while his stronger and more hardy companion, Jacobsen, was compelled, on account of the climate, to leave the expedition at Smyrna. Humboldt describes himself as being equally in his element in the region of the tropics as within the circle of the poles.

For some years prior to their departure for the University, the brothers resided more frequently at Berlin than at Tegel; for only in the capital could they procure the assistance of tutors capable of undertaking the various branches of their education and enjoy the advantages necessary for their preparation for college life. Concerning this period of his life, William von Humboldt relates the following particulars to his friend and correspondent Charlotte : [1]—' You ask where I was residing in 1786 and the following years. I lived at Berlin. My mother only resided there during the winter, but my younger brother and I continued there through the summer with a tutor, riding over to Tegel usually of a Sunday. This was my mode of life till the autumn of 1788, when, accompanied by the same tutor, I entered with my brother upon college life at the University then existing at Frankfort-on-the-Oder, where I remained till the Easter of 1789. I went soon after with my tutor to Göttingen, leaving my brother still at Frankfort. Once established at Göttingen, I bade farewell to my tutor, and from that moment, when twenty-two years of age, I was thrown upon my own responsibility. It was at this period that I made your acquaintance at Pyrmont in 1789. It was not till the Easter of 1790 that my brother joined me at Göttingen.'

By this system Kunth succeeded in completing the education of his pupils in the course of ten years, without the necessity of their attendance at any gymnasium or other public school. The friendship formed with his pupils remained unbroken during the ensuing forty years, till his death in 1829, and while they were winning for themselves positions of distinction, the one as a man of science, the other as an enlightened statesman, he continued to enjoy their familiar confidence and displayed towards them the same watchful care and devoted attachment he had ever manifested.

As early as the year 1782, a yearly pension of 400 gold florins was granted to Kunth by Frau von Humboldt, as the expression of a mother's gratitude ' for the faithful manner in which he has conducted the entire education of my two younger sons '—a gift which was confirmed to him, as a legacy, among the bequests of her will. As long as he lived, Kunth continued to act as administrator of Alexander's property. Even after entering the service of the State, he remained a member of the household of Frau von Humboldt till her death nine years afterwards, in 1796, and enjoyed a position of such friendship

[1] 'Briefe an eine Freundin,' p. 164.

and confidence, that on his death his remains were interred near the family vault of the Humboldts at Tegel.

Thus passed the childhood and early youth of Alexander von Humboldt. His lot was not that of many sons of genius, who are called to spend their energies in a constant struggle against poverty and adverse circumstances. His temptations were of another order; and it is due to his heaven-born nature, that notwithstanding the high position of his family, his aristocratic birth, and the manifold enticements to a life of luxury and ease offered by the possession of wealth, he evinced from his earliest years a thirst for knowledge and an aspiration towards all that was good and beautiful, which impelled him amid much bodily weakness to devote himself to study with unremitting application.

It will be desirable, before accompanying the brothers to the University, to take in review the general condition of Berlin at that time, since it forms, as it were, an historical background to the brilliant career of these distinguished men.

The august form of Frederick the Great was still the guiding star at Berlin during the childhood and early youth of William and Alexander von Humboldt. Alexander could lay claim to belong, as he himself expressed it on the celebration of the centenary of the great king's accession,[1] ' to the past generation, as one of those who, amid the earliest recollections of their boyhood, could still recall the image of the great monarch.' Although, since the days of Lessing and Mendelssohn, an enlightened tone of thought on religion, science, art, social life, and to some extent also on politics, had begun to evince itself at Berlin, though both Biester and Nikolai had since 1759 commenced in the ' Literaturbriefe ' a severe criticism against the courtly French muse, yet this pulsation of a higher life was so weak and intermittent, it was as yet so partial and spasmodic, that it cannot be justly regarded as a general elevation in taste. And as the number of ' new lights ' was small, so were also the social home circles, in which alone they could be appreciated. In every circle there were few among the

---

[1] Augsburg ' Allgemeine Zeitung ' of June 9, 1840, Supplement.

rising generation who gave promise of exceeding the prevailing mediocrity. It was not Berlin, but Königsberg, where Kant was inculcating his philosophy, that was destined to be the cradle of the new intellectual life of Germany.

At the court of Frederick the Great, the select circle immediately surrounding the monarch consisted exclusively of foreigners, for the most part French, who had grown old with him and been his companions through life. They freely indulged their taste for the high-flavoured dishes with which the French regaled them, such as the flimsy philosophy of Voltaire, the licentious paradoxes of La Metterie, while they took no cognisance of the movement of thought taking place in Germany. The circle that gathered round Prince Henry was distinguished for intellectual brilliancy and sarcastic wit, and was more captivating to younger men, but failed to prove a permanent attraction to superior minds.[1]

The higher officials, both civil and military, most of whom were of noble birth, avoided all society that was intellectual and elevating. Those of a lower grade, receiving small salaries, and overwhelmed with official work, found the calls of business and the necessities of their families too pressing to permit the cultivation of any elevation of sentiment. The wealthier commercial circles were noted only for the luxury in which their families were nurtured, but of true cultivation the appearance even was unattempted. Men of science remained hid in the restricted circles of their own families, and at most met together in the 'Montags-Club,' like the priests of the Samothracian mysteries, where the presence of ladies and strangers would have been regarded as a profanation of their sublime revelations.

It addition to this, the views held by the king on military, political, and administrative subjects were estranging him further and further from the living world around him, and even his vaunted philanthropic principles now showed a new phase, causing the severity of his government, which had not hitherto been burdensome, to be regarded as tyrannical and oppressive. There seemed, in short, to be a tone pervading every class of

[1] Pertz, ' Leben des Ministers Freiherrn von Stein,' vol. i. p. 21.

society, as if all the elements of an important epoch were on the point of dissolution.

This state of things accounts for the disagreeable impression produced upon George Forster by his visit to Berlin:—'I reached Berlin,' writes Forster [1] to Jacobi on April 23, 1779, 'at the close of January. I found I had greatly erred in my conceptions of this great city. The exterior is much more beautiful, the life within much more dark than I had pictured it. Berlin is certainly one of the handsomest cities in Europe. But the inhabitants! Hospitality and the pleasures of refined taste degenerated into luxury, high living, I might almost say gluttony; a bold and enlightened habit of thought perverted into licentious extravagance and unbridled scepticism. And then the rationalistic clergy, who, in their wisdom and out of the fulness of their virtue and moral perfection, would sweep from religion all that is incomprehensible, and reduce it to the ordinary level of the human understanding! I expected to find men of quite a superior order, pure, noble, and illuminated with light from Heaven, yet with the simplicity and modesty of children. I met, instead, with men of the ordinary type, and, still worse, with men filled with the pride and self-conceit of philosophers and theologians; . . . . more I need not say. As to the French Academicians!—let me shake off the dust from my feet and pass on. . . . In the five weeks that I was there, I dined or supped in fifty or sixty different houses, and each time I was obliged to listen to the same round of dull stories, to hear and answer the same questions, to amuse, in short, a thousand idle people, who, wishing to astound their neighbours by the wonderful extent of their knowledge, would put ten queries in a breath and recommence before the first was answered, only that they might captivate the weak and dull brains of the gaping listeners by the exuberance and rapid flow of their ideas, however foolish those might be. This kind of thing has tormented me almost to death, and Berlin swarms with people of this description. . . . . Of the fair sex I dare not even think. Were women ever entirely corrupt, they are so at Berlin, where self-love— that is to say, coquetry—is as universal as in Paris, where the

---

[1] 'Sämmtliche Schriften' (Leipzig, 1843), vol. vii. p. 112.

tone of good society is attuned to dull witticisms, insipid compliments, and the unceasing fabrication of so-called "*jolis riens*,"
where there is no thought, and, except in the greatest voluptuousness, no feeling. This is the state of society from court
circles down to the level of common life.' . . . .

There is no doubt, however, that during this visit to Berlin,
the private relationships of Forster had placed him in a most
uncomfortable and painful position. The applications he
had to make on his father's behalf compelled him, as he says
himself in the same letter, to mix in too many different circles,
and since 'the Berlinese exact from a stranger in everything
a pliability of character whereby he is rendered either a fool or
a knave,' he was frequently compelled to exercise a severe constraint upon himself, in order that no hindrance should arise
to the accomplishment of his father's plans.

Goethe also, who about the same time, in May 1778, was a
guest in the suite of his prince at the court of Berlin, complains
bitterly of the ' corrupt brood,' and of being forced ' to hear
the great king abused by his own miserable curs.'

Scarcely had King Frederick passed away (August 17, 1786),
when the fabric of the new enlightenment was completely broken
down, and all elements of corruption, all outgrowths of contracted official politics, of the overbearing patriotism of the bar-
rack-room, of political and ecclesiastical surveillance, of pietistic
hypocrisy and the deceptive mysticisms of illuminati, alchemists, and women of no reputation, of the censorship of the
press and restriction of speech, burst forth with audacious
effrontery, blunting and extinguishing every nobler aspiration.

King Frederick William II. was ambitious of being a *German* prince; he loved his native language, and was anxious that
it should be restored to universal use. Public buildings were
henceforth inscribed with German instead of Latin inscrip-
tions, and the lords and ladies of the court greeted one another
again with the simple ' Good morning' (' Guten Morgen '). In
striking contrast with the previous customs of the court, the
king and his suite attended divine service regularly; sometimes
at the Cathedral, to hear Sack, the preacher of the Reformed
Church; sometimes at the Marienkirche, to listen to Zöllner, a

Lutheran clergyman ; occasionally even at the Roman Catholic chapel, to hear the Bishop of Culm or the French preachers Ancillon, Erman, Dupasquet, and Reclam, but most frequently at the small Hospital Church, attracted by the preaching of Ambrosius, where the great world soon came in such crowds that no room could be found for the sisterhood in attendance at the institution.   This pious church-going, however, was accompanied, as is well known, by an extreme laxity of morals ; from ' gallant Saxony,' from the corrupt court of King Augustus, came many seductive syrens and abandoned women to the court of Frederick William, bringing with them the worst practices of heathen times.

A caustic description of the state of Berlin at this period is given by George Forster in the following letter to Sömmering, bearing date March 16, 1788 : [1]—' I could hold no converse with the all-powerful people, as you call them, much less even attempt to fathom them, without disclaiming the character of an honest man.   Had I found them to be a people who, like Cicero's augurs, laughed over the mysteries of their own trade, it might have been possible for me to have associated with them.   But to play the hypocrite, and call that high and honourable which I could never regard as such, is to me an impossibility.'   After a severe critique of Wöllner, Bischofswerder, Theden, and others, he says :—' What can be expected from such men ? '

And not only foreigners, but even residents describe the state of Berlin at that time as being in the highest degree sad and depressing.   On October 27, 1788, Professor Fischer, who has been already mentioned as one of the tutors of the two Humboldts, writes on this subject to the most distinguished mathematician of the day, Johann Friedrich Pfaff, at Helmstädt, as follows :—' Alas ! alas ! many and sad changes have taken place at Berlin since you left it.   I keep hoping, however, that the fermentation agitating the public mind will in the end, notwithstanding all counter-pressure, tend to further the progress of enlightenment ; for such a condition of things obliges all the friends of truth actively to bestir themselves.

[1] Wagner, ' Leben und Wirken Sömmering's,' vol. i. p. 266.

The *ecclesia triumphans* or *triumphare cupiens* has, meanwhile, to contend with all her might against a powerful opposition, and is occasionally defeated in her most cherished plans, as, for instance, in regard to a civil edict requiring that all enactments on the subject of religion should, whenever practicable, be issued by the crown. It is confidently reported that a new edict for the restriction of the liberty of the press has been lately rejected almost unanimously by the Council of State, two ministers only voting in its favour.[1] It may be regarded as an open question whether a victory of this kind is in reality a triumph of truth. For it is maintained by some that many of these liberal acts only originate from motives of financial policy, as, for instance, if the peasantry are not allowed to dance on a Sunday, the revenue from the music tax will be diminished, and so on. . . . Silberschlag[2] has recently delivered some lectures on the sun at the Academy of Sciences. The result of his reasoning, supposed to be incontrovertible, is as follows : —The sun is really a kitchen fire, and the spots are clouds of smoke and great heaps of soot ; consequently, where there is a kitchen fire, there must be meat to roast, such as godless people, Deists, Universalists, and Atheists, and the devil is the cook who turns the spit.'

Still greater absurdities had been committed at the Academy during the previous year. In 1787, the same year in which Humboldt matriculated at the University of Frankfort, Semler communicated to the Academy his discovery that gold was formed in a certain volatile salt when kept in a moist and warm condition. Klaproth tested this salt by order of the Academy, and actually found in it a small piece of gold-leaf— placed there by Semler's servant in order to cheer his credulous master in his labours.

Such was the intellectual and moral atmosphere of Berlin at the time when the two Humboldts were entering man's estate, and were therefore of an age to be most easily influenced by external

[1] The edict on religion was dated July 9, 1788 ; the edict on the censorship of the press December 19, 1788.
[2] Johann Esaias Silberschlag was principal preacher at the Church of the Trinity, Councillor of the Upper Consistory, and Privy Councillor of the Royal Commission for Public Buildings.

circumstances. What attraction, what stimulus could such a state of things offer to their gifted natures? To minds already fired by aspirations after a new world of thought, what congeniality could be found in society of this kind, where, notwithstanding the boast of high cultivation, Lessing was stigmatised as an innovator and a free-thinker?

There were but few, and those the disciples of Lessing and Kant, by whom the oriflamme of a higher intellectual life was unfurled; among these were Engel, Biester, Sack, Teller, Spalding, Meier-Otto, Mendelssohn, David Friedländer, Marcus Herz, and Zöllner, and a peculiar charm was lent to this circle from the presence of those gifted women by whom the tone of thought was powerfully influenced. Among these distinguished ladies may be mentioned the two daughters of Mendelssohn, the romantic enthusiast Dorothea Schlegel, and her sister Henriette Mendelssohn, to whom in later years was entrusted the education of the unfortunate Duchesse de Praslin —Fräulein von Briest, afterwards Frau von Rochow, and subsequently Madame de Fouqué—Henriette Herz, the friend of Schleiermacher and the two Humboldts, her sister Brenna, and the Sibyl Rahel, who was distinguished for peculiar mental acuteness and a subtilty of intellect worthy of Aristotle.

The brilliant circle that in after years gathered round Rahel, and formed an historical element in the intellectual society of Berlin, found its prototype in the select assemblage meeting at the house of the Jewish physician Marcus Herz, first attracted there by a course of lectures, on physics and philosophy, which he, an ardent disciple of Kant, had commenced at the advanced age of eighty years. The lectures on physics were exceedingly popular from the admirable experiments, remarkably elaborate for that time, with which they were illustrated.

It was through these lectures that, in the year 1785, William and Alexander von Humboldt first made the acquaintance of Herz and his household, and this acquaintance was further matured during several interviews that took place in reference to the erection of a lightning-conductor at Tegel, an appliance not yet in general use at Berlin.[1]

---

[1] The first lightning-conductors used in Berlin were erected in 1777, on the Royal Arsenal and on the barracks of the Pfuel Regiment at the

This occasional intercourse soon ripened into friendship of the closest intimacy, and the Humboldts were freely admitted to the small but interesting circle assembled at the house of Herz. In subsequent letters Alexander addresses Herz in terms of great affection, as his 'fatherly friend,' his 'dear teacher,' and, with feelings of gratitude and the modesty becoming a pupil, recounts to him the progress of his studies, while to the beautiful and gifted Frau Herz he not unfrequently wrote 'dreadfully long letters' in English, in order to merit her commendation for his industry. At Herz's house, the Humboldts made the acquaintance of Veit and Beer, a young physician with whom they formed an intimate friendship, and they here again encountered their friends Joseph and Nathan Mendelssohn, formerly their fellow-pupils under Fischer.

Owing to the high culture in which they had ever been nurtured, the two brothers, at the respective ages of sixteen and eighteen years, were, as described by Henriette Herz, 'vivacious and intelligent, and of distinguished manners, possessed of extensive information, and in every way estimable.' In the interest they felt in their fair companion there mingled, no doubt, some tinge of admiration for her great personal charms. Henriette Herz was the most noted beauty in Berlin. The universal homage she inspired is shown in the expression, which passed into a proverb, 'Whoever has not seen the Gendarmes Platz and Madame Herz, has not seen Berlin.' How far this feeling was experienced by William von Humboldt is evinced in his letters written while he was at the University, which were published in Varnhagen's 'Remains.'

Amidst his numerous occupations, the claims of society were not wholly forgotten, since we find that at this time Alexander bore the reputation of being a graceful dancer, and that he even instructed Frau Herz in the new 'Minuet à la Reine.' He also manifested a spirit of gallantry, and showed much keensightedness in reading the language of the affections, concerning which he often held opinions at variance with those around him. In the reminiscences of a lady who met him some years later [1] at

Köpniker Gate, according to the plans of Professor Sulzer and Privy-Councillor Gerhard.

[1] Frau Ilgen, wife of the well-known director of the educational institute at Schulpforta. See Laube, 'Moderne Charaktere,' vol. i. p. 366.

Jena, where he was often on a visit to his brother William, he is described as 'a wit, a diplomat, a philosopher, always busy with electrical machines and galvanic batteries,' and 'an amiable, good-looking man, undoubtedly the handsomer of the two brothers.'

It will be desirable here to direct particular attention to the fact that almost all the personages referred to in this review of Berlin were Jews; and it is worthy of notice that the Jewish element early formed an important constituent in the intellectual society at Berlin, and that, especially at the time now spoken of, the new phase of thought originated by Lessing was chiefly received among Jewish circles.

It is related by Henriette Herz in her biography, that, on account of the constraining influences prevailing in the Christian circles of the middle classes, men of thought gathered by preference around the centres of Jewish society.

While the men became earnest disciples of a severe school of philosophy, the women, with *naïve* originality and the fiery zeal of an eastern nature, threw themselves into the study of poetic literature, by which their youthful hearts were set in violent agitation and their souls inspired with hatred to all that was pedantic and obsolete. They read with delight the best works of French, English, and Italian writers, they admired the profound genius of Goethe, and indulged in sentimentality with Werther, they sympathised with the jubilant strains of Schiller, but their enthusiastic feeling reached a climax in their adoration of Lessing.

And as Lessing, the object of their highest admiration, had burst the bonds that oppressed the world of literature, and freed it from all conventionalities and the irksome restraints of established rules, so did this younger generation, in emulation of the spirit of their leader, seek to dissipate the depressing atmosphere that enveloped the social world around them, and banish from their social life the chilling influences which empty traditions and dead formalities had spread over them.

It appears that to Alexander von Humboldt, among others, this kind of society proved very congenial and attractive; for Henriette Herz writes of him as follows :—

'In those days, whenever Alexander von Humboldt wrote to

me or any other member of our intimate circle from the family
seat at Tegel, he usually dated his letter, "Castle of Boredom"
[Schloss Langweil]. This occurred chiefly in the letters he
wrote in the Hebrew character, in which I had given him and
his brother some instruction, and which, by the additional help
of other friends, they wrote very successfully. It was not to be
thought of, that a young nobleman should confess in letters
which could be read by everyone, that the society of Jewish
ladies was more entertaining to him than a visit to the mansion
of his ancestors.'

CHAPTER II.

COLLEGE LIFE.

The University of Frankfort—Studies in Finance and Philology—The
Winter of 1788 at Berlin — The University of Göttingen — Short
Journeys—'Observations on some Basalts of the Rhine '—Journey with
George Forster—School of Commerce at Hamburg—Application for
Official Employment—School of Mines at Freiberg.

ON the selfsame day, October 1, 1787, the two brothers,
William and Alexander von Humboldt, matriculated at the
Alma Viadrina, the University of Frankfort-on-the-Oder, at
that time under the rectorship of Johann Isaak Ludwig Causse,
Professor of Theology. In the college register the entry of
Alexander, who a fortnight before had completed his eighteenth
year, runs as follows :—

'Henricus Fridericus Alexander ab Humboldt, Berolinensis,
Cameralium Studiosus; pater meus jam mortuus est, mater
adhuc vivit ; domicilium Berolini.'

Although the youths were to be accompanied by Kunth, their
judicious and conscientious guardian, the University of Frankfort had been selected from the consideration, that they would
there be within reach of their mother's watchful care, and
that a comfortable home could be provided for them in the
house of Professor Löffler, their former tutor. The University,
however, though much frequented by the young nobles of
Pomerania and the Margraviate, was sadly deficient in auxilia-
ries for scientific instruction, there being no museum, no hall
of anatomy, no observatory nor botanic garden, and no well-
furnished library ; there was but one bookseller's shop, which
was very ill supplied, and a printing establishment exceedingly
inefficient.

It was his mother's wish that Alexander von Humboldt should devote himself to the study of finance, as a preparation for entering the service of the State.

The science of finance, or political economy, was at that time at its lowest ebb. The contempt in which it was held is evidenced by the expression 'He is studying finance' having become proverbial for any idler who was learning nothing. Beckmann, the most noted lecturer on political economy at Göttingen, brought before the consideration of the young students of finance in his principal course of lectures, a herbarium consisting of peas, onions, radishes, turnips, and other common vegetables. Nor were his lectures on mineralogy, technology, and manufactures much more instructive, for Leopold Krug,[1] as late as 1805, makes the following complaint:—'They learn to draw plans for a brandy-distillery, a tar-kiln, or a flour-mill, they learn the requisite number of threads in the warp and woof of linen and taffeta, they learn how to make cheese and to smelt iron, and how to destroy caterpillars and noxious insects ; but of the higher principles of political economy they have not the faintest conception.' Even so late as the year 1813 it was found desirable to issue an order in council, dated September 27, recommending that ' the students be disabused of the erroneous notion that the study of finance requires less strained application of the intellectual powers than that of theology, medicine, or jurisprudence.'

Among the professors connected with the University of Frankfort, there were none who exercised any permanent influence in the world of science. The names most worthy of notice are Schneider, known by his Greek and German lexicon, and Löffler, distinguished for his work on the Neo-Platonism of the early Fathers. Otto, the translator and editor of Buffon's works on natural history, was not till a later period professor at Frankfort.

There is but little to be gathered concerning the course of study followed by the two Humboldts : they attended few public lectures, and received most of their instruction privately —a plan which well suited their ability and zeal, and best corre-

---

[1] 'Betrachtungen über den Nationalreichthum des preussischen Staats' (Berlin, 1805), vol. i. Preface, p. 5.

sponded with their former method of study at Berlin. There they had been accustomed to finish exhaustively, in from six to eight weeks, a scientific course of study, which, according to the usual custom of the University, would scarcely be accomplished in a whole session. On this account William von Humboldt writes as follows to Henriette Herz [1] :—'It will be impossible for me to come to Berlin for Christmas, my dear friend. Kunth, as far as I can see, will not be going there, and I could hardly accomplish the journey alone. Besides, as we rarely attend the public lectures, our work goes on as usual during the holidays.' Other expressions connected with this period may here be given [2]:—' I sometimes wish I had my brother's temperament. It is true he complains of being dull, but on the whole he seems to amuse himself pretty well. He is for ever on the move, and continually joking; he is certainly never in the least sad, and says himself that he did not enjoy himself more at Berlin. You must not, however, suppose that he wastes all his time in this way: he is withal exceedingly industrious, and excels in many things. Moreover, we continue to live together as we used to do, and are always the best of friends, though rarely of the same mind; our characters differ too widely.' When at Göttingen, William wrote to Henriette Herz, [3] while Alexander remained in Berlin, as follows :—' People altogether misunderstand him, particularly when they suppose me to be so greatly his superior in ability and general knowledge. He is far more talented than I am, and considering that he is my junior, he possesses quite as much information, only it lies in other branches. Though he very often jokes me about you, this is partly to vex me, and partly because it is his way to joke everyone. Before others, he defends you with most energetic warmth. He has written me one of the most comical letters you can imagine ; it is commenced in Greek, continued in Latin, and concludes in German, with some Hebrew writing interspersed. What he has to say of you he writes in Greek, that Kunth may not understand it.' Again he writes to her

---

[1] ' Aus dem Nachlasse Varnhagen's. Briefe von Chamisso, Gneisenau, Haugwitz, W. von Humboldt, &c.,' vol. i. pp. 72, 79.

[2] Ibid. vol. i. p. 57.

[3] Ibid. vol. i. p. 98.

on this subject under date of February 14, 1789 :—' The news you give me of mon frère delights me. He is indeed a noble fellow, and will some day prove himself a useful man. Mischievous as he appears to be at times, he is really most excellent at heart. His chief failing is vanity, and a love of approbation, the cause of which lies in the fact that his nature has never been stirred by any deep or overwhelming interest.'

These traits of Alexander as thus sketched by William von Humboldt are clearly evinced in his own letters written about this time to his friends at Berlin, among whom must be reckoned David Friedländer, who has already been referred to in p. 23, and Beer, a medical student, who was an inmate in the household of Counsellor Herz, and who, in addition to his studies for the medical profession, devoted a good deal of time to philosophical subjects: eventually he practised as a physician at Glogau, where he died. After a fortnight's separation, Humboldt writes to Beer :—' I should have rejoiced to have heard from you sooner, and I would gladly have earlier fulfilled my promise of writing to you, if I had not been prevented by a thousand little hindrances. Now that we are somewhat settled, my dear friend, nothing can prevent my enjoying once more the pleasure of holding converse with you. Yet a letter is but a poor substitute for the enjoyment of personal intercourse, and the remembrance of an absent friend is always associated with a certain pang. But even in this pain of longing there lies concealed so sweet a feeling, that, without being sentimental, one is constrained to cherish it. Do not expect more from me to-day than these few lines. I dread every moment to hear the clock strike three, when we have to attend a lecture on jurisprudence ; so that I have only time to tell you that we are well, and, *quantum fieri potest*, leading a happy life. I shall write more another day. Remember me to the dear counsellor and his excellent wife, also to Veit, Levi, Herr Friedländer, and any others you may meet with who still bear me in remembrance.'

The following letter, written in November 1787, is of greater interest :—

' A thousand, thousand thanks, my dear friend, for the kind letter with which you lately gladdened me. I feel indeed half inclined to quarrel with you for believing so little in my excuses, only I cannot allow such a theme to engross my letter. Your kindness leads me to hope that you will have attributed my long (unintentional) silence rather to thoughtlessness than to any want of friendship. The Fair now going on here brings many people from Berlin, and of all my old acquaintances I have had most pleasure in meeting with Friedländer. We learnt from him that you are all well, and have not yet forgotten your absent friends. If you only knew how many unhappy hours have been sweetened to us here by the remembrance of you and the other kind friends whose society we enjoyed in your company! Whether I like this place, and whether I prefer my present position as a student to my former mode of life at Berlin, are questions which daily come before me, but to which as yet I can find no satisfactory answer. If it were not for the friendly intercourse we enjoy here so thoroughly, Frankfort would indeed be a dull place to both of us. It requires, however, but a very small amount of philosophy to be convinced that mankind is created for every spot upon the earth, consequently for the ice-bound bank of the Oder. What more exalted aim can the Goddess of Science (who certainly has no temple here) propose to herself than to make mankind contented! (I have been striving to appear before you in the character of a fine writer, but have not succeeded.)

'The number of students here is very small; there are only between 220 and 230, eight only of whom are studying medicine, while there is no university in Germany where so many Doctor's degrees in medicine are conferred. During the first five weeks of our residence here, no fewer than five candidates, of whom one only was a foreigner, disputed "ad summos in medicina honores legitime obtinendos." Such a manifestation of eagerness among physicians proves that good wares must be brought to market to attract so many customers. In Frankfort everything conspires to facilitate the acquisition of a Doctor's degree, since the disputation, if it can be so called, is nowhere easier. The præses must not merely write the disputation, but defend it in its simple acceptation. The respondents, who in most cases cannot put half-a-dozen words together in Latin, behave as if the objections of their opponents were quite irrelevant. They read out their compliments or addresses and patiently listen while the præses disputes with himself. As, however, it is quite possible to be a good physician without being able to speak Latin, I will not deny that among the new doctors there may not have been here and there a clever man. When we return to Berlin, I shall hope to bring you, my dear friend, a large store of disputations of this kind, which will at least possess some merit, since they are nearly all from the pen of Professor Hartmann. He is properly the lecturer on pathology, therapeutics, chemistry, and materiæ medicæ, but now that Meier is at Berlin, he constitutes the entire medical faculty. He is, moreover, a profound philologist and an agreeable Latin versifier. But enough and perhaps too much on these subjects. Remember me, as well as my brother and Herr Kunth, to the dear counsellor and his wife ; pray assure the latter that I shall not dare to write to her until I have received complete absolution, that is to say, until she can assure me that instead of feeling " a little vexed " with me, she is more than " a little pleased." She shall then have a dreadful letter in English from me. Remember me also to Veit, and Levi, and all my acquaintance.

' Ever yours,

' A. V. HUMBOLDT the younger.'

His correspondence with Beer seems to have been prosecuted with great activity, and was maintained at intervals even after his return from America. The following letters, belonging to this period, may here be introduced : they furnish evidence of the close friendship that existed between Kunth and his pupils, and were addressed to David Friedländer, who had been called suddenly away from Frankfort by a death in his family. The three letters are written on one sheet of paper, quarto size, and bear the same date, Frankfort, December 19, 1787.

The first letter is from William von Humboldt :—

' Sudden as was your departure from Frankfort, and much as it distressed me on every account, yet the receipt of your kind letter was almost as great a surprise to me, and, for more reasons than one, made me very happy. I ought indeed to have reckoned on this proof of your remembrance, but since doubts are so natural in a certain stage of love, why should they not also exist in a certain stage of friendship? I will not refer to the loss you have sustained, by which your heart has been so deeply wounded; but allow me to tell you how greatly I shall rejoice to see you comforted and cheerful again? That I do not apologise for the tardiness of my reply, you must consider to be more a token of bashfulness (for I have only been two months at Frankfort) than a want of politeness. Were excuses needed, I could find sufficient in the amount of work I have to do. Do not, however, say this to Engel; for by his kindness I have been entrusted for some time past with the task of reviewing for him the entire philosophic and scientific literature of this country, and if he were to hear that my increased labour did not permit me even to find a few minutes during several weeks in which to answer so kind a letter, it would perplex him to find me some other employment, and he would think himself obliged to create some new post for my benefit. Rather tell him something that will give him an idea of the contrary; he will then see that he is necessary to me, and on my return to Berlin will be all the more surely my devoted old friend. For the sake of the excellence of the motive, you must for once do violence to your love of truth. Do not forget at the same time to assure him of my grateful and unchangeable affection, and thus condone by a grand truth the evil of a slight untruth. I have received the " Mysterien " and " Guibert," and have read the former; many thanks for them: you shall have my opinion, if you care for it, when we meet. I have only left myself room to desire my affectionate remembrances to yourself and all your household, for the remainder of this sheet of paper has been appropriated by my brother and Herr Kunth.

‘ Farewell, my dearest friend. Ever yours,
‘ W. HUMBOLDT.’

Alexander von Humboldt then continues:—

‘ To be the youngest of the family has one disadvantage, which unfortunately intrudes itself even into this letter; for had the first page of this sheet fallen to my share, you would then have been able, my dear friend, according to the infinite progression of things (in the actual world, or only in the heads of philosophers ?), to rise from bad to better. But, as if my faults in style were not glaring enough, to give me the unfortunate middle place, and make use of me as a foil to bring out the light of surrounding objects, is quite as unwarrantable as your congratulating me upon my collection of satires. (I should be glad to dispense with collections of this sort, were it not that I am often asked on all sides for a contribution.) I do not tell you how sorry I was at your sudden departure and at the sad cause of your leaving, for I venture to trust from your friendship that you will credit me with these feelings. I ardently look forward to our next Fair, because one of my warmest wishes will then be gratified—the wish to express to you personally how I prize your affection above everything. Remember me to your family circle, and rest assured that my residence here, from whatever point of view it may be regarded by you, is only endurable to me from the thought that it is a necessary evil.

' Ever yours,

' HUMBOLDT the younger.

' P.S. The lyrical disorder which prevails in this letter must be ascribed in this instance not to me but to Herr Kunth, who stands behind, admonishing me to use despatch. I would rather be called stupid than a disorderly genius.'

Kunth concludes as follows :—

' Notwithstanding the modesty of which my predecessors, each in his own way, has said so much, they have left me so small a space on this sheet of paper that it stands in most unfair proportion to the desire I have to write to you and the time I have at my disposal. That you thereby may possibly be the gainer concerns me but little. Meanwhile I must cut my coat according to my cloth, and I shall be glad if you do not mistake greatly in thinking that this art of cutting is not wholly undeserving of merit, since one is after all at liberty to take a second sheet. Herewith I send you an acquittal *in forma probatissima* for Engel. If he should still entertain any doubt, I shall be happy to pay back the principal at the next

Fair. As for the interest, you will have to reckon it in the
pleasure this merry contest has afforded you. I think that in
this view of the matter you will be no loser. The bearer of
this letter is Herr Albinus, with whom you probably made
acquaintance during some of your walks here. He is companion to the young Count Dohna, and is a most worthy man.
He is taking advantage of the Christmas vacation to see Berlin.
If you can assist him in any way towards this object, or
render his stay more agreeable, pray do so—I will not say
because I ask this favour from you, but because he deserves it
so thoroughly. If you have anything further to tell me about
Engel, you may send me a few lines by him in return, for it
will be very long, unfortunately, before we meet again.

' Yours,

' KUNTH.'

Among the college friends of the Humboldts may be mentioned Count Alexander von Dohna-Schlobitten, who studied at
Frankfort between the years 1786 and 1788, and in 1808 was
appointed Minister of State in Prussia, at the same time as
William von Humboldt became Minister of Religion and Public
Instruction. They were also on friendly terms with Albinus,
the young count's companion, a well-informed, assiduous young
man. But of all his friends Alexander von Humboldt became
most strongly attached to Wegener, a young theologian, who
died in 1837 as superintendent of the Lutheran Church at
Züllichau. The letters still extant, addressed to him by Humboldt between the years 1788 and 1790, express the warmest
and most devoted friendship. They give evidence of the
nobleness and genuineness of the youth's nature, of his zeal for
science and love of knowledge, and throw a gleam of light upon
the events and persons of that time, the reflection of which
illuminates our own day. In these letters, which furnish
valuable details for portraying the history of the next two
years, are to be found the names of several other friends, Metzner, Keverberg, Herzberg, Bertram, Sartorius, and Fickert,
concerning whom nothing further is known. ' Could I only
look forward to the future with as much pleasure as I can
reflect upon the past,' writes Humboldt to this friend from

Berlin. 'The happy days spent at Frankfort are over. I can never think without emotion of that place ; we can never again enjoy such intimate communion. Yet who knows what pleasures may not be awaiting us ? . . . . If God only spare us, nothing can break the bond between two friends who are to each other more than brothers.' And again later :—' How quickly last winter passed away ; how long will this one seem ! How happy we used to be, talking for hours together in your old tattered chair by the stove. No day passed without our seeing each other once or twice at least during its course. How completely has the past disappeared ; where are all our old friends ! —Albinus in Silesia, Metzner and you in the New Mark, Herzberg in Halle, I in Berlin, and soon to be still farther away ! '

Although the instruction furnished by the University of Frankfort was of such a character as to fulfil the requirements at that time needed for the service of the State, and could thus accomplish the principal object contemplated by Frau von Humboldt in the choice of this university for her sons, yet it offered no facilities for satisfying their thirst for scientific knowledge. The youths remained, therefore, scarcely more than six months at Frankfort, and in Easter 1788, William went to the University of Göttingen, while Alexander returned to Berlin.

In his short autobiographical sketch,[1] Alexander von Humboldt remarks that he passed the next summer and the following winter at Berlin, 'in order to study the technology of manufactures, and to apply himself, in emulation of his industrious brother, more assiduously to Greek. At this time Humboldt formed an intimate friendship with Willldenow, already a distinguished botanist though young in years, and soon became an ardent student of cryptogamia.' During this residence at Berlin Humboldt maintained his connection with Frankfort, which he frequently visited, attracted apparently by sympathy with Reitemeier, who had just completed his prize essay on the ' History of the Mines and Smelting Works of the Ancients,' and was then publishing his ' Analecta ad historiam rei metall. veterum.' To his intimacy with Reitemeier may probably be ascribed the interest manifested by Humboldt in

---

[1] Brockhaus' ' Conversations-Lexikon' (10th ed.) ; ' Gegenwart ' (1853), vol. viii.

this branch of classic philology, for several of his early writings
are on kindred subjects, such as ' On the Basalt of ancient and
modern Authors ; ' ' On the Syenite of the Ancients ; ' ' On the
Basalt of Pliny and the Columnar Rocks of Strabo,' which were
incorporated into his work, ' Mineralogical Observations on
some Basalts of the Rhine.'

Concerning the course of his studies, even at this period,
nothing definite can be ascertained. The only information
on the subject is to be gathered from his letters to his friend
Wegener. In the beginning of May he tells him that he was receiving instructions in Greek from Bartholdi, and ' was still
struggling with prophets and otters [1] in the first declension.'
Yet within a month, in order to save his grammatical reputation from the charge of an error in declension, he writes
him a formal letter in Greek, without accents, but with this
preface :—' I must freely confess that I fear you will not understand a syllable of all that I have written, and then I shall
have to say with Sancho Panza : " Your worship does not
understand me ? No matter. God who knows all things
understands me. " '

The letter concludes in German :—' Thus much, and possibly
already too much for to-day, my dear friend. From these
few lines, which I have put together without a master (therefore the faults must all rest upon my own shoulders), you will
be able to judge of my progress in Greek. I work hard, and
continue to study it with pleasure. If I could only have
followed my own inclination, and had not always to contend
against a disagreeable mental constraint—a something that was
impossible to define, although I was conscious of its powerful influence—I should long since have prosecuted this study with
greater energy. The more I know of the Greek language, the
more am I confirmed in my preconceived opinion that it is the
true foundation for all the higher branches of learning. It was
certainly very ill contrived of me to build my house on mere
sand ; yet the foundations of so temporary a structure as mine
may easily be relaid, and therefore it does not distress me
that I am only learning to decline ἔχιδνα in my nineteenth
year.

[1] Προφήτης, ἔχιδνα.

'Engel was explaining to me the other day, in an admirable manner, how we should have been several centuries farther advanced in philosophy, enlightenment, taste, the arts, &c., if, in the great struggle between the bishops of the Greek and Latin Churches, the Western Church had not established its supremacy and introduced, in consequence, Latin instead of Greek literature, the copy instead of the original, an indifferent language in place of the most perfect and complete of all languages. What a vast amount of good must Rome accomplish before she can repair the mischief wrought by that Roman bishop!'

The theme on which Wegener was engaged in preparation for his examination, ' On the Gift of Tongues in Apostolic Times,' gave Humboldt an opportunity of expressing his views upon theological subjects. Whatever may be the value of these sentiments, it is at all events of great interest to notice the variety of his knowledge and the philosophical arrangement of his thoughts—a consequence, no doubt, of ' the logical education given by the Berlin professors.'

It was during this sojourn at Berlin that Humboldt commenced the study of botany; his remarks to Pictet on this subject have already been given in p. 26, but some additional particulars are furnished by the following letter to Wegener at Frankfort :—

'February 25 (1789).

'I have just come in from a solitary walk in the Thiergarten, where I have been seeking for mosses, lichens, and fungi, which are just now in perfection. How sad to wander about alone! And yet there is something attractive in this solitude when occupied with nature. To be in the complete enjoyment of the purest and most innocent pleasure, surrounded by thousands of creatures rejoicing in the mere fact of existence (precious thought of the philosophy of Leibnitz !), and with a heart lifted up to Him who, in the language of Petrarch, " muove le stelle e loro viaggio torto, e da vita alle erbe, ai musci, alle pietre " (*sic*) .... are circumstances, my dear friend, which tinge the reflections they inspire with a certain delicious melancholy! My friend Willdenow is the only one here who sympathises with me in these pursuits; but we are both so closely occupied that it is only occasionally that we are able to wander together, hand in hand, through the vast temple of nature. Will you believe that out of the 145,000 people in Berlin, there are scarcely four, besides ourselves, who cultivate this branch of physical science even as a secondary study, or as a recreation? And yet, in the ordinary exercise of their professional duties, ought not physicians, and especially all political economists, to be incited to follow such pursuits?—for the greater the increase in population, the higher the consequent rise in the price of provisions, and the heavier the burden which a corrupt system of finance must impose upon the people, the more important is it that fresh means of sustenance should be sought out to meet the ever-increasing deficiency. How many powers exist in nature of which no use is made, because they have hitherto been unrecognised, but which, could they only be developed, might furnish food and employment for thousands. Numbers of the very products which we bring from the farthest quarters of the globe are trodden beneath the foot in our own land; in course of centuries some accident reveals them, when by another chance the discovery is either lost, or, as much more rarely happens, is developed and widely propagated. Botany is regarded by most people as a science of very little use to any but members of the medical profession, except as an amusement, or, possibly, as a means for the subjective training of the mind—a use which is obvious only to a few, while to me it appears to be one of those studies from which the greatest benefit may be anticipated to the whole human race. What a mistake to suppose that the half-dozen plants which we cultivate (I say the half-dozen in opposition to the 20,000 different specimens covering the globe) contain all the powers with which the vegetable kingdom has been endowed by a beneficent providence for the satisfaction of our wants. It seems to me that similar misconceptions pervade the human mind in every department of thought: men think they have reached the truth in every direction, and they imagine that there is nothing left for them to improve, and nothing for them to discover. They shun investigation from the belief that everything is already investigated. The same spirit is apparent in religion, in politics, and in all subjects wherein the common herd are associated. My remarks upon botany are not, however, founded solely upon à *priori* conclusions; on the contrary, these thoughts have been awakened in me by the great discoveries I have found buried in the writings of the earliest botanists, and which have been confirmed in modern times by expert chemists and technologists. Of what avail is any discovery if there exist no means for its propagation? But I crave forgiveness, my dear friend, for wearying you with subjects in which you can have but little interest. They are of the highest importance to me, since I am collecting materials for a work on the various properties of plants, medicinal properties excepted; it is a work requiring such great research, and such a profound knowledge of botany, as to be far beyond my unassisted powers, and I am therefore endeavouring to enlist the co-operation of several of my friends. Thus far I have been working at it only for amusement, and in the course of my researches I am constantly meeting with things which (to speak familiarly) set me gaping with astonishment. I will send you further particulars of all my plans in course of time. Pray do not imagine that I am going to appear as an author forthwith; I do not intend that shall happen for the next ten years, and by that time I trust I shall have discovered something startlingly new and important.'

Soon afterwards he writes again as follows:—'I am delighted to hear that you are going to study botany in your hours of recreation. In your present loneliness you will find no more fascinating pursuit, nor one more capable of yielding you a pure and inexpensive enjoyment. Without talking sentiment, it may be said that plants become our friends, and among them we naturally find some more worthy of notice than others. Our walks are no longer solitary, since we are surrounded by the acquaintances we have made. What a pleasure it is when we occasionally meet several of our favourites together. The most insignificant places become invested with a peculiar interest, because we here first discovered some plant which had hitherto been unknown to us, because we there missed the flowers which had but lately been in bloom. The study of botany is a noble occupation for clergymen. An Englishman of the name of Ray (Raius), holding priest's orders in the

Church, was one of the greatest botanists of former times.
Jacquin, the distinguished botanist of Vienna, who is celebrated
for his travels in Jamaica and South America, published when
in his seventieth year a compendium of botany which is a complete and magnificent work. I venture to ask your acceptance
of a copy of this book, as a remembrance ; pray do not attempt
its perusal till the trees are in leaf, for the investigation of the
elements of botany indoors, without the opportunity of instituting a direct comparison with nature, is a dry and most wearisome undertaking.'

To this strong love of botany, his favourite pursuit at this
time, is due the publication of his first almost unknown literary
work, an anonymous treatise in French entitled, ' Sur le Bohun
Upas, par un jeune Gentilhomme de Berlin.' [1] It afterwards
escaped the memory of Humboldt, and only when his attention was drawn to the quotations he had made from it in
various publications (Crell's ' Chem. Ann.' (1795), vol. ii. p.
106, note ; his ' Subterranean Gases,' p. 376, note ; his ' Experiments in Galvanism,' vol. ii. p. 141, note), in which he expressly
refers to this treatise as his own, did he remember it and make
the remark that it was a translation of Thunberg's treatise,
' De Arbore Macassariensi,' [2] undertaken as a French exercise
for his tutor M. Le Bauld de Nans. The numerous notes added
to the work prove him to have been already possessed of
valuable stores of knowledge, and give evidence of that keen
observation by which his later works are so pre-eminently distinguished.[3]

It may well occasion some surprise to find the young finan-
cier attending lectures on technology from the lips of an
ecclesiastical dignitary, Zöllner, Provost and Counsellor of
the Consistory. After describing to Wegener Zöllner's versatility of talent and extensive theological acquirements, he
continues :—' It is a gross falsehood (and this you may boldly
maintain before everyone) that Zöllner possesses only a super-

[1] ' Gazette littér. de Berlin,' Nos. 1270 and 1271, for January 5 and 12,
1789.
[2] [The Upas-tree, or poison-tree of Macassar, in the island of Celebes.]
[3] This Upas-tree was again called to remembrance by the description of
the Manzanilla-tree in Meyerbeer's ' Afrikanerin.'

ficial acquaintance with the multifarious subjects he treats
of. I have had a good opportunity of judging of his scientific attainments during a course of lectures on technology,
which, as delivered by Zöllner, are well worth the cost of
100 ducats, for they are replete with information on a great
variety of subjects, such as mechanics, hydraulics, botany,
physics, chemistry, medicine, mineralogy, &c. Biester remarked very truly of him the other day: "What does Zöllner
not know?" His knowledge of medicine is so extensive that
some time ago he was desirous of undertaking a course of lectures on anatomy. I heard this from some of the physicians
here.'

Humboldt continued to cultivate the art of drawing in all its
branches. Besides free-hand drawing, he practised plan-draw-
ing, designing, and engineer-drawing, as well as the art of
etching. Among other specimens in our possession are two
heads ten inches by seven: the one inscribed ' Raphael pinx.'
' A. v. Humboldt fec. aqua forti, 1788,' is a study from the
' School of Athens,' and is the head of a figure in the right-
hand lower group at the extreme left of the picture; the
other represents the half-length figure of a man with a beard,
clothed in rich drapery and wearing a turban, and bears the
inscription ' Rembrant pinx.' ' A. v. Humboldt fec. aqua
forti, 1788.' The work of another hand is visible in many
places in this etching, and though neither of the heads possess
any especial merit as works of art, they are nevertheless of
undoubted interest as being the work of Humboldt when a
youth.

Under the tuition of Fischer, Humboldt had made considerable progress in applied mathematics, and he now devoted himself with marked success to the study of the practical sciences.
After remarking of his brother, then at Göttingen, ' he is
killing himself with study, has already read the whole of
Kant's works, and lives and moves in his system,' he continues :
' I expect to learn a great deal from him, for I have no time at
present to think of such subjects. I am so busily occupied
with practical matters, that speculation must needs be laid on
the shelf.'

These multifarious occupations prevented him attending the lectures delivered by Moritz on the Æsthetic in Art— lectures which at that time were setting the whole of Berlin in a furore for æsthetics—and he relates to Wegener in March 1789:—'Moritz commenced his course of lectures about three weeks ago, in the hall of the Academy of Arts, and they excite universal applause. Among his audience may be counted from fifteen to twenty of the most distinguished ladies of Berlin, and the minister Heinitz, Count Neale, and numbers of the court officials attend with the greatest regularity, never missing a lecture. They are certainly the most brilliant lectures that have ever been delivered in Germany. I attended one of them. His delivery is dignified and fluent, but somewhat too declamatory. As for the matter of the discourse, it was a strange mixture of brilliant errors, . . . for example: "One being passes into another; an inferior organisation, by becoming absorbed into a superior one, is raised to perfection. Animals live upon grass, mankind live upon animals; therefore vegetable life must rise gradually through animal life till it attain the highest form in human life." On hearing these words, one of the marshals of the court exclaimed, "Il est sublime!" What a mixture of materialism and monadology![1] An actual feast of monads! Again: "Nature created man in order to see her own perfections in him." But, notwithstanding all, he manifests much acuteness of intellect and some real flashes of genius. Little as I felt disposed to approve his propositions on the beautiful, I listened to Moritz with pleasure. His eloquence is fascinating, and is the secret of his popularity.'

In the active correspondence that ensued between Humboldt and Wegener, reference could not fail to be made occasionally to the events occurring in the capital, especially when those events were of a nature to mark the history of the advancement of civilisation. On one of these occasions he thus writes:—

---

[1] [The science of monads: an important part of the philosophy of Leibnitz, who taught that all bodies are compounded of certain primary constituents, to which, as simple substances without parts, he gave the name of monads.]

'Monday, September 29 (1789).

'The day before yesterday all Berlin was astir. To see Blanchard make his ascent was well worth the entrance-fee of two thalers. The sight of the enormous machine—twenty-six feet in width—the sight of the man who, with superhuman boldness, dared in such a contrivance to cross the ocean, the majestic movement of the balloon as it soared aloft, but chiefly the thought of the rapid march of human civilisation, whereby the third element had thus been brought into subjugation—all conspired to produce upon my mind a most powerful impression that stirred me to the heart. The balloon could be watched in its course for more than half an hour. It fell beyond Französisch-Buchholz. The parachute with the grappling-irons was a wonderful success; it was sixteen minutes in coming to the ground. But all this will be published in the newspapers. Blanchard is by no means the mean character he has been represented. I have met him at dinner at Herz's. He boasts but little of his achievements, and Herz assures me that he has a very fair acquaintance with physical science. He is certainly conceited, but in his position who would not be so? The stories told of him, that he gambles and has three wives, are complete fabrications; they are nevertheless believed by everyone in Berlin. This is the result of our intercourse with the philosophic Mirabeau! Madame Blanchard is a woman of refinement and of exemplary conduct, with whom I am well acquainted. Her husband has received most lavish presents, not only from the king, who gave him a snuff-box set with brilliants and 400 gold Fredericks, but also from many persons of the highest rank.'[1] . . .

It is certain that no discovery of a purely scientific nature had in the memory of man so completely aroused public attention as the invention of balloons. Great results were expected to ensue, not only in the interests of science, but in the furtherance of commerce and the useful arts, and all ranks in

[1] A munificence equally extraordinary was shown to the English physician Dr. Brown on the occasion of the inoculation of the prince. He received an autograph letter of thanks from the king, the title of Privy-Counsellor, a gift of 10,000 thalers, and a yearly pension of 600 thalers. (See the 'Berlin Almanac' for the year 1847.)

Berlin shared the enthusiasm which the new discovery had awakened. Blanchard was the most fêted hero of the day. The ascent was made an imposing spectacle, and the parade ground, at that time in front of the Brandenburg gate, was selected as the place of exhibition, where 2,000 troops of the garrison were stationed to keep the ground. The whole court was present. The queen and the princesses wore bonnets à la Blanchard, which were in the form of a balloon, with a parachute hung on the right side and a car with banners on the left. A royal carriage drawn by six horses conveyed the aeronaut back from the place where he alighted, the troops saluted him with the highest military honours, the public greeted him at the theatre with shouts of applause, and the king distinguished him by marks of personal favour. Besides the munificent gifts showered upon him by royal personages, he realised the sum of 12,000 thalers as the proceeds of his exhibitions.

The last leaf of the foregoing letter appears to have been purposely torn off, but from a small piece still remaining we gather that it contained some reference to Wöllner and his edict. Fortunately the following passage in another letter has been preserved :—' Now that Zöllner has a seat in the Consistory, I learn something of what is going on. You have no doubt heard the result of the debates upon the examination of candidates. With your extensive learning you have no cause for anxiety in the matter, but I confess I am alarmed for such men as Albanus, Köhler, Schüz, Israel, &c. . . . Wöllner's first proposition was that the examinations should be conducted throughout in Latin, and this is the principal alteration. Whoever, therefore, cannot speak Latin must be rejected. Wöllner insists further on being present at each examination, which so far (his zeal will soon wear itself out!) he has actually accomplished. Both of these regulations appear to me to be very injudicious. Many men may possess a good knowledge of Latin and yet be totally unable to speak it, and *vice versâ*. Of what use will it be to a country parson to be able to speak Latin? Would not the aim the examiners have in view be more successfully accomplished by deciding as to the usefulness of the intellectual gifts of the future preacher than by catechising him on the whole range of Christian ter-

minology, which is only another name for the trash of a dogmatic theology? All the candidates are filled with apprehension, for but few are competent to conduct a disputation
in Latin! And they have to be interrogated as to definitions, arrangement of subjects. . . . Another proposition of
Wöllner's was that every candidate should be rejected who was
not acquainted with Hebrew. This, I am thankful to say, has
been modified by the regulation that ignorance of Hebrew shall
only be made a ground of rejection in the case of those
candidates who are but ill prepared in other subjects, and
in this state it has passed. Lastly, a new form to serve as
a model for all examiners is to be drawn up, which is unfortunately to be made of universal application in the various sub-
consistories. The subject and the method of examination are
both to be prescribed. If this regulation should pass, it will
be drawn up by Zöllner, who will certainly see that it is not
too strict. What a mercy that it will not fall into the hands
of Silberschlag!'

On the restrictions that were imposed through the censorship of the press Humboldt remarks :—' Unger was fined ten
thalers the other day for printing a short poem—a wedding
ode—without licence. And lately, on the occasion of the
marriage of the Countess Lottum, I could not get the most
innocent of couplets printed on a pair of garters,[1] without
having the garters laid before the Court of Censorship. This
is my own experience!!!!! All that you have heard about
Wöllner's disgrace is a gross falsehood. The gossips here make
out that he is in and out of favour twice a day. He has really
nothing to fear.'

The last letter he wrote during this visit to Berlin is a farewell letter to Wegener, and is dated March 1789 :—

' My dearest Friend,—How can I find words to express the
pleasure your last letter gave me! The more I know of you,
the dearer you are to my heart ; the farther I am separated
from you, the more intensely do I long for your presence.
Those happy Frankfort days are gone for ever, for such
happiness can never be again. Nevertheless my fervent love

---

[1] In those days a customary wedding present.

and sincere friendship for you are as imperishable as the soul which gives them birth! I am now returning to my former career. My college life begins afresh. But my whole position is changed. I am now to take my first step in the world alone, and as a free agent; and critical as the circumstances may appear to be, I rejoice in the condition. After being kept as a child in leading-strings, man longs to set his pent-up powers in action according to the movement of his own will, and become as a free agent the architect of his personal happiness or misery. I contemplate my new position, however, with modest confidence. Limited as the sphere of my previous life appears to have been, I have yet found many opportunities for the study of human nature in observing the men by whom I have been surrounded. I am persuaded that no strong passion will ever sway me with an overwhelming power. Serious occupation and the calm induced by an absorbing study of nature will preserve me from the temptations of life. Of all my friends, my dear Wegener, you know me the best; therefore you can yourself pronounce a judgment as to whether I am strong enough to walk alone in this world's slippery path. How happy, how inexpressibly happy should I be, if I had a friend like you by my side! Göttingen is a perfect desert to me. I know I shall find there plenty of acquaintances; but my brother, Dohna, Stieglitz, Mecklenburg, Bing, and—oh! miserable me!—Keverberg—quod Dii avertant malum!—are all people with whom I have no sympathy. I doubt not that among 800 men there must be some with whom I could form a friendship, but how long is it often before we find each other out! Were not you and I acquainted for three months before we discovered how completely we were made one for the other? To be without a friend—what an existence! And where can I hope to find a friend whom I could place by your side in my affections!

'My departure is fixed for the 8th of April, and I have planned my journey through Magdeburg, Helmstädt, Brunswick, and Nordheim. Although I shall stop at each of these places, and remain for some days at Brunswick, where I must appear at court, I doubt if I shall be able to write to you on the journey. In any case, it will be my first business on reaching Göttingen to give you news of my arrival.'

In those days it was customary for travellers to expend much more time and pains than is usual now in seeking the acquaintance of distinguished men, and this was one of the motives that influenced Humboldt in the selection of this route to Göttingen. It will not, therefore, be out of place to insert here the letter of introduction from Professor Fischer, which Humboldt, on his way to the University of Göttingen, delivered to Johann Friedrich Pfaff, professor at the University of Helmstädt, and the most distinguished mathematician at that time in Germany: [1]—

'The bearer of this letter,' writes Fischer on April 5, 1789, 'is Herr von Humboldt, the younger of two brothers, in whose education, both as regards mathematics and the study of the ancient languages, I have for many years taken a not unimportant part. You may possibly recall hearing me mention the name when you were at Berlin. The elder brother is already at Göttingen, and the younger one is on his way to join him. He wishes to make your acquaintance, and I venture to think that the acquaintance will prove an agreeable one also to yourself. The two brothers are highly gifted both in head and heart, and have received an excellent education; therefore not one according to the present fashion of these times. The younger one is a student of political economy, in the various branches of which he has attained considerable proficiency. Had he been able to devote his attention exclusively or even partially to mathematics, I am convinced he would have become a distinguished mathematician; but I trust he will find the knowledge he already possesses of the subject sufficient for all practical purposes. I lose in him not only a good pupil, but also a friend whose society I shall miss.' . . . .

How highly this acquaintance was appreciated on both sides is apparent from the friendly and confidential letter which Humboldt, during the first few weeks of his residence at Göttingen, addressed to Pfaff.[2] The letter, which bears date May 11, 1789, contains interesting comments upon some of his studies,

---

[1] 'Sammlung von Briefen gewechselt zwischen Johann Friedrich Pfaff, u. s. w., herausgegeben von Dr. Karl Pfaff' (Leipzig, 1853), p. 170.

[2] Ibid. p. 231.

and may with propriety be inserted here, although somewhat interrupting the chronological order of events :—

'I am greatly indebted to you for your letter to Kästner. Your kindness makes me fear that you may have excited in him higher expectations than I with my limited attainments and slender abilities can possibly fulfil. Kästner has invariably received me with the greatest kindness. I have visited him repeatedly, and I find his society most instructive. Who could take offence at externals in a man so truly great?

'As I am destined to serve my country in various branches of political economy, I can only follow mathematics as an auxiliary science. Unfortunately, the career I have adopted, otherwise very attractive, requires an acquaintance with so great a variety of subjects—botany, mineralogy, chemistry, and statistics among others—that one is obliged to exert all one's powers to attain even to something very mediocre. Yet the study of mathematics, as well as that of mechanics, will ever remain the true basis of political economy. How little, however, can mechanics accomplish without the aid of a higher mode of analysis! Whoever is but slightly acquainted with the machinery used in various manufactures or in mines will not fail to discover, as well from the want of certain contrivances as from the mode of their application, the value of mathematics, and the mischief that might be occasioned through ignorance. Boulton's steam-engine and Höll's hydraulic press appear to me to be the best vindication of theoretical mechanics, if any were needed. From even the slight acquaintance I have with mathematics, I can well believe how important the study of this science must be to those engaged in the consideration of political economy: all the time, therefore, that I can spare from my numerous occupations I devote to the study of mathematics, and particularly to the differential calculus, in the use of which I feel myself very deficient. I am now going diligently through Tempelhof, a work I had commenced at Berlin. I continue also to prosecute machine drawing and the invention of new constructions, and though I am far from being vain enough to think that I shall ever arrive at any new combination, yet I have found these exercises of essential service in the close reasoning necessary to find the means for accomplishing a certain end. I have often laughed heartily with Fischer at the astonishment he evinced at the first sight of my designs, and his subsequent dismay on finding that, by the multiplicity of the combinations, power and weight had been brought to bear on *one and the same* point, and had thus become locked.

'But though I am compelled to renounce all claim to mechanical invention, I yet feel bound to acknowledge that I have made a discovery in another branch of mathematics, which (when is a young man ever dissatisfied with himself?) has afforded me some amount of pleasure. But however presumptuous it may seem to commence a correspondence with you, my esteemed friend, by a letter of such unusual length, I am yet going to trespass so far upon your time as will suffice for a preliminary explanation. The subject interests me so deeply. In my limited labours in algebra I keenly felt on one occasion the inconvenience in equations, where various sums and factors occur, of not being able to arrive at the precise value by means of logarithms. I thought over the possibility of remedying this evil, and discovered two ways for the accomplishment of my object—either to reduce all the sums and differences of two quantities into products, or to find a new kind of logarithm, which might be made available either for addition or subtraction.'

Humboldt adds some additional particulars in regard to the two methods, and begs his friend's permission to lay before him, at some future time, a more complete statement of his system of logarithms. Although neither method was completely successful in solving this problem, the solution of which was accomplished by Gauss twenty years later, yet it is evident from this communication that Alexander von Humboldt had already gained a great insight into the principles of mathematics, and had proposed to himself severer problems than were to be expected from a student in political economy, who was upon the point of entering only his second collegiate term.

The following letter to Wegener describes his journey to Göttingen:—

'We left Berlin, I think, on the 10th of April. I spent five

days at Magdeburg most agreeably, for I met there my friend
La Roche, a man in whom nature has for once condescended
to unite a noble intellect with a handsome and attractive exterior. What happy hours we spent together in our walks
along the secluded banks of the Elbe!'[1] . . . From Magdeburg
I visited the salt works of Schönebeck, Grossensalza, and
Frosen, as well as the new colony of the Moravians at Gnadau
in Saxony. High as my expectations had been raised concerning this institution, they were far surpassed by the reality.
The architecture of the houses, their cleanliness, their excellent
state of repair, the industry of the inhabitants, their consideration for the poor—the entire management of the colony in fact,
forms a complete ideal of a small, well-ordered state. Göttingen, a University, that is to say, an emporium of wisdom (where
wisdom is to be had for the fetching, there ought to be no lack
of it), where probably half a dozen lectures on physics are
going on at a time, yet leaves the college library without a
lightning conductor; while at Gnadau, a colony of superstitious
enthusiasts, there are no fewer than five conductors, though the
whole town consists only of some twenty houses!!! And there
is besides a lightning conductor to the church.'

At Helmstädt he was much interested in the celebrated
museum, formed by Professor Beireis, who, on account of his
extraordinary acquirements and peculiar habits, was called
the Alchemist of Helmstädt:—' Beireis does not himself know
the full extent of his treasures. At home he is always engaged
in prosecuting discoveries, and just now, as Crell assures me,
he spends sixteen hours a day in reading, on various subjects.
Besides the European languages, he speaks Egyptian, Chinese,
Japanese, as well as some of the dialects of Northern India,
and he read out to me, with facility, in German, some passages
from a Japanese book, yet many people venture to doubt
whether he knows Hebrew! He is in short a most extraordinary man, who with the most profound knowledge of chemistry
and numismatics, combines the charlatanry of the most cun-

---

[1] Humboldt here alludes in very favourable terms to Gurlitt in Kloster
Bergen, to Funke in Brunswick, and to other men of learning in Helmstädt, and then proceeds to matters of a personal nature which are irrelevant
to the present subject.

ning juggler. A number of little traits which I have collected of him can better be related by word of mouth. He tells one that he can make corn to grow, that he knows of a tree that bears truffles, that he lives without sleep, and in conversation says every minute that " he has thought upon that subject for six weeks together without eating or drinking."

' While I was at Helmstädt, I made an excursion to Harbke, where there is the oldest and most extensive plantation of American trees in Europe. The trees appear to grow as if they were wild, and the cedars, six feet in diameter, flourish as well on the Harz Mountains as upon Lebanon, where now they are becoming so scarce. At Brunswick I led a most unsettled life, for I went a great deal into society, and my attendance at the court, which, in comparison with other courts, I found very amusing, took up a great deal of my time. . . . I was delighted at the freedom with which certain changes at Berlin were discussed among court circles here, especially at the receptions of the Dowager Duchess. I met with all the learned men of Brunswick, and indeed it would be scarcely possible, except at Göttingen or Berlin, to find such a congeries of the patriarchs of German literature, the friends of Gellert, the venerable Gärtner, Schmidt, Ebert, Jerusalem, Eschenburg, and Campe—*semper idem*. . . .

' William, who sends you an affectionate greeting, came to meet me at Brunswick on his way from Hanover. Göttingen was empty when we first arrived, but the lectures commenced to-day. Last Sunday was the thanksgiving-day for the king's recovery, when there were great preparations for rejoicing, but no real joy. The proceedings consisted of—1. A sermon from Less. Heavens, what a sermon! Heyne says, Less gave thanks like a beggar-boy! 2. A levée at the English princes'—very crowded. 3. A ball. There were several songs in honour of the princes, which were sung with much bellowing. The princes shouted famously too. All the students, the princes included, wore favours inscribed with " Long live the king! " Oh, what folly! '

On April 25, 1789, Alexander von Humboldt was enrolled in the college register of the University of Göttingen, sub. No. 48, as follows:—

'Fridericus Alexander ab Humboldt, Berolinensis juris studiosus, ex academia Viadrina.'

It was in this year that the startling events of the French Revolution burst upon Europe, and without doubt the lightning flash of this terrific storm blazed above the horizon of the two Humboldts. William went immediately to Paris with Campe, 'to assist at the obsequies of French despotism,' while Alexander only paid a flying visit there a year later, when travelling with George Forster. During his stay at Göttingen, Humboldt took up his residence at No. 82, Weenderstrasse, and lived in the same house as the young Count, afterwards Prince Metternich.

The number of students at that time in the University was 812, of whom 210 were registered as students of theology, 405 of jurisprudence, 104 of medicine, and 93 of philosophy, under which were included mathematics, philology, agriculture, history, and æsthetics. Among the fellow-students of Humboldt were two of the English royal princes, Ernest Augustus, afterwards King of Hanover, and Adolphus Frederick, Duke of Sussex ;[1] also fourteen nobles of the rank of Count, of whom those who most distinguished themselves in after life were the Counts de Broglie and St.-Simon from Paris, Count von Einsiedel from Saxony, Count von Meerfeldt from Westphalia, and Count Metternich. Among other contemporaries of Humboldt the following names are also worthy of notice :—Von Vincke from Osnabrück, Von Nagler from Onolzbach, and Von Kamptz from Mecklenburg—all of whom afterwards rose to distinction and held some of the highest offices in the Prussian Government. In addition may be mentioned Oltmanns, a native of Friedland, who subsequently rendered much valuable assistance to Humboldt in the astronomical department of his work on America, and Van Geuns of Gröningen, a geologist, who accompanied Humboldt in his journey to England and the Lower Rhine.

The University of Göttingen was just then at the height of

---

[1] 'While there I received many distinguishing marks of favour from the English princes, whose governor, General Malortie, was strongly attached to our family, and would gladly have taken the surveillance of my brother and myself.'—*Alexander von Humboldt to Pictet, in 'Le Globe,'* vol. vii. p. 181.

its glory as a school of science. Next to science it was noted for the culture of classic philology, to which was united the study of political economy, here first launched into its vital element —publicity. To this is due the important influence exerted by Göttingen in the development of German thought.

For although Ernest Brandes, who in 1791 was appointed referee on the affairs of the University, thought he was entitled to boast that 'no disciple of Wolf or wild reformer in theology, no follower of Brown or other sectary in the school of medicine, no metaphysical prophet in any department of natural science, had ever occupied a professor's chair at Göttingen;' although his influence was always exerted in favour of the accomplishment of his own pious ejaculation, 'Heaven préserve us from the philosophy of the day ever being prevalent at Göttingen!' —yet notwithstanding all, the University of Göttingen at the time that Alexander von Humboldt entered as a student was reputed to be the first in Germany. For the men who had invested Leipzig and Halle with the glory of their celebrity had already passed away, and the splendour of Jena had scarcely as yet dawned.

The science of philology, hitherto restricted to the mere study of language, was developed at Göttingen by the genius of Heyne into a history of antiquity, and was brought to bear upon the ordinary concerns of life. The science of history had been entirely remodelled by Schlözer, in conjunction with Gatterer and Spittler, and the range of its subjects extended so as to include the study of politics, together with the history of inventions, the advancement of civilisation, and the constitution and legislative powers of a state in regard to the succession to the throne, a change of dynasty, or the events of war. Schlözer's 'Correspondence,' 'Göttinger Journal,' and 'Staatsanzeiger,' contained not only most valuable records of historical events, but constituted the highest political tribunal in the country, powerful enough at one time to induce Maria Theresa herself to pause in her projects with the thought, 'But what will Schlözer say?' In the faculty of jurisprudence, there gathered round Runde and Martens the 'clique of literati' ('gelehrte Eleganz') who had been enticed from Leipzig by Gebauer; and while Pütter attained a position of distinction as an expounder of German law, the youthful Hugo, appointed professor in 1789, excited universal attention as the founder of a new system in the study of jurisprudence.

But the pre-eminence of the University of Göttingen was undoubtedly due to the facilities it afforded to the study of mathematics, physical science, and medicine. The pursuit of these sciences had nothing in common with the philosophies of a revolutionary freedom, nor with the idle speculations of metaphysics; their object was the discovery and investigation of that only which was available and useful for human life. Kästner and Lichtenberg were not less distinguished for the pure science of their lectures on mathematics and physics, than for the classical wit and humour with which they were enlivened. Lichtenberg illustrated his lectures upon applied mathematics, the theory of the earth, meteorology, electricity, and physics, by means of a philosophical apparatus that was one of the most complete of the time. Among the professors of chemistry, Gmelin and Osiander are to this day mentioned with honour; the former distinguished for his work on ' The History of Chemistry,' the latter known as a successful accoucheur, and the founder of a museum of natural curiosities. Of all the professors at Göttingen, the most renowned was Blumenbach, ' who by his writings as well as by his animating lectures everywhere kindled a love for comparative anatomy, physiology, and natural philosophy—a love which he has carefully nurtured as a sacred flame for more than half a century.' [1] He was the first in Germany to raise natural history, hitherto regarded merely as a subject for the instruction of children, to the dignity of a science, and successfully demonstrated it to be in intimate connection with the history of the world and of mankind. His works have been translated into nearly all the European languages. He formed comparative anatomy into a distinct branch of study, and in 1785, long before Cuvier's time, he expounded in a complete course of lectures the system he had established.

Upon his arrival at Göttingen, Alexander von Humboldt found his brother already on terms of friendly intercourse with

---

[1] Alexander von Humboldt's ' Rede bei Eröffnung der Versammlung deutscher Naturforscher und Aerzte in Berlin am 18. Sept. 1828,' p. 6.

the distinguished men taking the lead in the circles of science
and literature, by all of whom he was welcomed with the most
gratifying cordiality. With Heyne and the other men of note
he at once entered upon terms of intimacy: nevertheless, he
found at first, as he complains to Pfaff in the letter above
quoted, that the stiff, unsocial society at Göttingen gave little
opportunity for free and unreserved intercourse—a formality,
however, which soon began to dissipate. He joined his brother
William in the study of philology at the seminary, and attended
a course of lectures by Heyne on archæology delivered in the
large hall of the University library, which was hung round with
engravings and casts from the antique; under Spittler he
studied the history of trade and commerce, and obtained
private instruction from Lichtenberg in certain branches of
physical science, such as light, heat, and electricity ; he
studied agriculture under Beckmann, and attended Heyne's lectures on the Iliad, to hear which an audience numbering about
fifty assembled. The most popular lectures were attended
by from 200 to 300 students. ' Heyne,' writes Humboldt in
a letter to Wegener, ' is undoubtedly the most clear-headed
man, and in certain branches of knowledge the most learned
professor, in Göttingen. His delivery is laboured and hesi-
tating, but he is in the highest degree philosophical in his turn
of mind, and logical in the sequence of his ideas.'

His sketch of the professors, ' although amid this posse of
great men, he could not know all equally well,' is marked by a
vein of humour and a tone of seriousness truly surprising in a
youth of scarcely twenty years of age, and seems to prove that
the maturity of mind which could thus early manifest so extensive an acquaintance with men and things had been acquired
more from association with refined society and men of superior
abilities than from any merely formal course of instruction.
We must, however, restrict ourselves to one or two of his
descriptions.

'. . . . Heyne is undoubtedly the man to whom this century
is the most deeply indebted; to him we owe the spread of religious enlightenment by means of the education and training
he has instituted for young village schoolmasters, to him is due
the introduction of a more liberal tone of thought, the esta-

blishment of a literary archæology, and the first association of
the principles of æsthetics with the study of philology. Yet
Heyne has never published a compendium upon any of the
twelve subjects upon which he lectures: these comprise the
classics, archæology, the antiquities of Greece and Rome, and
the ancient tragic poets, together with Aristophanes, Homer,
Virgil, Horace, Plautus, and Cicero. These lectures are always
delivered in a certain order, because they are in the first instance intended for the seminary, and therefore for the use of
those who, like my brother, are prosecuting their studies there.
Heyne's published lectures are so comprehensive and elaborate
that they are sold here at the price of from three to five Louis
d'or. Köppen's commentary upon Homer is in fact only a
dissertation of Heyne's surreptitiously reproduced. Heyne succeeded the renowned Gesner in the presidency of the seminary,
but where, in all Germany, can he find a fit successor? Schütz,
whom Heyne was so anxious to have near him during his
declining years, is much too inactive, and is besides fettered
with the literary journal.

' The seminary is in the most flourishing condition. Among
the students there are three, Mathiä, Kreis, and Woltmann,
who in a few years will be almost unrivalled in Germany for
the extent of their learning. People are astonished here in
Göttingen to find such extraordinary proficiency attained in so
short a time. Mathiä is the best Greek scholar, with the exception of Wolf, whom Heyne has ever had for a pupil; he is
familiar with all the most recent literature of England, Spain,
and Italy, and is intimately acquainted with Kant's philosophy.
The most agreeable and intellectual companions I find here
are among the collegians of the seminary. With Woltmann,
who has a wonderful talent for versification in German, and
indeed also in Greek, I have almost daily intercourse. He is
an excellent man, to whom I have been drawn by the striking
resemblance he bears in character to yourself. He has been
educated under the younger Stollberg, by whose influence he
has become imbued with a most extraordinary enthusiasm for
the writings of the ancients, and will certainly one day highly
distinguish himself. I usually spend from nine till eleven
o'clock every evening with him, when we read together Plautus

and Petronius. It is the *methodus vivendi* of Göttingen not to go out earlier in the day, for there is a great affectation of industry here.

'Spittler.—I attend his lectures on modern history; he possesses a fine understanding, with a stately delivery which passes with most people as the beau ideal of eloquence. It is too bombastic for me. Nations are " rapid torrents," the royal house of Prussia is " an ancient oak, under whose shadow a free German people delight to cast themselves." His mode of delineating history and his grouping of events is most masterly. It is much to be regretted that he does not take up the subject of church history, for such a course of lectures from him would be far more attractive than the disquisitions of Plank, who has nevertheless won great esteem by the impartiality of his views.

' Kästner has unfortunately a very indistinct delivery, owing to the loss of his teeth. He is very humorous, and is always saying something witty; but as he invariably laughs at his own wit before he has well finished, the humour of it is not always appreciated by his audience. He makes amends, however, to those who laugh with him by being polite enough sometimes to join in a laugh when nothing really witty has been said. The great drawback to this academy is the small attention paid to mathematics. Kästner is nevertheless the kindest and most agreeable of men; I see a great deal of him. He cannot help being very sarcastic, for which he is afterwards so remorseful that he invariably asks forgiveness.

' I have also attended the lectures on moral philosophy given by Less, and certainly never heard anything so miserable. In character, speech, and mode of thought, Less reminds me strongly of Fromm of Frankfort, though Fromm would be thought in comparison really eloquent. On one occasion Less enquired whether it was lawful for a Christian to put into the " Lotto di Genova," as he calls our lottery. Is not this lecturing on morals as a casuist ? It might as well be asked, ought a Christian to play cards or chess ? . . .

' The English princes are condemned to listen to this sort of trash for a couple of hours every day, and the unfortunate youths are obliged to write out each lecture—an exercise which is afterwards corrected by Less. Such is the folly actually demanded by that detestable English orthodoxy.'

Humboldt then proceeds to give some particulars regarding his brother and himself:—

'I am living here in an atmosphere of philology. Were my stay to be prolonged for a couple of years, I really believe, distasteful as it would be to me, I should end by devoting myself to Greek literature. My brother has been making an excellent use of his time. You cannot think how much general interest he is beginning to excite. By his confidential correspondence with Forster and Jacobi he has made quite a sensation both here and on the Rhine, and it is probable some of his letters will be published by Jacobi. I must confess that I am really beginning to wonder myself at my brother's extensive learning and high culture. Heyne says that he has not for long dismissed from his tuition so gifted a philologist. If to this be added his extensive knowledge of jurisprudence, history, and politics, his deep insight into Kant's philosophy (which Rehberg told me quite astonished him), and his acquaintance with English, French, and Italian, it must be acknowledged that, *ex professo*, there are few to equal him.

'Should you in course of time chance to meet with a small philological pamphlet, shortly to be published at Göttingen, and bearing on the title-page the words, " Edited with notes by Heyne," you may conclude it to be a production of mine. It is a dissertation upon the weaving-loom in use among the Greeks and Romans. The work is quite a prodigy of learning, and its compilation has been therefore most distasteful to me. I have discovered that the loom of the ancients was just the high-warp loom introduced by the Saracens into France—a fact capable of abundant proof from the bronzes of Herculaneum, the Onomasticon of Pollux, the writings of Isidorus, the Vatican MSS. of Virgil, the descriptions of Homer, &c. The proof is somewhat elaborate, from the number of authorities to be consulted. Heyne is delighted with the work. It is now easy from my interpretation to understand the meaning of the terms used by the ancient writers, *scapus, pecten, radius, insubulum,* &c.' . . .

In a postscript to this letter Humboldt mentions having made several excursions into Hesse, and through Lower Saxony, and describes a visit to Pyrmont, where he spent a week in daily companionship with Jacobi, Rehberg, Möser, Markard, Eschenburg, Mauvillon, &c., 'with whom, unfortunately, were included seven or eight princes.'

On the subject of this antiquarian treatise upon weaving, Humboldt writes to Sömmering, at the close of the year 1793, that it was a sort of commentary upon the Onomasticon of Pollux, and that the work had proved a great interruption to his ordinary occupations. The treatise was sent by William von Humboldt on March 8, 1794, to Friedrich August Wolf, for his revision, and Alexander availed himself of the opportunity to enclose to him the following letter : [1]—

'It is certainly very presumptuous in us to expect you to revise so youthful a production. What I ventured only to desire has been made by William into a request. For this he alone is responsible. I believe I am in a position not only to explain satisfactorily the meaning of the *radius* (κερκίς) [staff], but also of the *pecten* (ξάνιον) [comb], which hitherto has been confused with *plectrum* [rod], and which by modern commentators has frequently been confounded with *radius*, and sometimes even translated *Lade* [lay or batten]. . . . It would seem that in the ancient mode of manufacture the ἱστός [loom] stood upright, and that the weavers, especially when engaged upon the χιτὼν ἄρραφος [seamless garment], as they walked round the frame interwove the radius (a mere staff wound round with thread) into the warp arranged in a cylindrical form, making use of the *pecten* to drive together the threads of the weft. As it can be proved historically that the high-warp loom, which was introduced into Spain by the Saracens in the reign of Charles Martel, is originally from the same source as that in use among the ancient Greeks, as the *pecten* still employed in the East is of this form, and as by this hypothesis everything stated by Pollux on the subject of weaving may be easily explained, the theory I have propounded becomes at least a probable one.'

At the conclusion of the letter William adds the following

[1] Wilhelm von Humboldt, 'Gesammelte Werke,' vol. v. pp. 103, 106.

lines :—'My brother has also been occupied in some investigations regarding the linen cloth employed in the envelopment of mummies, and he intends at some future time to publish the result of his researches.'

Humboldt alludes to this treatise upon the method of weaving among the ancient Greeks as his first literary effort,[1] but this statement, in view of the pamphlet 'Sur le Bohun Upas,' mentioned in p. 58, stands in need of some modification. Unfortunately, the work has been lost.

The zeal with which Humboldt prosecuted these studies of a philological and antiquarian nature in no way lessened his interest in natural science, which was stimulated by the instruction and heightened by the social intercourse of such men as Blumenbach, Beckmann, Lichtenberg, &c. We may easily suppose that he was strongly attracted to Blumenbach, and influenced by him in a considerable degree, without the proof that Blumenbach was one of the first to whom he communicated the results of his experiments with galvanism upon the action of the muscles and nerves—results which were forthwith published in Grens' 'Neues Journal der Physik.'

In addition to the encouragement thus offered to the pursuit of science, Göttingen was the centre of all efforts which had for their object the development of the new principles of physical geography, and the establishment of a better system of map delineation. While Heyne devoted his attention to the ancient geography of the world, and Michaelis confined himself to that of the Holy Land, Blumenbach was occupied in illustrating and classifying the most recent discoveries in natural science, and at the time of Humboldt's residence at the University was elucidating Bruce's travels in search of the source of the Nile. Nor were the students of the university less active in their devotion to science than the professors. In the year 1789 Humboldt was the means of founding a society in conjunction with several of his friends, Seetzen, Link, Meyer, Van Geuns, Deimann, Kries, Kels, Schrader, Hofmann, and others, called the Philosophical Society,[2] which, by the valuable

[1] Brockhaus' 'Conversations-Lexikon,' 10th ed., art. 'Alexander von Humboldt.'

[2] Seetzen's 'Reisen,' edited by Kruse, vol. i. p. 5.

assistance afforded by the well-stored library, the ethnological
museum, and the museum of natural curiosities, rapidly attained
a state of useful activity.

After a long silence, Humboldt writes to Wegener from
Göttingen, on January 10, 1790 :—

' I was away from Göttingen for two months from September
24, spending the vacation in making a scientific tour with a
Herr van Geuns, a Dutchman with whom I became acquainted
through his writings on botanical subjects. Our route lay
through Cassel, Marburg, Giessen, Frankfort-on-the-Maine,
Darmstadt, along the Bergstrasse to Heidelberg, through
Speier, Bruchsal, and Philippsburg, to Mannheim, thence by
Alzei and Mörsfeld among the quicksilver mines of the Vosges
to Mayence, where we spent a week with Forster. From
Mayence we sailed down the Rhine to Bonn, whence we took
the road to Cologne and Düsseldorf, or rather Pempelfort, and
stayed a week with Jacobi; thence we returned to Göttingen
by Duisburg, Münster, Warendorf, Rittberg, Paderborn, and
Cassel. Of this interesting tour I can tell you nothing, because
in the first place this route has been described some hundreds
of times, and in the second place, amid this multitude of subjects
I should not know where to begin.[1] Your letter followed me
to Mannheim, where I spent three delightful days in the magnificent botanic garden of Counsellor Medicus.

' Amid the numberless distractions of the journey, which was
made sometimes on foot and sometimes by carriage, and with
the incessant occupation of packing up minerals and plants, I
was not very well able to write to you. I require to be in a

[1] Among the effects of Kunth, the botanist, was found Humboldt's copy
of Linnæus' manual on botany, ' Systema Plantarum sec. class. ordd. genn.
et sp. Edit. nova. cur. J. Reichard' (4 volumes in 1, 8vo., Frank. 1779-80).
The book is exceedingly interesting from the details of his early journeys
and studies, to be gathered from memoranda of conversations with Forster,
Banks, Willdenow, Sieveking, &c., and the frequent annotations (between 400
and 500) having reference to the time and place where the plants were
found. For example: 'Banks horto suo mihi monstravit 1790; vidi in
monte Mam-Tor versus Ax-edge, Jun. 1790; legi prope Helgoland 1790;
vidi in Hamburg, 1791; prope Cuxhaven, Dover, 1790; am Harz 1789;
prope Colberg, 1793; apud Calais 1790; prope Wittenberg 1790; praest.
spec. vidi in Harbke 1789 et in horto Kewensi 1790; legi prope Ostende et
Calais 1790; praest. spec. e Virginia v. in herb. Sieveking,' &c.

certain frame of mind to write to my friends: it is not a condition that will be enticed, therefore I must wait till it comes, however long that may be. I returned here at the beginning of November, and have since been occupied in a tissue of contrarieties. A host of lectures (I attend six) occupy the greatest part of the day, so that I can only visit the library, which I still need for my philological researches on weaving, on a Sunday, when I have express permission from Heyne to go there and lock myself in. Extra work, too, of a pressing character runs away with a good deal of my time; as, for instance, in occasional contributions to the "Zurich Botanical Magazine," for the last number of which I have written a Latin treatise.[1] I have also to write out the journal of my tour, and Forster expects me to furnish him with a mineralogical description of the basalts at Unkel, for the next number of his magazine, in which there is to appear an excellent paper by my brother, on the influence exerted on the morals of a nation by theism, atheism, and scepticism. My paper, however, grows so fast under my pen that I think it will probably have to be printed in a separate form. Do not think, my dear friend, that I wish to boast of my numerous occupations: the load is certainly not greater than I can bear, but in point of fact, I have but very little time left for my correspondence, seeing that I have always to write at least once and sometimes twice a week to my mother or Kunth.

'William, who has been in Paris with Campe, returned by way of Mayence, where he spent three weeks with young Forster, and thence made a tour through Switzerland, visiting Zurich, Schaffhausen, Kostnitz, Bern, the Grindelwald glacier, the pass of St. Gothard and Lausanne, down to Geneva, whence he returned to Mayence by Neufchâtel and Basle, and after a further sojourn there of a fortnight proceeded through Saxony to Berlin. This same William, who has caused me to construct this ungraceful period, asked me to meet him at Gothâ. Thither I went alone in the beginning of December, riding through Eichsfeld in the most detestable weather and along still more

---

[1] 'Observatio critica de Elymi Hystric. charactere, in Usteri,' 'Magazin für Botanik,' 1790, Part VII. p. 36; Part IX. p. 32.

detestable roads. The difficulties of the journey were willingly encountered for the sake of seeing a brother who had been an eye-witness of so many extraordinary events. We spent two days with Löffler, and it was delightful to me to renew the relationships of the old Frankfort days. What are all the sensations that inanimate nature can inspire, in comparison with those heart-stirring emotions elicited by the sympathy of friendship and the gratification of being loved by good men!' . . . .

The anonymous pamphlet, 'Mineralogical Observations on some Basalts of the Rhine'[1] was the result of this tour, undertaken by Humboldt when a youth of twenty-one years of age, and not, as is usually represented, of the journey he subsequently took to the Lower Rhine in company with George Forster: and it is necessary here to state distinctly that Humboldt was at this time self-taught both in mineralogy and geology.[2] His next works were two small treatises published in Crell's 'Chemische Annalen,' on 'The Aqueous Origin of Basalt,' and 'The Metallic Seams in the Basalts at Unkel.' In the chapter on 'Incidental Remarks on Basalt in ancient and modern Writers,' which precedes his 'Observations,' he proves with a great expenditure of philological learning, and in that lucid manner which subsequently developed into the full flower of his later genius, that the classic writers give no ground for supposing that the basalt of Pliny is identical with syenite, basanite, lapis lydius, or lapis æthiopicus; that it is erroneous to maintain, as hitherto has been the practice, that the basalt of our time is the same as that referred to by Pliny; that it is impossible now to ascertain what was the particular mineral rock that was termed by Pliny basalt; that the supposed basalt of Strabo is granite; and lastly, that it is quite uncertain whether the passage in Pliny has any connection with that in Strabo. The

---

[1] 'Mineralogische Beobachtungen über einige Basalte am Rhein.' Brunswick. 1790.

[2] Humboldt writes somewhat later to Freiesleben, 'I take the liberty of sending you my small pamphlet on the basalts, which (with many typographical errors) has appeared during my absence in England. I wrote it before I had received the benefit of any instructions in mineralogy, and I should never have ventured to have allowed it to be printed had I not been urged to do so by many considerations.'

chapter concludes with a dissertation on the lapis heraclius of the ancients.

In this work on basalt Humboldt displays rare powers of observation, a clear mode of description, and an extensive acquaintance with literature.   He is here discovered to be not merely a mineralogist and a geologist, but also a botanist. The plants which he found growing on the basaltic rocks of the Rhine are compared with those he had previously gathered from the basalt of the Meissner, and he enters into a searching criticism of the observations of Collini and De Luc.   'Everything,' writes Forster, 'upon which I had touched in few words concerning the supposed volcanoes on the Rhine, finds full confirmation in the two quartos of Dr. Nose, as well as in the condensed observations of our acute friend Alexander von Humboldt.'   His ingenuity was mainly directed to the support of the erroneous theory of the aqueous origin of basalt, at that time prevalent among geologists ; and the influence of the work was so considerable, that the book was often appealed to in support of those views long after the author himself had renounced them in favour of the volcanic theory.

The extravagant length to which the science of geology had been carried by these fantastic speculations on the nature of basalt is strikingly seen in a polemical treatise by the learned Professor Witte,[1] of Rostock, Counsellor of the Duchy of Mecklenburg.   He explained the Pyramids of Egypt to be the remains of a volcanic eruption, ' which had forced its way upwards with a slow and stately motion,' the hieroglyphics upon them as crystalline formations, the lake of Mœris as the sunken crater of an extinct volcano, the well or shaft of the great Pyramid as the air-hole of a volcano, the sarcophagus of Cheops as two pieces of lava, which, lying one over the other 'like a couple of biscuits,' before completely cooling, had taken the form of a coffin, &c.   Even the ruins of Persepolis, Balbec, Palmyra, the temple of Jupiter at Girgenti in Sicily, the two palaces of the Incas of Peru, at Lacatagua and Alkunkanjar, were supposed to be natural formations of basalt and lava.

[1] ' Ueber den Ursprung der Pyramiden in Aegypten und die Ruinen von Persepolis.' Leipzig. 1789.

Not less fantastic were the theories of the Abbé Giraud-Soulavie,[1] who thought he was able to prove that the geological formation of a country exercised an influence upon the physical condition and manners of the inhabitants. 'The inhabitants of basaltic regions,' he maintains, 'are difficult to govern, prone to insurrection, and irreligious. Basalt appears to have been an agent, though hitherto unacknowledged, in the rapid spread of the Reformation.' Humboldt, who in later life formed so just an estimate of the kind of influence exerted upon the inhabitants of a country by the natural conformation of the land, wrote even at this time :—'I need scarcely fear to be misunderstood, and be supposed to deny that the physical constitution of a country exerts an important influence upon the manners of a people. There can be no question that the inhabitants of a mountainous region differ very decidedly from the people dwelling in a plain; but to attempt to determine what particular influence upon the character is exerted by granite, porphyry, clay-slate, or basalt, must be regarded as a wanton trespass beyond the boundaries of our knowledge'— a proof of the caution he early displayed in the formation of his opinions and of the modesty which led him to avoid startling modes of expression.

The last few months of his stay at Göttingen, which he quitted after a year's sojourn in March 1790, passed without any remarkable event. Humboldt always preserved a grateful remembrance of the intellectual advantages that were here afforded him, and nearly half a century afterwards, on the occasion of the centenary of the Georgia Augusta (University of Göttingen), in September 1837, he expressed in grateful terms his acknowledgment that it was to this University that he was indebted for the most valuable part of his education.

It was at Göttingen, at the house of Heyne, that Humboldt first made the acquaintance of George Forster, Heyne's son-in-law, that remarkable genius who shone upon his youth like a guiding star, and was not only the friend with whom he enjoyed the deepest sympathy in all his tastes and pursuits, but was also the one who exercised the most powerful influence

---

[1] 'Histoire nat. de la France mérid.' vol. ii. p. 455.

upon his studies, his mode of thought, and the formation of his extensive scheme of life-long activity. In George Forster we have, in a certain sense, the prototype of Alexander von Humboldt.

George Forster, thirty-six years of age at the time of his introduction to Humboldt, and therefore his senior by only fifteen years, had already circumnavigated the globe as one of the expedition that accompanied Cook in his second voyage, and the masterly narrative he soon after published had justly raised him to fame. He had studied various branches of natural science, including physics and chemistry, he excelled as a draughtsman of plants and animals, he possessed an extensive acquaintance with philosophy, literature, and the fine arts, and had devoted himself enthusiastically to the study of geography, history, and politics; he wrote Latin and understood Greek, and he spoke and wrote with facility both French and English, he could read Dutch and Italian, and was to some extent acquainted with the Swedish, Spanish, Portuguese, Russian, and Polish languages. To these acquirements he united a disposition at once amiable and modest, which rendered him an agreeable as well as intellectual companion. Forster was a master in the art of portraying nature, his descriptions are such as to charm the artist no less than to instruct the enquirer, and while elevating by their poetic flights, enchanting by their picturesque adornment, they yet bring before the mind of the reader only the simple truth. But the delight with which his unrivalled books of travel can even now be read arises not so much from the extent of the information he displays upon the subjects he treats of, nor from the charm of the artistic colouring in which he portrays them, as from the intense human sympathy manifested throughout—a sympathy which led him always to select man, his condition, manners, and circumstances of life, as the chief object of his attention, which taught him to look with tender and loving interest beyond the feathers and the tattooing to the man himself, and enabled him under every form and in every position to acknowledge the right of reason.

At the close of an age characterised by a spirit of piracy and self-seeking which had even left its impress upon the grandest geographical discoveries of the century, Forster was the first to place the importance of peaceful aims and pure intellectual interests in their true light. In place of a mere love of adventure, he was led by a spirit of thoughtful investigation; instead of a search after worldly gain, he was moved by a desire to gratify the love of knowledge. The study of nature and political economy, the study of history and moral philosophy, as well as an acquaintance with the exact sciences, were all collectively regarded by him as the sole agents in moral elevation. From Forster, the world has learnt, in the fullest sense of the terms, both how to travel and how to describe travel.

Humboldt often mentions Forster's name with grateful remembrance and esteem. He speaks of him as a ' philosophe aimable,' and when at the height of his fame alludes to him in ' Cosmos '[1] as his ' distinguished teacher and friend, whose name I can never mention without a feeling of the most heartfelt gratitude,' and as one who powerfully influenced the literature of Germany by the introduction of the modern book of travel which affords a delightful contrast to the dramatic fashion of the middle ages. ' Through his influence there dawned a new era of scientific travel, having for its aim the comparison of various countries and peoples. Endowed with a keen susceptibility for the beautiful, and an imagination enriched by the lovely scenes presented to him by Tahiti and other peaceful islands of the Southern Seas, George Forster describes with a peculiar charm the varied glories of vegetation, the conditions of climate, the varieties of food in connection with the different habits of men, with reference to the races from whom they sprang, and the country whence they originated. In his works we see the reflection of a mind characterised by originality, a love of truth, and an observant thoughtfulness, a mind replete with images derived from a view of Nature in her exotic loveliness. Not only in his excellent account of Captain Cook's second voyage, but still more in his smaller works, is to be traced the germ of that greatness which he subsequently attained.'

Free from the obligation of earning his livelihood, unattracted by the honours of high station, and urged by no false

[1] ' Kosmos,' vol. ii. pp. 65, 72.

ambition, Alexander von Humboldt was possessed, through the
independence of his position, of sufficient means and leisure to
follow out his favourite pursuits, to gratify his love of travel,
to stimulate his mind with the contemplation of nature, and
prepare it by the habit of close observation for the most searching investigations.

During his stay with Forster at Mayence plans had been
laid for a journey to be taken in concert, the following spring,
to the Lower Rhine, Holland, Belgium, England, and France ;
for the old love of travel had been again aroused in Forster.
A visit to England promised to be of use in his father's affairs,
and he was anxious to gain some information on various
points connected with geography, natural history, and art.
How much of interest centred, too, in France, where the new
political *régime* had now been a year established !

Thus Humboldt, as if moved by an inspiration of his latent
genius, was led in the spring of 1790 to join with George Forster
and Van Geuns, his young Göttingen friend, in this projected
tour. It was as if the exciting guidance of this circumnavigator
of our globe was to become the preparatory training for those
vast expeditions which for extent and diversity of discovery
were to surpass all that had hitherto been accomplished.

His stay with Forster at Mayence was not long. On March
20, Forster wrote to Heyne that he was expecting Humboldt
to join him in the course of a day or two, and the first communication from the travellers was dated Boppart, March 24.

Forster has given an account of the journey in his classical
work, 'Sketches of the Lower Rhine,' but unfortunately the
reference to Humboldt is too occasional to afford evidence of
the impression the journey produced upon him, or the influence it
may have exerted upon his mind. That Humboldt kept, however, a copious record in his journal is evident from a portion
of the book still preserved bearing the inscription : 'Journey of
1790—England.'

The journey down the Rhine was happily accomplished; and
when a cloudy sky robbed the well-known scenery of any of its
accustomed charm, a book of travels placed them among the
wonders of Borneo, where the imagination was excited by the
glowing colours and gigantic growth of vegetation of that

tropical region, with which the wintry aspect of the Rhine district had nothing in common.

With Forster for a guide nothing could escape the closest scrutiny; nature and art, manufactures and churches, the past and the present, and all political matters, were alike viewed with interest; there was no literary celebrity, no park nor private pleasure-ground, no public institution, manufactory, nor piece of curious mechanism, there were no docks, mines, botanic gardens, nor observatories, that were left unvisited and uninvestigated.

One reflection made by Forster, referring to the Cathedral of Cologne, is of extreme interest. After describing the wonderful sublimity of the building he continues:—' My attention was arrested by a yet more engrossing object; before me stood a man of lively imagination and refined taste, who for the first time was experiencing beneath these majestic arches the impression of grandeur inspired by Gothic architecture, and who at the sight of this choir, rising more than a hundred feet above him, was riveted with admiration to the spot. Oh! it was glorious to see in this wrapt contemplation the grandeur of the temple repeated as it were by reflection. Ere we left the church, the shades of evening falling on the silent and deserted aisles, disturbed by no sound save the echo of our own footsteps as we walked among the graves of electors, bishops, and knights, whose effigies lay sculptured in stone, awoke in his imagination many sad images of bygone days. In real earnest, with his sensitive disposition and restless activity of imagination, I should have been sorry to have watched there alone with him through the night. . . . I hurried with him out of the Cathedral into the open air, and by the time we had reached our hotel he had once more regained that enviable state of mind through which, while displaying a keen appreciation of the charms of nature, he had so agreeably shortened the monotonous hours during our journey from Coblentz.'

The person here referred to was Alexander von Humboldt.

Forster further remarks:—'I have never yet been able to determine, whether it is most satisfactory to derive our ideas of real things direct from the world around us, or to receive them through the medium of an intelligent mind, which by selecting

and arranging an infinite variety of impressions has arrived
at an idealised conception more in harmony with our nature.
Each mode possesses its own peculiar advantage, and we have
fully enjoyed both during this journey. Immediate contact
with animated nature produces undoubtedly the most powerful
impression upon the mind ; the object, whatever it may be,
that exists without the co-operation of man, that is, and was,
and ever will be independent of him, impresses itself deeply
upon the mind with a clear and sharply defined image, in
which no detail is omitted. The most diverse ideas, on the
contrary, though gathered from all quarters of the globe, from
the past, and—may I add ?—from the future also, become,
when received through the apprehension of another mind, at
once associated with the present, with which they weave themselves into a drama that mocks reality.'

Thus, in the consecrated precincts of the Cathedral, Forster
saw with prophetic eye the greatness in store for Humboldt, as
if willing to read therein the fulfilment in a far wider sense of
his own destiny.

Two letters written by Humboldt during this journey to
Wegener have been preserved. The first, dated Castleton, High
Peak, Derbyshire, June 15, 1790, is as follows :—

'I must really ask your forgiveness for having allowed these
three or four months to slip by without sending you a line,
and for having even left the Continent without acquainting
you with my plans, my dearest, my best friend, to whom I am
indebted for the happiest hours of my life. But forgiveness is
easily obtained from one who is so uncomplaining, so rarely
vexed as yourself. Your last letter reached me when I was
laid up with a severe attack of influenza (*febris Göttingensis*),
which seized me before I had well recovered from the measles,
and which left me for some time in a state of great nervous
prostration. I was in no condition to reply to your letter then,
delighted as I was to hear from you. . . . I am much gratified,
my dear friend, to find that you are still so completely what
you used to be, so frank, so honest, so noble, so entirely unspoilt
that reason has enabled you to triumph over all the assaults of
dogmatic theology. Do not fear that I shall insult a position
which, though it may have been a torment to all mankind, has

yet been intended to further the highest human happiness. . . . We ought to be less disposed to grumble at the evils in the world, when we remember that their very existence furnishes opportunity for combating them and thereby securing a greater amount of good. The more you hear your confraternity preaching superstition, humbug, cant, as the spiritual virtues may be termed, so much the greater will be your delight in withstanding them. Your last letter contains a fine passage on this subject, so entirely the expression of a mind imbued with noble sentiments. Yes, my dear friend, I rejoice to see that you, with your zeal for investigation, your love of truth, your caution, your acquaintance with national and religious myths, are placed in this position, although it may cause you many sacrifices that may be painful to your heart. . . . Dogmatic theism is in my eyes far more dangerous than all the absurdities of the more positive system of faith, for even when it keeps the sword in the scabbard it commits spiritual slaughter upon reason. Nothing is so intolerable as those discerning princes who are determined to direct the thoughts of the rest of the world. The Berlin sophists appear to me in this light. What was more natural than that such a declension should result from an imposed form of religious faith? The substitution of Leibnitz for Luther is expected to cure the evil. And this is called freedom of thought! We are all groping in the dark. . . . . I am quite unstrung, and very tired, for I have spent most of the day below ground in the Peak Cavern, Eldon Hole, Poole's Hole, and among the mines. That it is quite possible to be fatigued among these hills you may learn from Moritz's travels.'

A few days later he writes again a further account of his journey :—

'Oxford: June 20, 1790.

'Do not expect from me, my dear friend, anything new about England. It is hard to say anything fresh of a country that has been so extensively visited; but I should like to give you my individual impressions if I had only time and quiet to write something reasonable. We shall therefore have all the more to say when we meet. Forster, my travelling companion, intends publishing an account of our journey, and I have read his descriptions piecemeal; they are well written, and I fancy the book will make a sensation. You must not, however, accept his opinions as mine, for we look at everything from a very different point of view. We could not have undertaken this journey at a more fortunate time. We travelled through all the provinces of Belgium, and were in the country during the late stirring events, the taking prisoner of General van der Mersch, the flight of the Duc d'Aremberg, the commencement of hostilities between Brabant and Flanders, and were actually present at the insurrection at Lille. Forster's name, which excites universal interest, his letters of introduction, &c., everywhere procured us access to the highest authorities.. At the Hague, in Amsterdam, and in Leyden, the attentions forced upon him were almost burdensome. And now at length we are here in England, finding, among other things to interest us, the trial of Hastings, the war with Spain, the music at Westminster Abbey, elections for a new Parliament,[1] various exhibitions of paintings, and innumerable collections in natural history and physical science. For the last fortnight we have been making an interesting tour through the country, visiting Reading, Bath, Bristol, Gloucester, Birmingham, and Buxton, thence among the hills to Castleton and Matlock, and by Derby, Stratford (Shakespeare's birthplace), and Blenheim to Oxford, where we have been the last three days. But the mere list of places that we visited can afford you little entertainment; you will doubtless be most interested in my individual impressions and feelings, and I can assure you that I have not only had a very agreeable journey, but one from which I shall derive much benefit and instruction. We shall start on our homeward journey in the course of a few weeks, and shall travel rapidly by way of Paris, where we shall make only a short stay, as by that time Forster will have exceeded his leave of absence of three months and a half. I do not intend returning to the University from Mayence, but shall probably go at once to Hamburg, to enter Büsching's School of Commerce, after which I shall perhaps study at the Gymnasium. I shall hope to

---

[1] He also attended some of the debates in Parliament. In conversation with Taylor, when quite an old man, he mentioned having heard 'Edmund Burke, Pitt, and Sheridan all speak the same night.'

remain at Hamburg till next spring, and then return to Berlin.
You will forgive my not having sent you my book, ("Mineralogical Observations on some Basalts, &c."), when I tell you
that I have not yet seen it myself. My physical health is
not so good as it was in the winter: you may easily imagine
how distressing this is to me just now, when it is impossible to
avoid fatigue. The journey has certainly done me good, but
the change has not benefited me as much as I expected. As
my stay in France and on the Rhine is uncertain, do not write
till you hear from me again. But I beseech you, dearest
Wegener, by all the affection which you know I bear you,
never to forget our brotherly love and friendship. You are
infinitely more to me than I can ever be to you. I have
now seen the most celebrated places in Germany, Holland, and
England—but, believe me, I have in seeing them never been
so happy as while sitting in Steinbart's arm-chair.

' ALEXANDER.'

The fragment of journal alluded to above, inscribed ' Journey
of 1790—England,' gives abundant evidence of the astonishing range of information possessed by the young traveller. It
contains observations upon mineralogy, botany, agriculture,
trade, technicalities, and the history of civilisation, together
with remarks of such various character that an abstract of them
can only find a suitable place in the Appendix.

The events transpiring at that time in France induced the
travellers to take the return journey through Paris, where
everything was still giving promise of great success. The universal enthusiasm for all that was pure and noble animating all
classes of the people in the preparations for the great national
festival, to be held in the Champ de Mars, was a gratifying
spectacle to all friends of humanity and lovers of freedom. The
sojourn of the travellers, however, was not prolonged beyond a
few days, and by July 11, Forster and Humboldt had again
reached Mayence.

Humboldt always referred to this journey as a time of
peculiar enjoyment. There can in truth be no greater happiness to a pupil in the school of knowledge than the opportunity of listening to the conceptions of the grand creative

faculties of a master mind. The reception of thoughts so suggestive, the outflow of a well-stored mind, warms and inflames the coldest, inspires the most indifferent, and imparts to the most retiring a sense of elevation. ' The companionship I enjoyed on this journey,' writes Humboldt, ' the kind interest shown me by Sir Joseph Banks, and the sudden passion which seized me for everything connected with the sea, and for visiting tropical lands, all exerted a most powerful influence in the formation of projects which, however, could not be carried out during the lifetime of my mother.'

Forster, ' whose noble, sympathetic, and sanguine temperament made him one of the happiest of men,' was always held by Humboldt, not only in respectful but in grateful remembrance. It is not generally known that the translation into German of Humboldt's ' Voyage aux Régions équinoxiales,' the ' Reise in die Aequinoctialgegenden des neuen Continents,' published in six volumes by Cotta, 1815–32, was executed by Frau Therese Forster, afterwards the wife of Huber; the work was entrusted to her as a means of pecuniary assistance, but unfortunately the translation was in many places so inaccurate that the undertaking proved a complete failure.

Even in the last year of his life [1] Humboldt thus expressed himself to Heinrich König, who had sent him a copy of his work ' George Forster at Home and Abroad ' :—' How can I adequately express to you my gratitude, for having followed the suggestion given you at Wilhelmsthal, in such a friendly spirit, by that noble and liberal-minded prince, the Grand-Duke ! You have furnished a biography of my departed friend which is characterised as much by the accuracy, fidelity, and impartiality of its representations as by the ability of its compilation. The perusal of your admirable work, in which so much penetration and good feeling are displayed, has kept me agreeably occupied for two long evenings. It has produced upon my mind many pleasing impressions, and many sad ones also. For the space of thirty years I have never known leisure but of an evening, and the half-century that I have spent in this ceaseless activity has been occupied in telling myself and others how much I owe my teacher

---

[1] On July 28, 1858.

and friend George Forster in the generalisation of my views
on nature and in the strengthening and development of that
which had already dawned in me before those happy days of
intimate friendship. On these two evenings, as the current
of my thoughts flowed back to the past, and I reflected with
sadness on the rapid deterioration of my powers, I was
more than ever reminded of the remarkable resemblances and
contrasts existing between Forster and myself; we held indeed the same political opinions, and mine, though in no
way derived from Forster, since they were formed long before my acquaintance with him, were yet strengthened and
matured under his influence; it was in company with him,
the circumnavigator of the globe, that I first beheld the sea
at a time when I had not the remotest expectation that
I should myself only twelve years later be sailing in the
Southern Ocean; my visit to London in company with him
during the lifetime of Cook's widow, when on the same occasion I, a mere youth of twenty-one, received so much kindness
at the hands of Sir Joseph Banks; in my expedition to Siberia
I trod the same shore of Samara, whence the elder Forster
had sent to Linnæus at Upsala the rare specimen of wheat
growing wild, I visiting the country in 1829, while Reinhold Forster, accompanied by his son George, then a boy,
was there in 1765, four years previous to my birth; I was invited by the Emperor Alexander through Count Rumanzoff, in
1812, to undertake an extensive scientific expedition through
the interior of Asia, in the same way as George Forster had
been requested by the Empress Catherine to undertake a
voyage round the world with Admiral Mulowski for purposes
of scientific investigation; similar disappointments awaited
our most cherished hopes, for both expeditions were abandoned
owing to the breaking out of war, in the one case with the
French, in the other with the Turks! How greatly have I
been excited by the awakening of these early reminiscences
through your valued gift! The whole of the sixth book is
masterly but very sad; the saddest of all to me are those
lines at the bottom of page 251, Part II.,[1] and yet such

[1] This passage refers to the part taken by Forster in the deliberations
against the citizens and officials of Mayence who would not take the new

words were necessary! With renewed expressions of heartfelt thanks and friendly esteem, I remain yours with the highest respect,

' A. VON HUMBOLDT.'

Humboldt remained a guest with Forster till the end of July.
It was at this time that he made the acquaintance of Sömmering,
with whom he at once entered into a lively correspondence ; his
celebrated work, ' Experiments upon the Excitability of the
Muscles and Nerves,' is dedicated ' with grateful respect and
affection to the distinguished anatomist Sömmering.' While
staying with Forster he wrote the following letter to Werner at
Freiberg, a noted professor of the science of mining and the
most distinguished geologist of the day :—

'Mayence : July 25, 1790.

' Sir,—I venture to send you a small pamphlet in which I
have propounded certain views upon the columnar basalts of
the Rhine. Although I can scarcely flatter myself that this
youthful production will gain your approval, I gladly avail
myself of this opportunity to express to you my profound
respect for your successful efforts in remodelling the science of
mineralogy. Unfortunately, I am not one of those who enjoy
the advantage of your personal instruction, and who can therefore observe under your own guidance the position of minerals
and fossils in their natural bed. Various circumstances have
prevented me hitherto from visiting the admirable institution
at Freiberg ; but perhaps I may yet be fortunate enough at
some future time to enroll myself as one of your pupils. I
have in the meanwhile, as far as my powers will allow me,
endeavoured so to familiarise myself with your system as to
adopt your ideas and your expressions as my own. How far I
have succeeded I must respectfully leave it to you to decide.

' In my journey to England, whence I have just returned, I
traversed for a second time the mountains of the Rhine. I found
nothing to necessitate the supposition of previous volcanic
agency, but on the contrary abundant evidence to prove the
aqueous formation of basalt. Your theory that a stratum of

oath of citizenship before March 30 ; while he himself had already set out on
his fatal journey to Paris on March 25.

basalt had once covered the whole earth, never seemed to me more reasonable or more obvious than at Linz and Unkel, where I noticed horizontal layers of basalt upon the highest summits. I shall be severely censured for this confession by many of our geologists, and my pamphlet (if it be not altogether overlooked) will be subjected to no tender criticism. But such considerations shall never prevent me from saying what I feel to be true, and I trust that I shall always maintain this resolution. I have resided for a considerable time in a region that has been called volcanic, and I have industriously traversed on foot the mountains of Hanover, Hesse, and the Rhine, as well as those in the neighbourhood of Zweibrücken; but I cannot accept the hypothesis so charmingly set forth by De Luc in his geological work, " Lettres physiques et morales."

' You have rendered to the science of mineralogy as great and important a service as that rendered to botany by Linnæus. In your work upon the external characteristics of minerals and your various other treatises, you have given us a philosophy of mineralogy. Through your investigations law has been discovered, confusion has been banished from the terms, and the rules for determining species and families laid down. Would that you could soon complete the work, and construct for us an entire system!

' I know how little my voice can do against the utterances of so many greater and more distinguished men ; yet I hope you will kindly pardon this rash expression of sentiment.

' I cannot flatter myself that I am known to you. My small botanical essays are too insignificant to have come under your notice, and though I was among the foremost of those who crowded round you on your last visit to Göttingen, I cannot suppose that you retain any remembrance of my name. I crave your kind indulgence both for this letter and the accompanying pamphlet. I am still very young, and though backward in many branches of knowledge, I am at least sensible of my deficiencies and anxious to repair them. With the assurance of the most profound respect and esteem, I remain, &c.

' I leave here to-morrow for Hamburg. Should you be disposed to favour me with a reply, my address will be : Herr von Humboldt, care of Professor Büsch, Hamburg.'

On Humboldt's departure from Mayence at the close of July 1790, Forster [1] furnished him with the following letter of introduction to Johannes von Müller, whose acquaintance Humboldt was anxious to make as he passed through Cassel:—

'I write to introduce to your notice Herr von Humboldt the younger, my travelling companion, a young man full of information and of a rare maturity of mind. He is well read in most branches of literature, but his particular province is finance and political economy. Should you have time to enter upon the subject with him, you will find him possessed of sound principles supported by a rich store of observations and a great amount of practical experience. His education has also included the study of fabrics and manufactures, in which he has made considerable progress. These varied acquirements of a practical nature, which are fitted to make him useful in an official career, are founded upon an excellent knowledge of philosophy and the classics, studies of which he has gathered the flowers without neglecting the less attractive portions. In a word, I think I shall be able to justify myself in introducing to you one worthy of being known, and one above all who merits your acquaintance. He is now on his way to Hamburg, whence he will return to Berlin. . . .' A singular letter of introduction certainly for a youth of one and twenty!

Humboldt's motive in going to Hamburg was to enter the School of Commerce, conducted by Messrs. Büsch and Ebeling, where he wished to attend a course of lectures upon the currency, to learn book-keeping, and to acquire some knowledge of the business routine of a merchant's office. With a mind deeply imbued with Forster's descriptions, and full of vivid impressions of England and her emporiums of commerce, the sight of Hamburg as the first sea-port in Germany must have excited his interest in a peculiar degree.

The School of Commerce at Hamburg had already attained a high reputation for sound instruction upon political economy, and it is an indisputable fact that the students in this science were enabled there to obtain instruction upon subjects, for the study of which the Universities at that time offered no facilities. The celebrity, too, acquired by Büsch as a jurist and a mathematician attracted to the Academy the attendance of young men destined for a political career.

That a number of foreigners availed themselves of the advantages afforded by this institution at Hamburg is apparent from the fact, that of the 159 students who received instruction at the Academy between the years 1767 and 1778, nearly half that number were from foreign parts, and included 25 Englishmen, 6 Frenchmen, 3 Danes, 4 Dutchmen, 2 Italians, 8 Russians, 6 Swedes, 14 Scotchmen, 2 Poles, 2 Portuguese, 3 Spaniards, 1 Norwegian, and 1 American—a proportion which has since increased year by year.

The concourse of so many youths from all parts of Europe furnished, in the words of Humboldt. a most favourable opportunity for learning in the best possible way the various living languages. With the object apparently of acquiring greater fluency in English, Humboldt shared his lodgings with a young Englishman, John Gill, whom he afterwards met at Barcelona in 1798, at that time a wealthy partner in the still prosperous mercantile house of that name. After the lapse of more than half a century, he still remembered this friend of his youth, and testified his gratitude for the hospitable reception he had received from him at Barcelona by the noble sympathy he tendered to one of the family who met his death at Berlin in 1848. Among other fellow-students of Humboldt were Speckter, the father of Otto Speckter, well known by his illustrations of Hey's fables; Wattenbach, the father of the distinguished historian of Heidelberg; Maclean, whose name became of the highest repute among the merchants of Dantzic; Böthling, a wealthy Russian from St. Petersburg, possessing a yearly income of 40,000 rubles, who at one time shared the same room with Humboldt, and who subsequently was desirous of accompanying him on his projected extensive journey.

A passage in a letter to Sömmering of January 28, 1791, furnishes the best insight into his studies and mode of life at Hamburg:—'. . . I am contented with my mode of life at Hamburg, but not happy, less happy even than at Göttingen, where the monotony of my existence was relieved by the society of one or two friends and the vicinity of some moss-grown mountains. I am, however, always contented, when I feel

that I am accomplishing the purpose I have in view. I am
learning a good deal here from the School of Commerce as well
as from personal intercourse with Büsch. Everything in the way
of mercantile knowledge was new to me, and I like the study
because I think it will be useful. I attend but few regular
lectures, therefore I work all the more industriously by myself.
Ebeling's extensive library is of great value to me; I find there
works on philology and history, and books of travel, while
Büsch furnishes me with various authorities on mathematics
and physics, and Reimarus with a very complete collection of
works on natural science. If to the unlimited use of these
means you further add the undisturbed use of a small room in
a secluded garden, with no interruption save the bell that rings
for dinner and supper, you will be compelled to admit, my
dear friend, that it is no less practicable to study at Hamburg
than at Göttingen. My leisure hours are occupied with geology and botany, both of which I am studying from books ! !
In addition, I have begun to learn Danish and Swedish, because I have a convenient opportunity for so doing. A life of
this kind may be supportable for seven or eight months, but
after that I shall begin to sigh for a more extended sphere of
action. As for society, which here means meeting at meal-
times, I am very well off. I visit in all circles, with citizens
as well as nobles, for the people here, after the praiseworthy
fashion of the Indian system of caste, have separated themselves into distinct classes. As card-playing is universal, I never
go into society before supper—a time when the pleasures of the
table are certainly very conspicuous. Much as the pride of birth
may be complained of on the Rhine, I am convinced that it
is far surpassed by the *hauteur* of the purse-proud circles here,
such as the Bentincks—not the Schimmelmanns. The common
sense of our western neighbours will triumph in this century,
while Germany will yet for long look on with astonishment, try,
prepare—and still postpone the decisive moment.'

In conformity with the practice he commenced at Göttingen,
Humboldt read Pliny's work upon the art of painting with some
of his fellow-students, and by this means excited in Hamburg
an unwonted interest in the study of philology. As a characteristic trait, we may notice the zeal with which he prosecuted

his natural taste for scientific investigation by going out on
the Elbe in stormy weather, in order to observe and measure the
motion of the waves.

Even in advanced old age he cherished the brightest recollections of the time he spent in Hamburg, especially of the
intercourse he enjoyed with the circle of friends assembling at
the house of Sieveking. This wealthy merchant was one of the
most distinguished men of the place ; he had filled some of the
highest offices in the senate, and stood in an important relationship with various political and learned men of note ; his wife,
a woman of great excellence and superior education, was a
granddaughter of Reimarus, celebrated in his day as the author of a work entitled ' Wolfenbüttler Fragmente,' edited
by Lessing ; her father was well known for his writings upon
lightning conductors, the instinct of animals, and various
branches of natural science. The most distinguished society
resorted to her house. It was there that Humboldt met with
Claudius Voss, the Stolbergs from the neighbouring province
of Holstein, and Voght, whose acquaintance must have afforded
him peculiar pleasure from being owner of the gardens at Flottbeck, noted for the extensive collection of rare plants.

Humboldt kept up an active correspondence with Forster
for a considerable time. On September 26, 1790, Forster
writes to Jacobi :—' The Humboldts are both prospering, but
in widely different ways. The elder is Counsellor of Legation,
and Assessor to the Supreme Court of Judicature at Berlin,
in which capacity he is serving his probation. When his time
has expired he will receive an appointment at Halberstadt, and
will then probably marry. His younger brother is with Büsch
in Hamburg, gaining a practical acquaintance with commercial
office routine ; he goes about a great deal among the various
eminent men of Hamburg, and has visited Christian Stolberg,
of whose praises he is quite full ; he makes expeditions too from
time to time to gather mosses, which flower during the winter,
and writes amusing letters full of lively wit, good nature, and
delicacy of feeling.'

These letters were afterwards returned to Humboldt by
Forster's executors between the years 1830 and 1840, and were
by him destroyed.

Greatly as this loss is to be regretted, there yet remain many other letters of this period which serve to throw some light upon Humboldt's studies and upon the circumstances by which he was then surrounded. They may here be given in chronological order, as they need no comment.

To His Excellency the Minister Von Heinitz at Berlin.

'Hamburg: September 10, 1790.

' Sir,—I venture to flatter myself that your Excellency will pardon the liberty I am taking in presenting you with the first-fruits of my studies (" Observations upon some Basalts, &c."). I do not aspire to the glory of ranking myself among the learned geologists of my country; for though I have had grand models before me, it yet requires a bold hand to catch the style of these masters. Men of noble and distinguished qualities have ever been characterised by a spirit of forbearance. I venture, therefore, to throw myself upon your kind indulgence. I am still very young, yet what little I do know, I know thoroughly, and I hope through greater industry to render myself eventually useful to my country, and to merit at some future period the favourable notice of your Excellency.

' With sentiments of the deepest respect and esteem, I have the honour to remain, &c.'

To Wegener.

'Hamburg: September 23, 1790.

' At length a few lines once more to you, my dear friend! You are not angry with me, dear William ?—nay, anger is impossible between you and me! It is indeed very long since I wrote to you, but you well know that the frequency with which we exchange letters is no thermometer of our friendship. You know the warmth and sincerity of my attachment to yourself, you are aware, my dear friend, how greatly I am indebted to you—nothing can possibly estrange us.

' I trust you safely received my letter from London. You probably expected me to write oftener while I was in England, but I must tell you that with the exception of my own family and Willdenow, you were the only person to whom I wrote during my absence. Only think how constantly I have been on the move for the last nine or ten months! On my return from France I stayed a month at Mayence, and thence made a tour to Aschaffenburg among the Vogelsgebirge, and through the district of the Rhone.[1] On my way here I visited Göttingen and Hanover. I am now a pupil at the School of Commerce under Professor Büsch, and see nothing all day but ledgers and account-books, so that I find it best to forget my plants and stones. I had scarcely been a week in Hamburg when I met with some natural curiosities from the island of Heligoland. I was immediately seized with the desire of collecting some, so I put myself on board a vessel and in eight days accomplished a stormy voyage of two hundred miles. In future I must content myself with the sight of the ships in the harbour, for the next time that I trust myself to the mercies of the elements . . . . If I can carry out my wishes, I shall visit England again in the course of a couple of years. I should now find a residence there very agreeable.

' I am surprised when I think how much I have seen since I left Berlin, of the variety of experiences I have passed through, and the number of interesting men with whom I have made acquaintance. I am disposed to be contented here, but I cannot say I feel very happy. I have made considerable progress in general information, and I am beginning to be somewhat more satisfied with my attainments. I worked very hard at Göttingen, but all I have learned makes me feel only the more keenly how much remains still to know. My health

---

[1] Further particulars of this part of the journey are given in the following passage from the letter to Sömmering of January 28, 1791, part of which has been already quoted at p. 98 :—' I left Aschaffenburg (which became endeared to me by the intellectual conversation of Müller and the unaffected good humour of Gallizin) with the determination of unburdening myself to you, of all that was in my heart, immediately on my arrival at Hamburg I fancied I had seen so much out of which to forecast a glowing future for myself, and I believed I should enjoy it all the more intensely by discussing it with a sympathising friend. An unfortunate tour which I soon afterwards made among the Vogelsgebirge and through the district of the Rhone, partly on foot and partly by carriage, in most unfavourable weather, introduced an entire change in the current of my thoughts. The minerals I had collected had to be arranged and several remarks appended, and you know how imperative these small matters appear at my age!' . . . .

suffered severely, but improved somewhat during my journey with Forster, yet even here I continue so closely occupied that I find it difficult to spare myself. There is an eager impulse within me which often carries me, I fear, over the bounds of reason; and yet such impetuosity is always necessary to ensure success.

'I send you herewith my book on the basalts. I shall esteem it a favour if you will read the first half, which is entirely philological; the remainder will be of small interest to you, though it has already created some sensation on account of certain paradoxes which it contains. An article in fulsome praise of the book has appeared in the "Hamburger Correspondent." You may easily imagine how little I am flattered by such vulgar trumpeting. I have no doubt it originated through Büsch, whose institution is rather declining, and who would therefore gladly proclaim to the world that men of such learning resort to his academy. Par intérêt !! I was much gratified by the criticism in the "Göttinger Anzeigen." If you feel disappointed in any way with my book, pray remember that while engaged upon it I daily attended five lectures, and was twice laid aside by serious illness, and that I wrote it entirely without philological assistance, for though Heyne encouraged me to undertake the work, he never saw the MS.

'This desultory letter must suffice for to-day. I will write more at length another time. Let me hear from you soon. My heart rejoices at the mere thought of you. With brotherly affection and grateful esteem, I am, &c.,

'HUMBOLDT.'

To Werner at Freiberg.

'Hamburg: School of Commerce, December 13, 1790.

'Sir,—I trust you will pardon the liberty I am taking in venturing to address you without having the pleasure of your personal acquaintance, and in being so bold as to trouble you with an enquiry. The motive by which I am actuated, namely, the strong desire to complete my education under your auspices, will, I hope, appear to you a sufficient justification.

'It is now nearly two years since I commenced the study of geology. My enthusiasm for the science has been recently

increased by my residence at Göttingen, my botanical wanderings among the mountains of Germany, and my delightful but
sadly too hurried journey through the Peak district in Derbyshire, which I made in company with your friend George
Forster. I have read as many books on the subject as my time
would allow, and have always been a close student of nature ;
but I am conscious that though up to the present time I have
acquired a great variety of information on mineralogy, my
knowledge is very disconnected and is of a somewhat superficial
character. I have long felt an ardent desire to go to Freiberg,
and enrol myself among the number of your pupils, but circumstances have hitherto rendered it impracticable. At length
these obstacles have been overcome. I have already attended
a course of jurisprudence and political economy at two universities, and I am now at the School of Commerce here, for the
sake of acquiring a knowledge of business routine ; but on
leaving this institution I shall still have six months at my
disposal—the summer of 1791—before entering on official
employment. I am, unfortunately, only too well aware of the
insufficiency of six months for passing through a complete
course of instruction in mining. I shall endeavour, however, to
content myself with the privilege of your valuable instructions,
for a limited period, rather than be deprived entirely of so
great an advantage. I hope, as the will is not wanting, to apply
myself to the subject with so much energy and zeal as to learn
a great deal even in six months.

'I leave this Institution at Easter, and a few weeks after shall
be at liberty to enter upon a course of study at Freiberg. I
therefore make so bold as to enquire if you could receive me
for so short a term as six months, and whether I and my servant
could be accommodated in the School of Mines, or whether I
should have to seek for lodgings in the town. I should be
greatly obliged if you would send me a few lines at once in
reply. I should have requested Herr Rosenstiel, Chief Counsellor of Mines, or the Assessor, Herr Karsten, to write to you
on my behalf, had I not ventured to think that I might address
you in this more direct way.

' Pray accept the assurance of my highest esteem, &c,

The following is from the letter to Sömmering of January 28, 1791, from which extracts have been already given:—

' I had written thus far about three weeks ago, and have since been prevented by illness from finishing my letter.

' To-day I am going to add to this already lengthy epistle a drawing of what has been called a petrified child's hand, which was found in the slate beds at Riegelsdorf. The phalanges show clearly enough that it is the paw of an animal, and could never have formed part of any human being. Perhaps you may be better able to determine its classification; is it likely to be a species of otter? Such a thing is not altogether improbable, for thirty fathoms below the surface at Riegelsdorf there is an extensive bed of limestone, considerably contorted, containing fossil fish. Pray assure Forster of my heartfelt esteem, and tell him that I have frequently repeated his experiment of the phosphorescence of potatoes with great success. If, in connection with Fourcroy's discovery of the existence of albumen in many plants, we call to mind the fact that animal lime is contained in the cereals, and volatile alkali [1] in the tetradynamia —the luminosity seen in potatoes is also to be observed in beef and salmon—we shall begin to see something of the affinity between animal and vegetable life. Yet another question, my dear friend. Where could I meet with some comprehensive treatise on the formation of animal bone?—which, though really composed of limestone, is impregnated with phosphoric acid, and constitutes that which Werner has designated uncrystallised phosphate of lime. I am at a loss to know how the calcareous earth can be evolved from the slight nourishment taken by an infant.'

There are still extant some letters written about this date to Dr. Girtaner, whose acquaintance Humboldt made in London. The subject of one of these letters is an unpleasant misunderstanding with Usteri, the editor of the ' Magazin für Botanik ; ' while in another letter Humboldt mentions that he was to leave Hamburg at the end of April, and proposed to spend a fortnight at Berlin with his mother and ' excellent ' brother, on his way to Freiberg, where it had been arranged that he should pass the summer.

[1] [An old name of ammonia.]

To the historian Wattenbach at Heidelberg we owe the publication [1] of some letters received by his father from Humboldt about this time. This gentleman was related to Büsch, acted as his secretary, and, together with Humboldt, was an inmate of his house. These letters contain the names of several of his friends at that period, and they evince the warm and lively recollection which Humboldt long preserved of the friendly relationships he maintained with his companions at Hamburg. They are full of the 'lively wit' to which Forster refers, while the unpretending good nature which is manifested throughout keeps in the background that overwhelming superiority of which Humboldt must even then have been conscious when in the society of his youthful companions.

The few weeks spent in Berlin, from the end of April till June 11, were occupied by Humboldt in the pursuit of his favourite study, botany, and in experimenting with Hermbstädt in his chemical laboratory, but his attention was chiefly given to acquiring a more intimate acquaintance with mineralogy in preparation for his visit to the School of Mines at Freiberg.

As a matter of courtesy more than of necessity, Humboldt sought the permission of the minister Von Heinitz, Head of the Department for the Regulation of the Mining and Smelting Works, in the following terms :—[2]

'Berlin : May 14, 1791.

'Sir,—The unbounded confidence inspired in me by the estimation in which the character of your Excellency is universally held, combined with the remarkable kindness with which you received my small work upon the nature of the basalts of the ancients and the basalts of the Rhine, leads me to hope that your Excellency will forgive my boldness in venturing to lay before you a *sketch of my future public life.*

'I have now attained an age when I cannot but wish to enter upon some fixed sphere of labour, and devote to the service of my country such limited powers as I possess. The

[1] 'Preussische Jahrbücher,' vol. xvi. pp. 139–148.

[2] The original is preserved among the papers of the Royal Mining Department at Berlin.

wish to complete my education by securing an express commission from your Excellency to labour in the various departments under your charge, arises not only from the decided taste I possess for geological investigations, and the strong interest I feel in all the operations connected with mines and salt works, but still more from the flattering hope of some time being able to associate myself with you in the accomplishment of those grand and beneficent plans, by which your Excellency has for a long series of years not only succeeded in opening up to the State new sources of national wealth, but has shown how wealth could be best employed, in accordance with the unalterable principles of philosophy and sound political economy.

' My studies have hitherto been principally directed to finance and the science of political economy, the pursuit of which has been the main object of my residence at Frankfort-on-the-Oder and at Göttingen, and of my travels through Germany, the Netherlands, Holland, and England, as well as of my sojourn at Hamburg. In order to complete my scientific education, and acquire a practical acquaintance with mining and the machinery employed in its various operations, I am anxious to spend six months at the School of Mines at Freiberg. It would, however, be a great relief to my mind could I arrange something definite about my future career in life before taking further steps towards the accomplishment of this design; I venture, therefore, humbly to request of your Excellency to dispose of me as you may see fit, to grant me access on my return to the reports in your department, and to permit me at once to hold some appointment in the administration of the mines and smelting works.

' I remain, with the profoundest respect,
' Your Excellency's most obedient servant,
' A. VON HUMBOLDT.'

In little more than a fortnight after this date, Alexander von Humboldt received a reply on the thirty-first of the same month couched in the following flattering terms:—'That his Excellency, out of consideration of the attainments already possessed by Herr von Humboldt, and on account of the additional knowledge he proposes to acquire during his stay at Freiberg, as

well as in consideration of the zeal evinced by his offer of
service to the State, is willing to grant him employment in the
various departments presided over by his Excellency, and therefore makes him the preliminary promise that immediately upon
his return from Freiberg next winter he shall not only be
commissioned to draw up reports on the salt works and mines
in the province of Westphalia, but that, in order that he may
become practically acquainted with the management of the
general correspondence, book-keeping, &c., he shall also be
appointed assessor *cum voto* of the mines, smelting works, and
turf-fields. At the same time he is informed that after having
visited the salt works at Schönebeck and Halle next spring,
and become familiar with their mode of management, he will
then be deputed to inspect the salt works of neighbouring
countries.'

In the journey to Freiberg Humboldt passed through Dresden, where he spent several cheerful, happy days in the family
of Herr Neumann, Secretary of War to the Elector of Saxony,
to whom he thus writes on his arrival at Freiberg:—' During the
last year or eighteen months that I have been wandering about
at my own disposal, I can scarcely recall any time in which I
have experienced more intellectual and æsthetic enjoyment
than during the few days I spent with you and your family.
If thanks could be given in return for love and proffered
friendship, you should have mine, but the best and purest
thanks I can render to one of your nature must be the deep
appreciation of your affection.' Farther on he continues:—' If
it be a gratifying spectacle to observe the harmony that reigns
among the greatest diversities of inanimate nature, it is still
more delightful to see good men, closely united by the ties of
affection, striving towards one common object of the highest
intellectual development, and to see this object near its attainment. Such an enjoyment was granted to me in the intercourse
I had with you and your family.'

The School of Mines, established by Heinitz at Freiberg in
Saxony in 1766, enjoyed at this time the highest repute in consequence of Werner's celebrity. Werner was regarded as the
first geologist of the day and the founder of that science.
No one could approach him in his knowledge of minerals,

and the authority of Linnæus in the science of botany was never higher than was that of Werner in the science of geology. Under his banner the upholders of the aqueous theory had gained a decisive victory over those contending for volcanic agency in the incessant strife concerning the history of the formation of the earth's crust. The weight of his authority, supported by personal examination of the Erzgebirge, completely set aside the theory of the upheaval of mountains, although the strongest evidence of its truth could be adduced in other districts. The strength of his influence may be inferred from the theories he introduced into the science Geology was at this time passing through a phase by no means unusual in a new science, a phase in which it seems necessary to follow to their ultimate consequences the principles of a theory before its weakness is revealed and truth discovered. Thus the native genius of his gifted pupil was insufficient to break the ban under which the teachings of the master had laid him, till through his extensive travels he had contemplated Nature in her widest aspects.

'Werner,' writes Alexander von Humboldt,[1] thirty years later, 'the father of the science of geology, recognised with surprising acuteness the salient points to which attention must be directed in observing the separate formations occurring in the several classes of primitive, transition, and secondary rocks. He pointed out not only what was to be observed, but also what was essential to know; in regions he had never himself examined he anticipated some of the later discoveries—it might almost be said that in some instances he had a presage of the facts which geology was hereafter to reveal. As geological formations are independent of latitude and the vicissitudes of climate, a comparatively small extent of the earth's crust in any quarter of the globe, a region even of a few square miles in which several distinct formations are exhibited, may suffice, like the true microcosm of the ancient philosophers, to awaken in the mind of an experienced observer many just conceptions concerning the fundamental truths of geology. Thus most of the early conclusions of Werner, even those to which he had arrived

---

[1] 'Essai géognostique sur le gisement des roches dans les deux hémisphères,' translated into German by Leonhard: 'Geognostischer Versuch über die Lagerung der Gebirgsarten in beiden Erdhälften' (1823). p. 67.

previous to the year 1790, are characterised by an accuracy which still calls for admiration.'

Freiberg became the resort of mineralogists, geologists, and miners from all quarters of the globe—from Sweden, Denmark, Russia, Poland, Transylvania, Italy, England, France, Spain, India, and America. On July 14, 1791, Alexander von Humboldt entered the Academy as the 357th pupil. His reputation as 'an interesting young student of science' had already preceded him, and his 'Observations upon the Basalts of the Rhine' secured him a warm reception from Werner.

On June 15, the day after his arrival, he took his first lesson in practical mining by a descent into the 'Elector' with Karl Freiesleben, a fellow-student, who had been appointed by Werner to be his guide during a tour of inspection through the mines—an expedition which interested him so much, that in the following week the two young men undertook a journey through the central chain of the Bohemian mountains, the results of which appeared in the 'Bergmännisches Journal.' [1]

Humboldt was disappointed in obtaining accommodation in the buildings of the Institution, under the same roof with Werner, but apartments were prepared for him in the first floor of a private house,[2] at the corner of the Burggasse and Weingasse, since distinguished by a tablet commemorating the circumstance. From many of his expressions it may be gathered that personally he was not strongly attracted to Werner, whereas in the house of Freiesleben he felt as a beloved member of the family, everyone being at pains to further the object of his stay. It was, however, to Karl Freiesleben, his junior only by two years, that he was pre-eminently attracted, and to him he became devotedly attached. It was one of Humboldt's characteristics, evinced early in life, to select, wherever he might be, one particular friend upon whom to lavish the full force of his affections: thus in Berlin it was Beer, later Willdenow, at Frankfort Wegener, and at Hamburg Wattenbach, with whom by turns he was accustomed, like his distinguished brother, to revel in sentimental friendship and

[1] 1729, vol. i.
[2] Scherer, 'Theorie und Praxis (Freiberg, 1867), p. 143.

indulge in a correspondence which, from its enthusiastic character, recalls the extravagant devotion of the Hainbund,[1] a
passionate sentimentality of which we cold Epigoni have now
no comprehension. Of all these friends, however, none exercised so decided and permanent an influence over him as
young Freiesleben, who became his inseparable companion,
ever ready with advice and enthusiastic sympathy in the
various branches of his mining studies. It may be confidently
asserted that in the moulding of his thoughts in this direction, Humboldt felt more indebted to young Freiesleben
than to any of the instructors under whom he studied at
Freiberg—a statement that receives ample confirmation in the
long and connected series of letters addressed to him by Humboldt between the years 1792 and 1799, as well as by various
occasional letters written up to the latest years of his life.

The subjects at this time embraced in the college training
may be thus sketched. Werner gave lectures on the art of
mining, the classification of minerals, geology, and the smelting
of iron, besides conducting classes for the working out of
problems in mining. Charpentier, who in the first instance had
lectured on particular branches of mining operations, especially
on the machinery employed, had in 1784 been succeeded by his
pupil Lempe, who, in addition to these subjects, gave instruction on pure and applied mathematics. Köhler lectured on
the existing laws of the country relating to mining, Klotzsch
on the art of essaying, while Freiesleben, the father of Humboldt's youthful friend, gave instruction in practical survey-
ing as applied to mines. A lectureship in chemistry was first
added in 1794 under the tuition of Lampadius.

With so many opportunities for acquiring an intimate acquaintance with the various branches of this science, Humboldt
was able to write as follows to his friend at Dresden:—'I am
living very happily here in Freiberg, although alone. I shall
be able to accomplish the various scientific purposes I wished to
attain by a residence here. I entered upon my labours, which
are certainly of a multifarious character, immediately upon my
arrival. I spend nearly every morning from seven till twelve in
the mines, in the afternoon I attend lectures, and of an evening

---

[1] [A society formed among the students at Göttingen.]

I go moss hunting, as Forster calls it. Werner shows me much kindness, and for the cordial reception I have met with at the house of Charpentier I am indebted to you, my dear friend.'

Under Werner and Charpentier, with both of whom he was admitted to familiar intercourse, Humboldt devoted himself with true enthusiasm to the study of the art of mining, not only as a science but also in its practical details. His ' Flora subterranea Fribergensis ' testifies to the wide circuit of his wanderings in the extensive labyrinths of the mines around Freiberg—expeditions which he took according to the plan laid down by Werner, in company with Freiesleben, impelled by the excitement so often kindled in young and ardent natures by the dangers of subterranean explorations.

Various other studies of a different nature also strongly excited his interest at this time, and were carried on during his leisure hours. As no official instruction was then given in chemistry, he devoted himself to the study of the works of the French chemists, Guyton de Moreau, Fourcroy, Lavoisier, and Berthollet—a pursuit in which he enjoyed the sympathetic companionship of Franz Baader of Munich, who had already been three years in Freiberg, and had made himself a name by his treatise on the nature of heat, and reviews of scientific works. It was while engaged in the study of fossils, during these wanderings in the vast subterranean passages, that he conceived the happy idea of turning his attention to the vegetation of that lower world from which the light of day is ever excluded. His ' Experiments and Observations on the Green Colour of Subterraneous Vegetation,' the result of his researches in his ' small subterraneous garden,' which could not be visited by a single ray of sunshine, and could at most be illuminated by the meagre, unproductive light of a miner's lantern, is closely allied to the investigations of Bonnet, Priestley, Ingenhouss, and Sennebier upon the influence of sunlight on vegetable organisms, and was a preparation for his comprehensive work on the physiology of plants, ' Flora subterranea Fribergensis.' The phenomenon that vegetation, even when wholly excluded from light, was yet tinged with various shades of green, regarded at that time as a very extraordinary fact, was explained by Humboldt to arise from the disengagement of oxygen, in which the luminous matter constituting the basis of noxious gases exerts a powerful influence. He requests, however, with the unassuming love of truth by which he was even then so honourably characterised, that the facts educed might not be confounded with the explanation he had offered concerning them, and quotes the too often neglected words of Spinoza : ' Videmus enim omnes rationes, quibus natura explicari solet, modos esse tantummodo imaginandi, nec nullius rei naturam, sed tantum imaginationis constitutionem indicare.'

Humboldt thus alludes to his literary occupations in a letter to Wattenbach, written shortly before he left Freiberg, and dated February 18, 1792 :—

' If the few letters which I find time to write now did not all begin with excuses, I should willingly find some for you, my dear Wattenbach. But I am heartily tired of making the everlasting excuse of want of time. I am indeed quite distressed not to have sooner answered your kind and affectionate letter of November 14, but if you only knew how I am situated here, I am sure you and Hülsenbeck and all my other friends would excuse me. If you will bear in mind that during the nine months that I have been here, I have travelled nearly 700 miles on foot and by carriage through Bohemia, Thuringia, Mansfeld, &c.—that I am daily in the mines from six till twelve o'clock (nearly two hours are occupied in the transit, which in the snow is very fatiguing)—that as many as six lectures are crowded into an afternoon—you will then be in a condition yourself to pronounce my sentence. I have never been so much occupied in all my life. My health has suffered in consequence, although I have not been laid up with any serious illness. Nevertheless I am on the whole very happy. I follow a profession which to be enjoyed must be followed passionately ; I have acquired an immense amount of information, and I never worked with so much facility as I do now.

' Everlastingly about myself ! You have been ill, poor fellow ! I heard of it from Maclean. I was heartily sorry for you. Illness in itself is no misfortune, but the monotony of the life and the commiseration of others is insupportable. The death of Pepin and Metzer of Embden will be a loss to the Academy. Pray take occasion to mention wherever you go that Von

Heinitz, Chief Director of Mines, may possibly send his son to Büsch. I hear a whisper of it occasionally, but I doubt if it ever comes to pass; however, pray speak of it whenever you can. Madame will be delighted at the mere thought of such a thing. I was intensely amused by what you told me of Giseke and Flottbeck. Please ask Gill for his address in Amsterdam. I think I must still owe you some money, but I have not the slightest idea how much. I have also to send some money to Arendt, and I will remit it to you both from Berlin, where I expect to be in a week, as the postage from here is so heavy. I was delighted with Böthlingk's letter. I love that man intensely, for he is by no means so cold-hearted as he forces himself to appear; I consider him in every way worthy of esteem. What has become of Losh? You must send me a complete list of all the academicians with their separate histories. Tell me something of yourself, the studies at present engaging your attention, and the prospects you have for the future ; you know how deeply all this will interest me. By my own fault, I am so completely cut off from all correspondence that I have not had a line from Forster for six months. In what part of Switzerland is Speckter now? As for Hülsenbeck, I cannot for shame write to him. I don't know what he must think of me, for he wrote me an exceedingly warm-hearted and affectionate letter, asking me to do him a trifling service; and I am ashamed to say I never answered him by a word. Steps, however, have been taken on behalf of Herr Christ. Mund, partly by my own exertions and partly through my brother's influence ; but owing to the dilatoriness of Counsellor Klein in the Court of Judicature, nothing further has been heard of his affairs. Please explain all this to Hülsenbeck and make my peace with him. I am sure that neither of you think I would be really disobliging to anyone, least of all to such friends as you both are to me—friends to whom I am indebted for so many agreeable hours. I shall write to Hülsenbeck at once on my arrival at Berlin.

'I have now a small favour to ask. Will you kindly give the enclosed pamphlet to Brodhagen, and by help of a little flattery persuade him to review it in one of the papers? You will know how to do this without in the least compromising me or my dignity. To blow one's own trumpet is part of an author's trade, therefore I hold reviews to be of some importance. Brodhagen might avail himself of the opportunity afforded by the new impulse lately given to the journal by the advent of Herr Hoffmann, who since January has joined Herr Köhler in the management of the magazine.

'I have been living lately very much among the printers. Only think of the following articles being all either actually printed or else going to press :—

'For the "Annalen der Botanik."

'1. On the Motion of the Filaments of the Parnassia Palustris.
'2. On the Double Prolification of the Cardamine Pratensis.
'3. Diss. de plantis subterraneis Fribergensibus.

'For Gren's "Journal der Physik."

'Experiments upon the Green Colour of Subterraneous Vegetation.

'For Crell's "Annalen."

'Table of the Conducting Powers of Heat in various Bodies, calculated according to Maier's Formulæ.

'Do not in any case omit to read my essays on the "Theory of Evaporation," and on some "Experiments on the Decomposition of Common Salt." Both are new.

'So much discovered and observed! Nos poma natamus!

'HUMBOLDT.'

Among Humboldt's fellow-students may be mentioned Leopold von Buch, who subsequently attained so high a position in science; Esmark, a Dane, who died in 1840 as Professor of Mineralogy in Christiania; Andrada, a Portuguese; and Del Rio, a Spaniard, whom he afterwards met twelve years later in Mexico as a teacher in the Colegio de Mineria.

Humboldt's departure from the School of Mines was signalised by a formal celebration given him by his friends on February 26, on which occasion he was addressed in two poetical effusions, one in the German and the other in the Latin tongue.

In a spirit of grateful remembrance Humboldt alludes to this period of his life in a letter he addressed to his friend Fischer, on February 8, 1847, when congratulating him on attaining the jubilee of his degree of Doctor : [1]—' Allow me to offer you my heartiest and most fervent congratulations, for I was privileged, together with our departed friend Freiesleben, to be among the first to recognise the greatness of your talents and the amiability of your character. Can you still recall to mind the garden behind the church at Freiberg, the sojourn at Dresden with Reinhard von Haften, the residence at Paris, where Caroline von Humboldt was your pupil, and where you received so many gratifying marks of esteem, both from my brother and from Cuvier ? These are reminiscences from the world of shadows which are to me most precious and affecting ! '

In his address at the festival in commemoration of the centenary of Werner's birth, on September 25, 1850, Humboldt expresses his deep sense of the obligations he was under to the institution at Freiberg and the powerful influence which his sojourn there exercised upon the whole course of his life. He therein states that he was indebted to the comprehensive grasp and methodising power of Werner's genius for an important part of his mental culture and for the direction which had been given to his efforts ; that it was his constant endeavour to honour the name of Werner and to elevate his works, in modern times so often misunderstood, to their right position ; that he had devoted his energies for several years exclusively to the practical art of mining ; that he felt proud of having held the office of Superintendent of the Franconian Mines in the Fichtelgebirge ; that the happiest recollections of his youth were associated with all those advantages, for which he felt indebted to that excellent institution, the School of Mines at Freiberg, which had, especially during the brilliant period of Werner's administration, exercised so powerful an influence not only upon the rest of Europe, but even upon Spanish and Portuguese America ; and finally, how much he owed to the encouraging kindness of the various officials in the mines in Saxony, and to

---

[1] ' Séance extraord. de la Société impér. des Naturalistes de Moscou du 22 févr. 1847, à l'occasion du jubilé semi-séculaire de S. Exc. M. Fischer de Waldheim,' p. 7.

the instructive companionship of his fellow-pupil and co-worker Karl Freiesleben.

The completion of his studies at Freiberg brought the college life of Humboldt to a close. But neither to mineralogy nor to mining, neither to botany nor to physics, nor yet to chemistry, had he exclusively devoted his attention; his interest had been excited much more to ascertain the conditions of organic life which he had already sought to discover even in the darkest and deepest recesses of the mines. The laws regulating the growth of plants, the rapid germination of seeds in diluted oxydated muriatic acid,[1] the movement of the delicate filaments of the Parnassia Palustris, the cause of the production of green in the most intense darkness, were all only preparatory studies for his later, more comprehensive researches.

The nobility of Humboldt's frank and ingenuous character is strikingly portrayed in the following extracts.

In a letter addressed to his friend Neumann at Dresden, a week after his arrival at Freiberg, on June 23, he thus speaks of himself:—

'You have seen me, my dear Neumann, as I wish to appear before my friends. Warmth of heart and frankness of disposition are the only excellencies to which I venture to lay claim. It is these qualities which have gained for me the friendship of Jacobi and the confidence of our mutual friend Forster, and since they have now procured for me an affectionate interest in your heart, their value in my eyes has been trebled. I know I am hasty and inconsiderate in my judgments, an impatience you must forgive on account of my youth and the peculiar circumstances of my early education. Though powerfully influenced by the dictates of reason, I am only bewildered by the suggestions of the imagination—in a word, it cannot have escaped your own observation, nor that of your wife, how completely unfinished my character yet is, and how much there is in me still needing to be developed.'

The following sketch of Humboldt's character at the time of his departure from Freiberg is given by Freiesleben, afterwards

---

[1] [In the language of modern chemistry, a weak solution of chlorine.]

Counsellor of Mines, who at this period was his intimate friend
and the companion of his studies: [1]—

‘The salient points of his attractive character lay in his
imperturbable good nature, his benevolence and charity, his
remarkable and unselfish amiability, his susceptibility of friendship, and appreciation of nature ; simplicity, candour, and the
absence of all pretension characterised his whole being; he
possessed conversational powers that made him always lively
and entertaining, together with a degree of wit and humour
that led him sometimes to waggishness. It was these admirable qualities, which in later years enabled him to soften and
attach to himself the untutored savages among whom he dwelt
for months at a time, which obtained for him in the civilised
world admiration and sympathy wherever he went, and which
gained for him while a mere student the esteem and devotion of
all classes at Freiberg. He was kindly disposed towards everyone, and knew how to make himself useful and entertaining in
every circle of society ; and it was only against every species of
inhumanity and coarseness, against every kind of insolence,
injustice, or cruelty, that he ever manifested either scorn or
indignation, while to pedantry and sentimentality, or, as he
called it, “the sloppiness of feeling” [Breiikeit des Gemüths],
he invariably showed the greatest indignation.’

[1] From an earlier biography of Alexander von Humboldt, published in
the ‘Zeitgenossen’ (Leipzig, F. A. Brockhaus), 3rd Series, vol. ii. Part I.
p. 67.

CHAPTER III.

OFFICIAL EMPLOYMENT.

Assessor of Mines—Animus of Official Administration—Official Employment only a Stepping-stone to more important Scientific Schemes—
Commission in Franconia—Appointment as Superintendent of Mines—
Extension of Commission to January 1793—Visit to Berlin—' Flora
Fribergensis '—Experiments on Sensitive Organisms—Commencement
of Official Employment in Franconia, May 1793—Condition of Fran-
conia—Free Schools for Miners—Practical Undertakings—Scientific
Labours—Appointment as Counsellor of Mines, 1794—Commission
in Southern Prussia—Diplomatic Service under Möllendorf—Repeated
Offers to become Director of the Silesian or Westphalian Mines
and Salt Works—Refusal—Tour in Switzerland, 1795—Return and
well-directed Industry—Dangerous Experiments—Visit to Berlin—
Diplomatic Mission to Moreau, 1796—Proposals for further Official
Service—Death of his Mother.

THOUGH the residence of Alexander von Humboldt at the School
of Mines at Freiberg scarcely extended beyond eight months,
from June 14, 1791, to February 26, 1792, yet even this limited
period sufficed to enable him to acquire every qualification then
necessary for official employment in the Mining Department of
the State. A farewell fête was given him by his fellow-students
at Freiberg on February 26, and three days afterwards, on the
29th of the same month, a ministerial rescript, in fulfilment of
the promise made by Heinitz, was issued from Berlin, appointing
' Alexander von Humboldt Assessor *cum voto* in the Administrative Department of Mines and Smelting Works,' an appointment
conferred upon him, in the words of the official document, ' on
account of the valuable knowledge, both theoretical and practical, possessed by him in mathematics, physics, natural history,
chemistry, technology, the arts of mining and smelting, and
the general routine of business.' It was further enacted that

before being attached to any special department, the Assessor von Humboldt should devote himself to the internal regulation of the service, such as the mode of conducting the correspondence, the drawing up of reports, &c., in furtherance of which he received a commission to inspect the various processes employed in the smelting establishments, in the lime quarries and kilns, and in the peat works.

We will now take a glance at the official practices and mode of administration prevalent in those days.

In the various boards of administration, the inseparable evils arising from bureaucracy, red-tapeism, and idleness prevailed to a very large extent. The subalterns by sheer drudgery accumulated a vast amount of so-called valuable information, according to the caprice or perverted notions of their superiors. Scientific education among the officials was quite the exception, and any participation in literary work was as good as forbidden. Von Hippel, subsequently holding the position of President, dared not write under his own signature. A candidate for office who had given a conditional affirmative to the question, whether scientific pursuits comported with an official position? had his papers, which were in themselves excellent, returned to him by the presiding examiner, with the significant remark, that opinions of that nature were inadmissible. Stein was accustomed to relate of the minister Count von Hagen, that on one occasion, when his subalterns came to congratulate him on his birthday, he received them with the greatest cordiality; but when they were about to present him with a printed copy of their congratulations, the minister stopped them somewhat harshly with the remark: 'You know I read nothing in print; give it me in manuscript!' The most able officials and statesmen had alike fallen into a state of literary and scientific stagnation. According to a communication from the President von Schön[1] to the Burgrave von Brünneck, even Stein himself up to the year 1808 had not read a line of Goethe.

These and a number of other evils, equally great, had not escaped the observation of Kunth.[2] He knew the sacrifices that were required, in comfort, in expense, and in health, before

---

[1] From a private letter. MS.
[2] Stein's 'Leben von Pertz,' vol. vi. p. 75.

a conscientious official occupying a position on a provincial board, whether administrative or judicial, could rise, after a service of more than twenty years, to a salary of 600 thalers, or could obtain after thirty or forty years spent in constant exertion, in a dependent position, even a salary of 1,500 or 1,800 thalers. 'How many tradesmen and artisans do I not know,' said he, 'who, as regards their position of affluence and independence, would ridicule the offer of exchanging their occupation for the most lucrative post in the ministerial council!' Since the publication of Wöllner's edicts, and since the maladministration of such men as Görne, Struensee, &c., the whole system of official life had sunk into a state of unheard-of corruption and depravity. In acknowledging with sadness such a condition of things, Von Vincke, afterwards President, thus expresses himself about that time (1793), when still a youth:— 'If I, with the requisite qualifications for serving my country, could not obtain official employment without first becoming a Rosicrucian, a visionary, an alchemist, a hypocrite, or an intriguer, I would rather be a merchant than be thus obliged to submit to the unreasonableness of prejudice and self-interest.'

When it is remembered that William von Humboldt voluntarily resigned his office in the service of the State after practical experience during a year and a half of this deplorable condition of official life, it may well be conceived that the eagerness displayed by Alexander von Humboldt to devote himself to public life must not be interpreted as a wish to enter upon an ordinary official career. The service of the State was from the first regarded by him merely as a stepping-stone to the service of science. To this position of independence may be ascribed his exceptional preferment over the highest of his superiors, to this may be traced his fearless judgments of persons and things, and his indifference, not to say ironical disregard, of all public recognitions and preferments, as well as of the most flattering proposals for continued employment.

The department of the public service in which Humboldt entered formed an honourable and happy exception to the corruption so rife in the administration of all other official boards. Heinitz, the minister of this department, was one of the most excellent men of his time. The principal traits of his estimable character were genuine religious feeling, an earnest, persevering endeavour to cultivate the purest affections, an absence of all self-seeking, a susceptibility to everything noble and beautiful, and inexhaustible kindness and gentleness. He was distinguished for the lively interest he took in the subject of education, for the readiness with which he acknowledged merit in his subordinates, and the care with which he ever selected only worthy and capable men for the appointments he had to bestow. While in the service of the Brunswick-Hanoverian Government he had brought the mining works in the Harz Mountains to a flourishing condition, later in 1766, when in the employ of the Elector of Saxony, he founded the School of Mines at Freiberg, and since 1777 he had devoted himself to the task of improving the condition of the Prussian mines and smelting works, which were at that time at a very low ebb.

Such was the state of the public service when Alexander von Humboldt, at the age of twenty-two, entered upon official employment. The path to honour and dignity in the State lay open before him, and, without waiting for any proof of his ability, he was received with the highest expectations and in the most flattering manner.

The enjoyment experienced by Humboldt in the execution of the commissions entrusted to him whereby his interest in science and love of travel found constant gratification, may best be gathered from his letters to his friends and recent fellow-students, especially from his correspondence with Freiesleben; and these letters, together with some legal documents still preserved in the Prussian and Bavarian archives, furnish the best means of tracing the course of events in his official life. As early as March 2, 1792, he wrote to Freiesleben from Berlin :—

'It was impossible for me to write sooner, as the mails for Saxony only leave here on Tuesdays and Saturdays, and I arrived so late on Monday evening that the post-office was already closed. . . . It is still quite uncertain how long I remain here ; it must be decided now in the course of a few days. My memoir on the salt works and their mode of management, and my other literary efforts, have produced a sensation. The minister has loaded me with encomiums. It seems as if everything were conspiring to meet my wishes, and yet I regard the whole thing as I should the issue of a game at chess—that is to say, almost with indifference. The olfactory nerves are rendered at length insensitive, from the incessant offering of incense so unmerited!'

Only a few days later, on March 7, he writes to the same friend:—'I received yesterday my commission as Mining Assessor *cum voto* in the mining and smelting departments. I felt quite ashamed of myself for being elated by such a trifle. I have, however, taken no steps to gain this post. It seems unfair to make me at once assessor over the heads of a whole troop of cadets and former pupils, &c., since my literary merits can yield neither ore nor water-power, certainly not the latter. I have openly expressed this opinion here, but I have been told that in this department no one had a better claim than myself, and this is very likely true. I shall be sworn into office next week, and introduced to the board. I shall certainly not remain long in Berlin, for I have expressly begged to be relieved from a residence there, since to my mind Berlin is no more suitable for an administrative board of mines than for a board of admiralty. I shall have first to go to Halle, Rothenburg, &c., and thence?—the minister will have to decide. Count Rheden gave me to understand, in our first interview, that he thought I had studied the practical details too closely, that technicalities should be left to the old hands, and that a man in my position was certainly not born to be a common foreman. This did not in the least disconcert me; I told him that I considered that everything depended upon an accurate knowledge of technicalities, for establishments that deal only in generalities accomplish little. I mention these trifles merely as an instructive comment on the philosophy of life. The very man who hates all scientific study is the one to reproach me for having educated myself to be a practical miner. How consistent! Rheden is particularly kind and pleasant to me now. He suffers much in health, and great allowance must be made for him on this score. I observe a great change in Karsten, much to his advantage. I believe him to be truly kind-hearted, and his manner towards me is wholly free from assumption. His style when speaking is quite different from his letters or books. He has a great respect for you. I do not see much of him, nor indeed of anyone but Willdenow, since I am still engrossed with botany. He is the first to make me rightly appreciate the value of my " Flora Fribergensis." He thinks it all new, and exceedingly remarkable, and strongly urges me to prepare a more complete edition.'

About six weeks later, he writes to the same friend on May 19 :—' I keep well and work a great deal at night. As yet my official post has given me but little to do. My course is still undecided ; it is certain, however, that I do not remain here—I shall probably first go to Thuringia, and then to Westphalia. For with practical mining *I will* have to do.'

He thus again writes to his friend on June 4, 1792 :—' In five or six days I am to leave for Linum, where the extensive peat-cutting works are, then to Zehdenick, to see the smelting furnaces, and afterwards to Rheinsberg, where I have received a commission to inspect the porcelain manufactory. Suitable occupations for a miner !! There will, however, soon be an improvement. I am delighted at the prospect, which I must communicate to you in half-a-dozen lines, of going to Bayreuth and the Fichtelgebirge, in about three weeks. I have been honoured by a commission to investigate the geological structure and mineralogical constitution of the two Margraviates. Eight weeks only have been granted me at first, that I may merely travel through the country and furnish a general report to the minister. What I shall do then, whether I shall remain there altogether (and become Overseer of Mines !!) or go to Silesia, is quite uncertain at present. I am quite delighted at the thought of seeing a new mountain range, and so many different kinds of mines, and to be once more in your neighbourhood. It will be impossible for me to pass through Freiberg, for I am obliged to go by way of Erfurt and Saalfeld. My route is rigidly prescribed me—irreparabile fatum ! ' [1]

[1] Humboldt's official commission included a visit of inspection to the Royal Porcelain manufactory, where he greatly interested himself in the erection of the first steam-engine, or ' fire-engine,' as it was then termed. He often referred to this 'pre-ogygian' activity. So lately as October 12, 1857, he wrote to the proprietor of the extensive porcelain factory at Herend, near Veszprim, in Hungary :—' I recall with pleasure that when I was twenty-two years of age I was appointed with Klaproth, the celebrated chemist, on some technical business connected with the Royal Porcelain Manufactory

The next letter is dated ' Gräfenthal, July 11, 1792.'

' I am so tired, dear Freiesleben, with going about and descending mines and writing reports, that I can scarcely keep
up. Yet I must write you a few lines to tell you I am well. I
must solace myself in these late hours of the night by the
affectionate remembrance of yourself, and alas ! of those happy
hours we have spent together. . . . " Where have I been ? "

' Ask rather where I have *not* been. My last expedition was to
Saalfeld, Kuhnsdorf, &c. I have enjoyed myself immensely,
amid this constant bustle.   Only think what I did in *one*
day ; I walked to Saalfeld and back, and in the most frightful
heat spent from four o'clock in the morning till six o'clock in
the evening in visiting the several mines—the Pelican, the
New Joy, the Unexpected Pleasure, the Iron John, and the
Twilight.   I am quite foot-sore with so much running about,
but that will soon be better.   How many things crowd upon
me that I should like to talk over with you!   At present I
shall only throw out hints of the subjects about which we
shall have to correspond. . . . . Thus much for to-day. To-
morrow I leave for Naila.'

The Commission to the Franconian Principalities, to which
Humboldt refers in the letter to Freiesleben above quoted, was
undertaken at the instigation of the minister Von Heinitz, who
was anxious to institute some important improvements in the
management of the various mines and smelting works in those
provinces, as well as in the porcelain manufactories and salt
works of Gerabronn.   While undergoing a course of mineral
waters at Carlsbad, Heinitz resolved, as a preliminary step, to
send Humboldt, armed with a commission, to inspect the chief
establishments in the principalities, and on his return to meet
him at Bayreuth with a full report of their condition.   For
this purpose he addressed a communication bearing date July
23, 1792, to the Government at Berlin, and the minister,
Hardenberg, with whom he had often conversed of Humboldt's
extensive information and solid acquirements, furnished all
necessary official assistance for the prosecution of this tour of
inspection.

at Berlin, and that I have even made experiments in the process of rolling
the clay.'

Towards the end of August, the minister Von Heinitz arrived at Bayreuth, and Humboldt there reported to him that in the execution of the commission, he had not only made an inspection of the mines, and investigated the general formation of the mountains, examining particularly the beds of ore, but had collected statistical and general information concerning the former mode of working the mines and smelting works. He then enlarged upon the general character of the Franconian mountains and the condition of the three mining stations, Wunsiedel, Goldkronach, and Naila, their development and decay, and upon the means of working the mines in such a manner as to ensure their being self-supporting, and to allow funds for repairs and improvements—in short, he dwelt upon all that had reference to the administration and technical arrangement of these mines and smelting works. Entering upon his favourite subject, he discussed in a supplementary report the nature of iron, and the formation of sulphuric acid in the manufacture of alum and vitriol, besides reporting on the salt works at Gerabronn and Schwäbisch-Hall, on the porcelain manufactory at Bruckberg, on the vitriol works at Schwefelloch near Gräfenthal, &c.

By this preliminary verbal report Humboldt at once gained the warmest approbation of both ministers, and when he afterwards presented it in the form of nearly 150 sheets of manuscript, the acknowledgment was expressly made ' that . . . von Humboldt had not only displayed therein a further proof of his praiseworthy and indefatigable activity, but had with sound judgment and penetration pointed out how by good management, and a judicious application of scientific principles, the mines, smelting furnaces, and salt works of the Franconian Principalities might be raised to a condition of prosperity, and even considerably extended, concluding with the recommendation that the Report ' should be circulated among all the officials of the department, in order that they might become accurately acquainted with its details.'

Such an acknowledgment, so honourable to all the parties concerned, was not likely to remain long without results.

Accordingly, on August 27, Humboldt writes to his friend:—

' Just two glad words, my dear friend, to tell you, on con-

dition of profoundest secrecy, that I was yesterday appointed
Superintendent of Mines in the two Franconian Duchies. I
have won so much renown by my report upon mines, that
the sole direction of the practical working of the mines in
the three districts of Naila, Wunsiedel, and Goldkronach has
been committed to me. All my wishes, my dear Freiesleben,
are now fulfilled. I shall henceforth devote myself entirely
to practical mining and mineralogy. I shall live among the
high mountains, at Steben and Arzberg, two villages in the Fichtelgebirge, where the geological formations are peculiarly interesting, and I shall be near you and able to see you at least
once, and perhaps many times in the course of the year. . . . .
I am quite intoxicated with joy. I shall probably not be able
to see you this autumn, but I hope we may meet some time
during the winter, or in the spring at Leipzig. Do not feel
anxious about my health; I shall take care not to overexert
myself, and after the first the work will not be heavy. I cannot conclude without acknowledging that it is again to you
that I am indebted for this happiness; indeed I feel it only
too keenly. What knowledge have I, dear Freiesleben, that
has not been taught me by you! It is only a year ago since I
was asking you what a winze [Gesenk] meant, and now I am
Superintendent of Mines. What wonderful progress I have
made! It is very impudent of me to undertake such a post.
However, I did not seek it, I even remonstrated against it; but
it was pressed upon me, and the thought of living in your
neighbourhood made me yield. Think how much I shall learn
in such a position! *None of our plans* will be disturbed by
this arrangement. Heinitz told Hardenberg that he could only
spare me for a year or two, and assured me personally that the
journeys already planned should suffer no interruption. I think,
therefore, dear Freiesleben, that you will rejoice with me.'

It was while Heinitz was still at Bayreuth that Humboldt
received the appointment of Superintendent of Mines in the
Franconian Principalities—an appointment bestowed upon him
only six months after his entrance on official life. But remarkable as was the mere fact of this rapid promotion, the manner
of its accomplishment, as shown in the transactions between
the two ministers, is still more remarkable.

On September 6 an official communication was addressed by Hardenberg, Governor of the Principalities, to Heinitz, requesting that Assessor von Humboldt might be appointed for a few years to the office of Superintendent of Mines in the district round Bayreuth, since no more suitable person could be found at Berlin for a position involving so much responsibility. The reply of Heinitz, granting his entire concurrence to Hardenberg's request, is dated even the same day. He accompanies his consent with the proviso, highly flattering to Humboldt, that he should report periodically to the head of the department at Berlin—with whom he would retain his connection throughout this temporary appointment—on the progress of his work, the improvements he introduced, and the discoveries he might make of any geological phenomena. He then proceeds to suggest to Hardenberg the propriety of committing to Humboldt a search for salt-springs throughout the provinces, as well as the inspection of the vitriol and alum works at Crailsheim and the porcelain manufactory of Bruchberg. He concludes his letter by informing Hardenberg that Humboldt would not be able to enter upon the duties of his new position before March in the approaching year, since he would be occupied till then in completing a commission in Bavaria and Silesia on salt works and methods of evaporation.

The official notification of his appointment as Superintendent of Mines was at once communicated to Humboldt, and it is worthy of remark that in the rough draft of this document, preserved among the official papers, there appear some words in Humboldt's own handwriting. There is no allusion in the documents to the amount of salary attached to this office, but from other sources it is certain that it could not have been more than 400 thalers.

On the selfsame day, September 6, 1792, he writes very fully to Freiesleben in a most joyous strain, and 'pours out his whole heart to his friend.' The events of the last two months pass before him as in a magic mirror, his imagination is filled with most pleasing impressions of the present, and with most delightful plans for the future; to work side by side with this friend, to travel in company with him, is all his desire. Before all other considerations come the assurances of tenderest friend-

ship, and expressions of deepest thankfulness for the variety of
instruction so happily conveyed. 'How sweet is the thought
to me, dear Freiesleben, that it is to you that I owe all this;
it seems as if it bound me closer to you, as if I carried something about me that had been planted within me and cultivated by yourself. Do not write a syllable in reply to this. I
can easily imagine what your modesty would dictate, but leave
me in the enjoyment of my feelings, for I rejoice in them
inexpressibly.'

In the midst of the daily routine of his new duties, Humboldt's bent for historical research was soon apparent. 'In
addition to all this,' he writes, 'I am busily engaged in examining the ancient records of the mines. You know how much
I am set upon this. I have had three chests of mining documents belonging to the sixteenth century sent to me from the
fortress of Plassenburg, which I must read *ex officio*, since they
contain records of official inspections. They will make glorious
reading in the damp autumn weather of that wild region. As
I have already devised a plan for reworking the prince's mine
at Goldkronach, where we have come upon some gold in a seam
of antimony, I am anxious to make myself acquainted with the
past history of this mine, which has been abandoned since 1421.
I have already been fortunate enough to come upon traces of
an old gallery which had not hitherto been suspected. Do not
feel anxious about my health, dear Freiesleben. I have kept
extremely well throughout the summer, and my delicacy seems
to be leaving me. I consider, as I used often to tell you at
Freiberg, that the improvement in my health is entirely due
to my mining occupations; and though you were apprehensive
that my daily descents into the mines might prove injurious, I
am convinced that they were in fact highly beneficial.'

Humboldt had in reality good grounds for satisfaction.
His services had met with the warmest recognition from
both ministers. In his official employment he encountered
no opposition to his plans or wishes, but was, on the contrary, solicited to undertake commissions of a most flattering
nature, in which his love of travel found gratification. Even
while the minister Von Heinitz was still at Bayreuth, the
commission originally planned to Franconia was, as he inti-

mated in his letter to Hardenberg, so far extended as to include expeditions through Upper Bavaria, Salzburg, the Government salt works in Austria, Galizia and Upper Silesia, for the purpose of investigating the rock-salt mines and the processes of evaporation.

For details of these official journeys, which extended from June 1792 to the end of January 1793, we are again indebted to the letters to Freiesleben. Humboldt thus writes from Traunstein, in the district of Salzburg, on October 4, 1792:—
'. . . I am moreover very well, but more dull than I ought to be in this exceedingly interesting region. The weather is detestable, everything is covered with snow and hidden from investigation. Hence I go to Reichenhall, to visit the salt springs, and on to Hallein, Berchtesgaden, and Passau, whence somewhere about the 16th of October I go to Vienna. The country about here is sublime. I feel as if I had never seen a mountain before, everything here wears a new aspect. Real Alpine mountains rise one above another, pile upon pile. The Appenzeller (?) Alps lie before me as if I could lay my hand upon them. We must see them together some day, dear Freiesleben. You might easily accomplish it during a vacation tour. In three or four days you could join me at Bayreuth, and we should be able to reach here in the course of five or six days. The whole expedition from Leipzig and back would only require four or five weeks at the furthest.'

From Vienna, Humboldt writes on November 2:—'I arrived at Vienna on the 21st of October. Notwithstanding the heavy snow and considerable fatigue, I have accomplished an exceedingly interesting tour through the mountains of Salzburg and Berchtesgaden and the Austrian Alps. I visited Kessenberg (?), Hallein, Berchtesgaden, &c., districts where rock-salt abounds, and I paid a most instructive visit to Reichenhall, where I spent twelve days investigating the salt springs with Von Claiss, the director of the works. I regard him as undoubtedly the first authority on all subjects connected with the management of salt works, whether practical or theoretic. He possesses an extensive acquaintance with physics and mathematics, and has spent seven years in England, where he worked much with Franklin, he has also been for a long time in France; he is the owner of some salt works in Savoy and of some sulphuric acid works in Winterthur, and has besides the direction of all the salt works in Bavaria. I plied him with questions from morning till night, and I scarcely know anyone whose conversation I have found so instructive. He wrote of his own accord to tell me that he considered my treatise the best that has been written upon the manufacture of salt. I now see the subject in a clearer light, and I mean to postpone the publication till the autumn. I have gained a great deal of information from Claiss, and have got possession of some of Franklin's manuscripts upon methods of heating ; I also wish to complete my chart, showing the connection between all the salt springs of Germany. I don't think I have mentioned to you anything about this map. It originated from an essay appended to my report from Bayreuth, " On the Method of Boring for Brine." The leading ideas are that the mountains of Franconia, Suabia, and Thuringia have one main position of strata, that they are connected by a valley extending from twenty to thirty miles between Eisenach in the mountains of the Thuringian Forest and Osterode among the isolated Hartz Mountains, that all the brine in Franconia and Suabia flows in the upper gypsum, that all the salt springs in Germany lie in one given direction, that it is possible to draw lines upon the map, by following which salt springs may be found mile after mile, that these salt streams follow the general slope of the land, which throughout Germany is from the south-west to the north-east, and flow round the primitive rocks wherever these project above the surface.'

Even in later years Freiesleben frequently expressed regret that this map, showing the course of the salt-streams of Germany, and the treatise on the method of boring for brine, had never been printed and were unfortunately lost.

The last of the few letters still preserved of this journey is dated Buchwald, January 14, 1793 :—

' I was three weeks,' writes Humboldt, ' in Breslau, and the rest of the time at Waldenburg and Kupferberg among the Riesengebirge. Never longer than two days in a place, travelling in the midst of extreme cold and far into the night, that I might at least visit the principal mines, I found no possible opportunity of writing to you. At Breslau I stayed three weeks with Count Rheden. If ever I was industrious, I was so there. I wrote out my report on the salt works of Traunstein and Reichenhall, which occupied twenty sheets, and which I illustrated by drawings on twenty-one large sheets of royal paper. I have been more engaged than usual with drawing this year. I practise it unremittingly, especially plan drawing, which in my present position I find of great importance. I am just now engaged in some researches among the mountains with Count Rheden. I came here yesterday from his estate at the foot of the Schneekoppe in a sleigh, and to-morrow I go with him *recta via* to Berlin.'

During his stay at Breslau, Humboldt was elected associate of the Leopold-Charles Imperial Academy of Natural Science, in a manner that was peculiarly flattering to him. By the laws of the Academy, it is required that no one shall be eligible for membership who has not previously received the degree of Doctor from one of the Universities. The president, however, is empowered to summon at any time a sitting of the faculty in any of the Universities throughout the German Empire, for the examination of any candidate he may present to them, and for conferring upon him, 'examine vili superato,' a Doctor's degree by imperial authority; Von Schreber, therefore, the president at that time, who seems to have prophetically beheld, even at this early period (1793), the future achievements of the proposed member, proceeded without delay to arrange for Humboldt's reception on June 20. The following address was presented at his election :—

'Esto igitur, ex merito, nunc quoque noster! Esto Academiæ Cæsareæ Naturæ Curiosorum decus et augmentum, macte virtute Tua et industria, et accipe, in signum nostri Ordinis, cui Te nunc adscribo, ex antiqua nostra consuetudine cognomen *Timœus Locrensis*, quo collegam amicissimum Te hodie primum salutamus.'[1]

[1] The surname *Timœus* bears a flattering reference to the Pythagorean philosopher of that name, described by Plato in one of his dialogues, who by birth and fortune ranked among the first citizens of Locri, was invested with the highest offices of the town, and was the author of a work still extant on the nature and the soul of the world (περὶ ψυχᾶς κόσμου καὶ φύσεως). Sixty-two years later, in the year 1855, the President of the Academy, Nees von Esenbeck, renewed the homage of this honorary memorial. The

He remained at Berlin till the end of April, and, besides his official labours connected with the salt-works, was much occupied in the publication of the 'Flora Fribergensis,' and in various chemical and galvanic experiments upon plants and animals. He also entered upon some investigations in Hermbstädt's laboratory on the absorptive properties of sponge.[1]

Humboldt writes from Berlin to Wattenbach, on February 9, 1793, as follows:—' Since June I have travelled nearly 3,000 miles, and remained at no one place for any length of time. . . . For the last fortnight I have been at Berlin, where I intend remaining quietly till April, while I see my long-announced work " Flora Fribergensis " through the press. In April I leave for the Fichtelgebirge, as I have been appointed Superintendent of the Mines in the Franconian Principalities. During the summer I visited Suabia, Bavaria, the Tyrol, Vienna, Moravia, and Silesia, returning through Poland to visit the rock-salt works. Man is indeed a wandering being, yet ever glad to find himself once more among old familiar friends, for he does not easily forget the pleasures of social intercourse.' . . .

Three days later, on February 12, he writes to Girtaner:— ' My course of life is so far removed from your political career, that it has been scarcely possible for me to follow your successes, notwithstanding the sensation they have created throughout the whole of Southern Germany; but I have been profoundly interested in the discoveries you have made in chemical physiology. Your paper, " Sur le Principe de l'Irritabilité," which has been zealously supported by men of such ability as Sömmering, Scherer, Planck, and Herz, afforded me an opportunity of becoming acquainted with the antiphlogistic system, or rather, with the antiphlogistic truths. I at once commenced making experiments, and for the last two years have exerted myself to the utmost to study everything that bore the

value placed by Humboldt upon such tokens of respect is shown by some expressions addressed to Bonpland in a letter from Rome, June 10, 1805 :— ' If you wish it, I can procure your admission into the Arcadian Academy. It will cost you forty francs, and they will assign you a Greek name and a cottage in Greece or Asia Minor. They have given me the name of Megasthenes of Ephesus, and a piece of land close to the Temple of Diana.' (De la Roquette, 'Humboldt, Correspondance, etc.' vol. i. p. 179.)

[1] 'Aphorism. ex doctr. physiol. chem. plant.' Fischer's German translation, p. 109.

slightest relation to the subject. As a result, I am now as firmly convinced of the fact of oxygen being the principle of vital power (notwithstanding the mysterious galvanic fluid, which is certainly neither magnetic nor electric), as you were when you first told me about it in Green Park. Since then, nothing that you have written has appeared to me of greater value than the admirable section *on vegetation* in your " First Principles of Antiphlogistic Chemistry," for which I hope soon to have an opportunity of publicly expressing my deep obligations. . . . All my leisure now is devoted to chemistry, especially to the chemistry of the physiology of plants, for the study of which I am provided with efficient apparatus. I intend some day to devote more time to this subject. I have instituted a series of experiments on the germination and growth of plants in various substances supposed to be unfavourable to vegetable life.' An account of these experiments appeared in various periodicals.

In April Humboldt went to Schönebeck, to plan the construction of some salt works, and shortly after his return to Berlin, upon the completion of the publication of the ' Flora Fribergensis,' started in May for Franconia. On his way he paid a visit to his brother and his family at Erfurt, whence he wrote to his friend Freiesleben on May 26, 1793 :—

' I leave to-night for Bayreuth. I am now officially engaged in practical mining, of which I was so ignorant only two years ago as to be learning from you what a lode [Spatgang] was. . . . . I possess a certain amount of vanity, and am willing to confess it; but I know the power of my own will, and I feel that whatever I set myself to do, I shall do well.' Then a feeling of diffidence passes over him. ' But little,' he remarks, ' is expected from a Counsellor or Overseer of Mines, one is too much accustomed to incompetency; but a Superintendent! However, there is no help for it.'

In order to follow him intelligently in the prosecution of this commission in Franconia, it will be well to take a glance at the political condition of the principalities, and to pass in review the men who were associated with him in his official labours.

Upon the annexation by Prussia of the margraviate of

Anspach-Bayreuth, in the year 1791, Hardenberg, who had in the previous year relinquished the service of Brunswick to enter that of the margraviate, was, on the occupation of the country by the Prussians, appointed on December 2, 1791, Governor of the Principalities. About the same time Count von Hoym was nominated Governor of Silesia, to which was soon after added the provinces acquired by the partition of Poland. While Hoym distributed the places he had to bestow among personal favourites, and, instead of order and justice, introduced into the country only confusion, dishonesty, and self-interest, by which he aroused the bitterest feelings of the inhabitants, Hardenberg, in the appointments he made, admitted at first only two Prussians to any official post, his private secretary Koch, and his librarian Albrecht, and always observed the strictest forbearance, and the most judicious circumspection, in the introduction of the new form of government. Subsequently, when he had occasion to introduce Prussian officials from the older provinces, the selections were always made exclusively upon grounds of personal merit. As coadjutors with Alexander von Humboldt stand the names of Langermann, Hänlein, Schuckmann, Nagler, and Altenstein; the last two were at this period assessors, and, with Schuckmann, afterwards attained the position of Ministers of State. Up to this time the Mining Department had been presided over by Tornesi, Counsellor of the Upper Court of Mines, who was a good-natured man, but quite unfit for his position, and who, according to a letter of Ludwig Tieck, a student of Würzburg, then travelling through the district, was at the same time Master of the Hotel and Governor of the Lunatic Asylum.[1]

On June 10, 1793, Humboldt writes to Freiesleben:—'I have just come from the mines. I have ridden nine miles, and spent three hours in the prince's mine; therefore do not be surprised if my letter should show signs of confusion. I get on faster than I expected with my operations. The preliminary organisation is nearly complete; the office of administration is open, every arrangement made for the miners' relief fund, and now there only remains the filling up of the various appoint-

[1] 'Aus Varnhagen's Nachlass. Briefe von Chamisso, Gneisenau, &c.' vol. i. p. 204.

ments. I have been here some days, for the purpose of erect-ing new works at the prince's mine, at Sibald's mine, near Langendorf (for coal), &c. The universal confidence shown me by the miners makes me enjoy my occupation, otherwise my position is strange enough ; I am really doing the work of a foreman, not of a Superintendent of Mines. I will not attempt to give you any details of my mode of procedure. The heat is unbearable, and the atmosphere of the mines ener-vating.'

On July 19 he continues:—'I have ridden quite alone in this great heat, 146 miles in five days. My object was to inspect a new vein of ore, and apparently it will prove only useless rubbish. I am afraid you will gather from this, dear Freies-leben, that I am in a bad humour; but I can assure you I am not. The last four days I have been spending quietly at Steben, in the district of Naila; I am daily in the mines from half-past four till ten o'clock, and everything is making satisfactory progress. I possess the confidence of the men, who think I must at least have four arms and eight legs, which is pretty well in my position among so many lazy officials.'

In the midst of these various employments, his benevolent spirit was actively directed towards the improvement of all classes among the miners, not only in providing means for the education of the young, but also for the instruction of adults. The valuable efforts of the present day for diffusing a better education among the lower classes, especially workmen and artisans, were in fact originated by Humboldt at this period ; for while closely occupied with official duties, and in the midst of various scientific researches, he yet found time to labour, even at the sacrifice of his own means, for the improvement of the miners. When he first came to Naila, he complained that the most appalling ignorance prevailed among the lower orders of the miners, that they were full of prejudice in re-gard to choke-damp and the search for ore, and were quite unable to recognise the commonest minerals. He therefore established, without appealing for official authority, a free school for miners in the village of Steben, during the winter of 1793, and committed it to the charge of a young foreman named Spörl, whose salary he paid out of his own purse. This was the act of a true philanthropist, and of one who had a keen perception of the value of education. The hours of instruction were at first restricted to the afternoon and evening of every Wednesday and Saturday; but the interest excited both in teacher and pupils soon became so great, that the lessons were continued up to eleven o'clock at night.

It may easily be understood that Steben should have become so endeared to Humboldt that in after life he thus expressed himself:—'Steben has exercised so powerful an influence upon my mode of thought, I there projected so many of my greatest plans, I there abandoned myself so completely to feeling, that I almost dread the impression it would make upon me were I to see it again. During my stay there, especially in the autumn and winter of 1793, I was kept in a constant state of such nervous tension, that I could never see the lights of the cottages at Spitzberg shining through the evening mist without emotion. On this side the ocean no place would ever seem to me its equal!'

It was not till March 13, 1794, that Humboldt transmitted a report to the minister Von Heinitz upon the free mining schools which he had established; almost immediately afterwards a similar school was set on foot at Wunsiedel, and both were maintained for many years with good results.

While expending so much care upon the intellectual wants of the miners, he was by no means unmindful of their material interests. A proof of his unselfish consideration for the subordinate officials is given in the following passage from a subsequent letter to the minister Von Heinitz, dated May 21, 1795. The minister had sent him a present of a sum of money in recognition of his services, but Humboldt declined the gift and continued:—'So far I have done nothing to deserve it. It would be laying myself open to the charge of pecuniary motives, from which I am quite free. . . . . I would beg very humbly of your Excellency to distribute this sum during the winter. Such men as Birnbaum and Barrisch have a greater claim upon your consideration than I have. Another request which I also venture to make interests me much more. I have summoned young Sievert from Wettin to Arzberg as foreman. To him alone is due the impulse which the mines have received;

he has done much more than I have. He has (. . . . (?) illegible); no one could do more. I know no one so efficient in minéralogical research, and at the same time so thoroughly acquainted with practical detail. He has a good knowledge of geometry, is a good draughtsman, and has had considerable experience in smelting and in the carpentry of the mines. . . . . Your Excellency is aware how badly off we are here for pensions. I had written to Herr Veltheim to beg him not to abandon the idea altogether . . . . he promised he would do what he could, but alas! he has left, and I am obliged to trouble your Excellency with my very humble petition.' . . . . In a similar manner he frequently presénted the warmest appeals on behalf of others.

Meanwhile the 'Flora Fribergensis' had met with a brilliant and flattering reception; princes and learned men vied with each other in acknowledging its worth. The Elector of Saxony honoured the author by sending him 'an enormous gold medal,' accompanied by a letter, 'to serve as a public testimony of the pleasure your work has afforded me.' The Swedish botanist Vahl distinguished the youthful author by naming in his honour a magnificent species of an East Indian laurel, the *laurifolia Humboldtia*, 'in honorem botanici eximii F. A. Humboldt, auctoris præstantissimæ Floræ Fribergensis'—tokens of homage which, as is well known, were subsequently endlessly repeated.[1]

Thus passed the year 1793, occupied in multifarious undertakings, and on January 20, 1794, Humboldt writes to his friend at Freiberg, after some complaints as to the state in which he found the mines, in the following strain :—

'Upon the whole the mining works are now progressing rapidly. At Goldkronach I have been more successful than I

---

[1] Critical notices of Humboldt's works are out of place here, but it is a mere matter of history to state that Herr Mayer on two occasions read a paper upon the 'Flora Fribergensis' before the Berlin Academy of Sciences. At the commencement of his discourse, he thus speaks of the author of the work :—'Well versed in the necessary studies, both preliminary and accessary, and furnished with a large amount of erudition, he has far outstripped, even by his first steps in a literary career, the achievements of all our men of science.' ('Histoire de l'Acad. roy. des Sciences, etc., 1794 et 1795,' pp. 11-26.)

had ever dared to hope. The documents of the sixteenth century, recently discovered, which I have been studying with the greatest assiduity, have given me the right clue. My various predecessors in the superintendence of these mines have all failed for want of this source of information. An expenditure of 14,000 florins in eight years has scarcely produced 150 tons of gold ore, while I have procured from this mine alone in one year, and with the labour of only nine men, 125 tons of gold ore at a cost of barely 700 florins. Some experiments that have been made show that it will readily amalgamate. The mining commission of Berlin assured Hardenberg last year that a hundredweight of gold ore is scarcely worth three kreutzers, while this year I have reported the value of it to be twenty-four kreutzers. You see, my dear friend, that I am becoming quite a boaster. But I speak in this strain only to you. The district of Naila is progressing as rapidly as that of Hammsdorf is declining. The mines of the latter furnish from 100 to 112 tons of iron-stone, while ours produce 188. They employ five or six men only in the workings, whereas we have twenty men, and, in the case of one mine, as many as forty men at work. Our results this year are in iron to the value of 163,000 florins, in vitriol to 28,000 florins, and in cobalt, tin, antimony, copper, grey copper ore, and alum conjointly, to the amount of 300,000 florins. This is certainly an ample return with only 350 miners. At Steben I have at length commenced the working of the Frederick-William gallery, the preparations for which kept me busy the whole summer. I made out a very elaborate estimate of the cost, amounting to 20,000 florins, in which everything is included, down to the plank-nails ; it is an *opus operatum* which I must send you some day, together with a history of the most recent copper-mine at Steben. The new copper workings are everywhere improving, and I am sure that with the Frederick-William gallery, which can also be made navigable, they will in time again yield from 100 to 150 tons of pure copper. But enough of this boasting.'[1]

An attack of intermittent fever, caught by exposure to the

[1] As a means of comparison, it will be well here to give a passage from a treatise by Heinitz on the mineral products of the Prussian States, ' Ab-

damp of the mines, and in frequent journeys in the wild country of the Fichtelgebirge, was not allowed during its three weeks' duration to interfere with his zealous activity. 'You will doubtless scold me, my dear friend,' he writes further, 'yet how can I fulfil my duty here without exposing myself to such risks? You ask if I am engaged on any literary work at present. Yes, I am, and, as usual, on many subjects at once; but that which interests me most deeply, "Experiments on the Excitability of the Nerves and Muscles," is too elaborate for me to give it you in detail. I am also engaged upon an important work on geology, for which, however, I have as yet no title. I may perhaps call it "Results of my Observations," or "Results of my Travels in Germany and Elsewhere." My idea of the subject is this:—Geological descriptions of entire districts, accompanied with well-arranged flora, are merely vehicles for bringing before the world personal observations; such works always contain much unimportant information, which is not sufficiently accurate for the purposes of a mineralogical geography. The hurried nature of my journeys renders it quite as impossible for me to give a complete flora as a good geological description. I therefore remark only what is new, and state it in short aphorisms of half-a-dozen lines, after the manner of the following notes:—Granite Boulders.—I found some recently six feet in diameter; stratified granite is everywhere the oldest form of granite; compass-observations as to the dip of the strata; relative age of Franconian and Bohemian syenite; beds of alum in almond stone. . . . You shall certainly see the manuscript before it is published. However rhapsodical it may appear, I take the trouble to work out every entry with the accuracy I should if it were a monograph.

'You are aware that I am quite mad enough to be engaged

handlung über die Producte des Mineralreichs in den königl. preussischen Staaten' (Berlin, 1786), where in p. 110 he thus speaks of the margraviates of Anspach and Bayreuth :—' Formerly mines were worked here successfully, both for gold and copper. The mines, however, were deserted, because it was thought a greater profit could be obtained from the direct sale of the timber. In those days there existed in the two margraviates thirteen smelting furnaces, where in one campaign of thirty-nine weeks 3,042 tons of excellent pig-iron were produced,' upon three books at once, therefore I may as well tell you that I am going to bring out, at Schreber's suggestion, a magnificent " Flora Subterranea," in folio, illustrated by steel engravings. I have besides discovered several new lichens. I have also been occupied upon the history of the weaving of the ancients. I am carrying on these works simultaneously, just as I have leisure or inclination, so that I hope they will all be published about the same time, probably during the winter.' Neither of the last-named works, however, ever saw the light.

Notwithstanding these labours in various branches of science, Humboldt never for a moment neglected his official duties. ' I have been extremely busy,' he writes to Freiesleben on April 2, 1794, ' in arranging for everything to go on without me, as I am leaving the works for three months. I was at Gold-kronach from the 17th to the 26th of February, at Kaulsdorf and Jena from the 26th of February to the 13th of March, at Naila from the 15th to the 26th of March, and at Wunsiedel from the 26th to the 31st of March. Is not this being in a whirl? . . . . My head is quite distracted with all I have to attend to—mining, banking, manufacturing, and organising; . . . . the mines, however, are prospering. . . .

' I am promoted to be Counsellor of Mines at Berlin, with a salary probably of 1,500 thalers (here I have 400), and after remaining there a few months I shall most likely be appointed Director of Mines, either in Westphalia or Rothenburg, and receive from 2,000 to 3,000 thalers. I tell you everything, my dear Karl, and open my heart to you. I am just setting off on a commission to the shores of the Baltic and the Polish mountains, but I return here again as Superintendent. My former plans remain undisturbed; I shall resign my post in two years, and go to Russia, Siberia, and I know not whither.'

The object of this new commission was the investigation of salt works, and for this purpose he travelled through Colberg, Thorn, in the province recently annexed to Southern Prussia, and along the left bank of the Vistula, to the districts of Slonsk, Nieszawa, Racionzek, Woliszewo, and Ciechoczinek, which had lately come into notice from its salt springs, thence westwards back to Bayreuth, through Lenczic, Inowraclaw, Strzelno, where numerous saltpetre works existed, Gnesen, Posen, Glogau,

Prague, and Eger. Elaborate reports upon Colberg, and upon some boring experiments at Ciechoczinek, dated from Goldkronach towards the end of June, 1794, are still extant.

Soon after his return, he was summoned, by the unexpected course of political events, to take part in some diplomatic negotiations connected with the army on the Rhine, occupying Munzernheim, Mayence, and Wesel—a service which detained him about four months, until October 1794.

In the hope of rendering essential service in the settlement of the dispute with France, the minister Von Hardenberg arrived in June, 1794, at Frankfort-on-the-Maine, where the king, who took a personal part in the war, was holding his head-quarters. The deeper the significance attached to Hardenberg's arrival at the royal camp, inasmuch as he came upon his own responsibility, and the more unpromising the position of affairs at that time appeared to be, so much the more flattering to Humboldt was the distinction shown him by the minister, by whom had been effected the breach of the contract of subsidy with the Hague, in selecting him as his most intimate and confidential companion. As early as October 21 the army recrossed the Rhine, and a separate peace was concluded at Basle.

In what manner Humboldt was employed in these negotiations is not known, for all that remains to us of this period is the following fragment of a letter, dated September 10, 1794, from the head-quarters at Ueden, in Brabant :—

'My life was never so changeful as it is now. I have for some time been removed from my own department to undertake some work connected with the diplomatic mission of the minister Von Hardenberg, and though I am nominally attached to the suite of Field-Marshal von Möllendorf, I am just now under orders here in camp. I leave Ueden on the 14th for Altenkirchen, to make a general inspection of the mines of that district, and thence I return to camp by Kreuznach and Frankfort. Something of this kind is always going on: it does not afford me any particular pleasure, and yet it amuses me sometimes, for the constant travelling through interesting mineralogical districts has greatly helped me in my work upon strata and stratifications. I have now ascertained with accuracy the geological distribution of the whole of Western Germany, I have descended several mines and described the veins in detail, so that I intend next winter to employ myself systematically on an extensive mineralogical work, a kind of geological aspect of Germany.'

In the midst of labours so arduous and so greatly diversified, he received on February 3, 1795, a communication from Heinitz, soliciting him, with many flattering inducements, to accept the appointment of Director of Mines, smelting works, and salt works in Silesia—a post likely soon to become vacant by the removal of Count von Rheden to Berlin. However gratifying such a proposal might be, it was nevertheless declined by Humboldt. In his reply, dated from Steindorf, Fichtelgebirge, on February 27, 1795, he writes:—

' I am contemplating a complete change in my mode of life, and I intend to withdraw from any official connection with the State. A few years ago I had the honour of respectfully submitting to your Excellency the plan I had formed of preparing myself for a scientific expedition by a practical employment among the mines. As I have a deep conviction that such an expedition is highly important for increasing our knowledge of geology and physical science, I am exceedingly anxious to devote my energies at once to the execution of this design ; the more so as the sad conviction has been gradually forcing itself upon me that my physical frame, in consequence of premature exertions, will feel the effects of age at an earlier period than I used to think possible. Grateful as it would therefore be to me, on account of my small annual income, to improve my pecuniary position, it would yet be indefensible in me to accept an appointment which I should have to resign almost immediately. Under such circumstances, I must beg most respectfully to decline the post of Director of Mines in Silesia, which your Excellency has done me the honour to offer me.' . . .

Notwithstanding this refusal, the proposal was renewed in almost pressing terms on April 7, 1795, by Heinitz, who employed every means in his power to induce him to change his determination, but with no better result.

At the solicitation of Hardenberg, who was most anxious to retain his valuable services, Humboldt was appointed on May 1, 1795, to the office of Actual Counsellor of the Upper Court of Mines, with the offer of retaining his salary, and with permission to prosecute his various foreign travels as opportunity occurred. But flattering and alluring as such proposals would have been to most men, they yet possessed no charm for Humboldt, and his resolution remained unalterable of leaving the service of the State, in order that he might accomplish his long-cherished plans of undertaking extensive foreign travel in the furtherance of science. A letter about this time to Freiesleben gives additional particulars of his plans:—

'I think that I now possess a more than ordinary acquaintance with practical mining. Whether after travelling for five or six years I shall again enter the public service, in either Saxony, Austria, Russia, or Spain, (you see I expect my fame as a miner to have greatly increased by that time,) I cannot now determine.

'At present, my plans are to spend from July (1795) till October or November in Switzerland, to pass the winter in Germany, and in the spring of 1796 to start for Sweden and Norway. I wish particularly to go to Sweden for the sake of visiting Lapland, *botanices causa*—an expedition in which there would be no danger. One of my heartfelt wishes, my dear Karl, is to take you with me, not only to Switzerland but to Sweden. I shall relieve you from all expense in either journey, as I have 1,000 thalers at my disposal; I depend absolutely upon you to accompany me. Your wishes shall be to me as commands, and you shall not repent going. . . . . In the tour through Switzerland, which is to include Salzburg and the Tyrol, I wish to introduce an element which will, I trust, not prove an insurmountable obstacle to you. You must, if you please, consent to make one in a trio with me and a friend of mine to whom you are a stranger. I will try to give you an accurate description of him.' . . .

The companion referred to was Lieutenant Reinhard von Haften, of Westphalia, an officer in the Grevenitz Infantry Regiment, at that time garrisoned at Bayreuth: he was a

young man of winning manners and finished education, gifted
with a noble character, and devoted to the pursuit of science.
Humboldt accorded him his most intimate friendship, often
residing with him under the same roof, and writing whole
pages full of enthusiastic expressions in his praise.

In conclusion, he says of him:—'I learnt to appreciate
him from his conduct in taking my part, . . . circumstances
of great moment in social life,—my gratitude therefore,
my ' . . . .

The following lines, torn away by some discreet hand, probably contained the confession of an attachment, which, according to a distinct and circumstantial statement of Kunth
the botanist, existed between Humboldt and the sister of
his friend Von Haften, and which, though faithfully cherished
for upwards of ten years, was never consummated by the union
so ardently desired.

This glimpse into a part of the inner subjective life of Humboldt is no reckless exposure of one of the deepest secrets of his
heart; it is here referred to merely as a means of shedding a
fresh gleam of light upon the true humanity of the character
of this remarkable man. Humboldt was not placed by nature
beyond the reach of ordinary human sensibilities, but the unconquerable desire for a course of many years' adventurous
travel, the necessity for being in a state of continual readiness
for a change of residence, and his unreserved devotion to
science, all exacted from him the ruthless sacrifice of the
comforts of a stationary home, and the sweet happiness of
domestic ties. Such was the sacrifice of the affections made
by Humboldt to the shrine of science.

On one occasion, when Humboldt's affectionate disposition
had evinced itself in an enthusiastic apostrophe to his friend
Freiesleben, he suddenly broke off with the expression: 'Haften,
who is looking over my shoulder, asks in astonishment how it
is possible we can still go on addressing each other as *Sie*
[you]. I could give him no satisfactory reason, so I laid the
blame on you. Let us therefore at once renounce this piece of
formality, though we cannot thereby become more like brothers;' . . . and so the confidential *Du* [thou] ever after took
the place of *Sie*.

Von Haften succeeded in obtaining leave of absence, and Humboldt wrote to Freiesleben repeating the invitation to join the party: the route was finally arranged through Munich, Innsbruck, Halle, and Treviso to Venice, where on Haften's account they were to spend a fortnight, thence through Vicenza and Verona to Milan, and by way of the Lago Maggiore and the Pass of St. Gothard into Switzerland and on to Schaffhausen, the main object of the journey being to observe the connection between the various mountain ranges of the Tyrolese, Italian, and Swiss Alps. 'This tour is planned more for Haften than for myself; I am, as you are aware, under some obligation to him, therefore I would rather neglect some of my scientific aims than not be entirely at his disposal during the first part of the journey. The only question now remaining is, where and when shall I meet you?'

In a long letter to the minister Von Heinitz, dated Schwarzenbach-am-Walde, May 29, 1795, he furnishes a detailed report of the results of his official and scientific labours; and while reminding him of the necessity of appointing a successor, communicates his plans for extensive travel—all expressed in terms of the most respectful devotion, but yet with unmistakable evidence of the full consciousness of his independent position, and the unalterable nature of his determination.

The fame which Humboldt had earned for himself by his official labours, and the rumour of the comprehensive plans he had laid for extensive foreign travel, had spread far and wide among all circles of his acquaintance. With reference to this period, David Veit wrote as follows to Rahel on June 15, 1795:[1]—'Alexander has been appointed Counsellor of the Upper Court of Mines; he has erected works at Bayreuth at an extremely small cost, and has carried them on with such an extraordinary amount of ability and honesty of purpose that the mines now yield as much in one year as they formerly did in fourteen years, while he has put the whole under such admirable management that the works may be conducted by any ordinary mining engineer. He receives no salary, and therefore can leave at any time; next summer he spends in Switzerland,

[1] Varnhagen, 'Galerie von Charakteren aus Rahel's Umgang,' vol. i. p. 51.

VOL. I.   L

and in the following summer he intends to travel in either Lapland or Hungary, in prosecution of scientific discovery.'

They started at length upon their tour on July 17, 1795, and Humboldt thus writes to Willdenow, his 'dear friend and companion,' who had written to announce the birth of a son, and to ask him to stand godfather at the baptism:—'I leave to-day for Venice, by way of the Tyrol, thence by Vicenza and the Venetian Alps to Milan and Switzerland.' The letter is mainly of a congratulatory character, and it reveals Humboldt's feelings with reference to 'human interests' of this description. 'I cannot tell you,' he writes, 'with what intense sympathy I read both your letters. There is no one in the world to whom I am so strongly attached as yourself, no one who lies so near my heart. I rejoice with you most truly in the fulfilment of your ardent wishes. How completely I can realise the joy that you and your dear wife must experience—you in becoming a father, she in being a noble-hearted devoted mother; and how can I sufficiently thank you for allowing your poor friend, buried in the subterranean regions of the wild Fichtelgebirge, to participate in your joy. A boy withal! a strong and healthy child! . . . Next winter I shall hope to embrace you all, and hold your son in my arms.'

On July 28 Humboldt writes to Freiesleben from Trieste:— 'You are aware, my dear Karl, that the main object of my journey is to investigate the connection between the several ranges of the Tyrolese, Venetian (*montes euganei*), Lombardian, and Swiss Alps. Every arrangement is made with reference to this object. I am also making a collection of plants, therefore I am busy enough. I have met with such a great variety of new specimens, both in the Tyrol and here among these Venetian mountains, that I must refer you to my journal—I write it up most conscientiously—and to my published work on strata.' . . . . After reference to a projected tour through Hungary and Greece, he continues :—' How striking the contrast between the wild districts of the Tyrol, where two days ago I was knee deep in snow, and the sunny plains of Italy around Bolzano and Solmino, where, fanned by Italian breezes, fig-trees flourish in the open air!' With yet greater delight he describes a long stay at Venice, whence he proceeded on August 9, through

Padua and the plains of Lombardy to Vicenza, Verona, Parma, and Milan, so as to reach Switzerland on September 1, and join his friend Freiesleben on the 20th of that month at the Hotel Krone, at Schaffhausen.

From Schaffhausen, whence Herr von Haften, on the expiration of his leave of absence, returned to Bayreuth, Humboldt continued his journey, in company with Freiesleben, through the Jura and the Alps of Savoy and Switzerland ; in the course of their travels, which lasted from September 20 to the beginning of November, they visited De Luc, Pictet, Saussure, and other notable men of science, and the intercourse thus commenced proved a source of mutual gratification and benefit.

'Throughout all these journeys,' relates Freiesleben,[1] 'he was chiefly occupied in observing the connection between the flora and the stratification of the mountains. At the same time no other subject having any reference to the physical constitution of the earth, the atmosphere, or any point of natural history, was allowed to escape his attention. And when I remember that within the short space of seven or eight weeks we visited, chiefly on foot, the mountains surrounding Schaffhausen, Zurich, and Bern, descending as far as the Valley of Chamouni, passing afterwards from Altdorf over the St. Gothard to Airolo, I am inclined to congratulate myself even now on the good use we made of our time—an art in which Humboldt is certainly a master. His zeal for science and his unexampled industry have led him from boyhood to employ every moment in some useful or instructive occupation. Even his night's repose was never allowed to extend over more than a limited number of hours.'

On the homeward journey, Humboldt visited Rastatt, where the Congress, which was exciting universal attention, was then sitting, his motive being not so much to see the diplomatists as to meet with Faujas, the French mineralogist. 'Certainly,' narrates Herr von Lang, the satirical diplomatist of that assembly by which the world was to be reorganised, 'certainly Humboldt was never so panic-stricken by any storm at sea as was Count Goerz, the Prussian Plenipotentiary at the Peace Con-

[1] From an earlier life of Alexander von Humboldt in 'Zeitgenossen,' 3rd Ser. vol. i. p. 71 (Leipzig, 1828).

gress (*Reichsfriedenspacificationsverhandlungstractate*), while
standing at the table, by the entrance among the diplomatic
dignitaries of Herr von Humboldt—the invited guest, a whole
hour behind his time, heated with haste, in boots and travelling
dress, fresh from a tour among the mountains of Baden. The
Count however put the assembly completely *au fait* by a shrug
of the shoulders and the whispered apology, "A philosopher."[1]

On his return home in November, 1795, Humboldt resumed
his occupations among the mountains at Steben, Lauenstein,
Goldkronach, and Arzberg near Wunsiedel, till the spring of
1796, and amid the various distractions that beset him, some
of an annoying, some of a pleasing character, (among the latter
was a grand ball given by him at the old castle on the marriage
of his friend Von Haften), 'there poured in upon him from the
minister a flood of work of all kinds,' which took him for a
fortnight to Anspach in company with Herr von Schuckmann.
It was about this time that he received the distressing news of
the painful and incurable illness of his mother,

In the midst of these multifarious occupations, he laboured
unremittingly on two elaborate works widely different in character, the one on geology, the other on physiology.

Some particulars on the subject of the first-named work may
be gathered rom his unpublished letters; in one of these
Humboldt writes to Werner at Freiberg as follows:—' I am
working uninterruptedly at a great geological work which is to
appear under the title " On the Formation of the Crust of the
Earth in Central Europe, especially with regard to the Disposition of Strata in Mountain Masses." I intend to furnish a
general description of the stratifications from the Light-house
at Genoa to Warsaw and Segeberg (?) in one direction, and from
the Forest of Ardennes and Chalons to Ojców in another. I
wish to show that the bearing and dip of the strata have no
reference to the direction or declivity of a mountain range, nor
to the waste of its material, but are connected with something
of much wider significance, and that in all the mountains comprising the great European chain from Savoy to the Tyrol
which I have traversed on foot, although the direction 3—4
and declivity are from north to west, the bearing and dip

---

[1] Ritter von Lang, 'Memoiren,' vol. i. p. 329.

of the stratifications run from south to east. I have been for three years collecting observations for this work, and it is not from laziness, but from a wish to produce something really valuable, that it has been kept so long from the public.'

In a similar strain he writes to the minister Von Heinitz on February 3, 1796 :—' I have succeeded in arranging a general system for the inclination of the strata with respect to the horizon, as well as for the order in which the strata are superposed. It is a very remarkable phenomenon, which has hitherto escaped the observation of our physicists. My work will be published in the course of the summer, and I shall consider myself well repaid for the many journeys I have undertaken on foot, and the fatigue I have undergone, if this endeavour to establish the laws of geology should be deemed worthy of the support of your Excellency.' . . .

This work, however, was never published ; the information collected for it was incorporated into a later work, entitled, ' Essai géognostique sur le Gisement des Roches dans les deux Hémisphères.'

The second work on which he was engaged was 'Experiments on the Excitability of the Fibres of the Muscles and Nerves, with Conjectures on the Chemical Process of Life in the Animal and Vegetable Kingdoms.' As early as 1792, during his first visit to Vienna, Humboldt became acquainted with the discoveries of Galvani and Volta ;[1] from that time he watched the discussions of physicists as to the nature and cause of animal electricity, and undertook repeated experiments and counter-experiments, not confining himself merely to the frog— hitherto the favourite subject of all such investigations—but experimenting upon his own person, with an amount of self-sacrifice so extreme as to produce a permanent derangement of the nervous system. Detailed accounts of the experiments he instituted, and records of the progress of his work, are contained in numerous letters, both published and unpublished, to Blumenbach,[2] Sömmering, Herz, Reil, Girtaner, Willdenow, Marc-Aug.

---

[1] 'Versuche über die gereizte Muskel- und Nervenfaser, nebst Vermuthungen über den Chemischen Process des Lebens in der Thier- und Pflanzenwelt.' Preface, p. 3.

[2] 'Blumenbach,' writes Humboldt to Freiesleben, ' may possibly add

Pictet, Van Mons, Fourcroy, Loder, &c., as well as in various treatises in Gren's, Crell's, Millin's, and other periodicals.

In June, 1795, Humboldt writes to Blumenbach:—'I can only describe to you here one experiment. I applied two blisters to my back, each of the size of a crown-piece, and covering respectively the trapezius and deltoid muscles: I lay meanwhile flat upon my stomach. When the blisters were cut, and contact made with zinc and silver, I experienced a sharp pain, which was so severe that the trapezius muscle swelled considerably, and the quivering was communicated upwards to the base of the skull and the spinous processes of the vertebræ. Contact with silver produced three or four single throbbings which I could clearly separate. Frogs placed upon my back were observed to hop; when the nerve was not in immediate contact with the zinc, but separated from it by half an inch, and the silver only in contact, my wound served as a conductor, and I then felt nothing. Hitherto my right shoulder was the one principally affected. It gave me considerable pain, and the lymphatic serous humour produced in some quantity by the irritation was of a red colour, and, as in the case of bad sores, was so acrid as to produce excoriation in places where it ran down the back. The phenomenon was of too striking a nature not to be repeated. The wound on my left shoulder was still filled with a colourless watery discharge, and I caused the nerves to be strongly excited in that wound also by the action of the metals. Four minutes sufficed to produce a similar amount of pain and inflammation with the same redness and excoriation of the parts. After being washed, the back looked for many hours like that of a man who had been running the gauntlet.'

On his return from Italy and Switzerland, Humboldt instituted in December, 1795, a repetition of his former experiments, and supplemented them by an additional series, in consequence of his interviews with Volta and Scarpa, both of whom he had visited when at Pavia and on Lake Como.

He thus describes them :—' Dr. Schallern experimented upon my back for fully three-quarters of an hour. The painful pro-

some notes to my book. He is so eager about it, that he has already sent translations of certain portions to Banks in London.'

cess of galvanizing the wound by means of zinc and silver had scarcely commenced, when the serous humour poured out copiously; it rapidly deepened in colour, and in a few seconds inflamed my back in blood-red streamers as it ran down. I incautiously applied cold water to cleanse the excoriated places, when they instantly increased so manifestly in size and intensity of colour, that both the doctor and I became alarmed, and we bathed the wounds with luke-warm milk, without however producing much effect. The experiment was thus incontrovertibly confirmed to my own mind.'

Many experiments were in a similar way repeated upon various wounds in the hands, and also in the cavity left in the jaw by an extracted tooth; the attempt to carry on the process so far as to benumb the irritated nerve was unsuccessful, since the pain became too violent.

It is not in accordance with the systematic arrangement of this work to enter in this chapter on the details of Humboldt's scientific labours; it will be well, however, to mention here some facts which may serve to illustrate his personal heroism and his self-sacrificing devotion in the pursuit of science. With this intention, therefore, the following proof may be adduced of the zeal with which he engaged in scientific investigation.

It is well known that the existence of fire-damp and noxious gases, besides a variety of other injurious influences, render working in the mines an employment fraught with extreme danger. It is found, however, that the sudden and murderous explosions are, on the whole, less destructive to human life than the insidious diseases to which the miners in their underground labours become a prey, such as asthma, disease of the bone, jaundice, induration of the glands, palsy, and skin diseases. Delighted as was Humboldt to enlarge the boundaries of scientific knowledge by means of his discoveries, ' he experienced a still deeper gratification in devising methods for the preservation of life and health among an industrious class of men.'

Equipped with all the apparatus then at the command of science, Humboldt made, ' at the cost of months of exertion and self-sacrifice,' chemical analyses of the various gases to be found in mines, and investigated the causes of their peculiar localisation, thus originating a new system of subterranean meteorology.

These labours led to the invention of a kind of respirator, and of four lamps of different construction suitable for employment in various circumstances.

The respirator was to prevent the inhaling of injurious gases, and to supply the miner with good air; the lamps were constructed to burn in the most inflammable kind of fire-damp without igniting the gas. They were the forerunners of Davy's later invention, and were frequently made use of by the miners, who availed themselves also to some extent of the respirator. The experiments, however, for testing these inventions were not unfrequently connected with considerable personal risk, and an instance of one such experiment, which occurred at the alum works at Berneck, on October 13, 1796, will be found in the following extract:—'The air in the cross-cut was so impure,' relates Humboldt, ' as to extinguish every light as completely as if immersed in water. The arm-lamp, though usually one of the most effectual, burned only with difficulty, while the ring-lamp remained as bright as in the purest upper air. In order to ascertain whether it would be possible to extinguish the flame, I crept through an opening in the screen shutting off the passage which had been left for the purpose of allowing access without removing the screen. I went in alone. The air was so loaded with carbonic acid gas that I could not obtain a light, nor ignite paper even for an instant at my lantern. I penetrated to the distance of six or eight fathoms, over ground strewed with the traces of burnt sulphur, and as I stood surrounded with rotten wood, my ring-lamp continued to burn as brightly as at first. I placed it upon the ground, in order to observe its behaviour in the lowest stratum of air, when the quantity of carburetted hydrogen gas suddenly deprived me of consciousness. I became exhausted, and at last sank down insensible near the lamp. Fortunately I had just before summoned the attendance of Bauer, the foreman of the mine, and he and Herr Killinger hastened to my assistance and pulled me quickly back by the feet: I soon revived in the fresher air of the mine. I had the gratification, on regaining my senses, of seeing the lamp still burning.'

Humboldt thus learned, from his own experience, how brightly his lamp would continue to burn in gases which were too noxious for human life.

The illness of Frau von Humboldt summoned both her sons to Berlin in the middle of February, 1796. Alexander von Humboldt passed there six weeks in a state of great mental depression, as we may gather from his journal of this date, where an entry occurs that on March 26, his will was deposited in the Municipal Court of Justice.[1] 'The condition of my poor mother is beyond measure distressing,' he writes to Freiesleben; ' she suffers fearfully from a cancer in the breast, and her case is not only hopeless, but beyond the reach of any alleviation. I do not think it possible she can survive till the autumn, and I shall therefore remain at Bayreuth through the summer.'

On his return to Bayreuth, which he reached on April 16, he was laid aside for several weeks with a severe attack of influenza and nettle-rash; scarcely had his recovery permitted him to resume his labours on the subterranean gases, respirators, and experiments in galvanism (for the results of which he received from Decker, a publisher in Berlin, the sum of three gold Fredericks per sheet[2]), when his appointment in July on a diplomatic commission connected with the war which had recently broken out, caused a long interruption to his scientific undertakings.

The unexpected invasion of the Duchy of Wurtemberg by the French under Moreau, and the flight of the Duke, caused some alarm in Prussia from the fear that the territory of the Prince of Hohenlohe, where Mirabeau, brother of the distinguished statesman, had raised an emigrant legion in 1791, might in

---

[1] This notification is to be found in vol. v. of the 'Tagebücher.'

[2] The 'Versuche über die gereizte Muskel- und Nervenfaser' was published by Rottmann, son-in-law of Decker, who was at that time the proprietor of the business carried on by Rottmann. It may be interesting to give here some particulars of the sums paid to authors at this period. Gellert received thirty florins for his Fables, and Lessing absolutely nothing for his ' Minna von Barnhelm.' Goethe and Merck shared in common the cost of printing 'Götz von Berlichingen,' and had not even paid for the paper at a time when Goethe's name had attained celebrity; Mylius of Berlin offered twenty thalers for ' Stella,' and Goethe received a tea and coffee service of Berlin china from Himburg for permission to reprint in Berlin an edition of his works in four vols. Schiller, too, was obliged to pay fifty florins for the publication of the ' Robbers,' since no publisher would undertake the work, and he had to start it at his own cost. (Dr. Aug. Potthast, *Geschichte der Familie von Decker und ihrer Königl. Geh. Ober-Hofbuchdruckerei*, p. 293, note 222.)

revenge suffer plunder and injustice at the hands of the invading
army.   It was hoped, however, that in consideration of the
friendly relationships which had existed since the peace of Basle,
it would be possible to protect the threatened territory as an
enclosure within Prussian domain, and to make its neutrality,
as well as that of the Franconian principalities, respected by
the invaders, and Humboldt received unexpectedly the commission to make the necessary diplomatic arrangements. Guided by
his accustomed simplicity of purpose, he greatly distinguished
himself in the execution of his task, and the few letters still
preserved of this date contain valuable sketches of persons and
events.   He thus writes from Ingelfinden, July 17, to Baron
von Schuckmann, his coadjutor in Franconia :—

'I am become a person of great importance,. and if I
should end by being appointed major domo to the Prince
d'Oeringen, who is expected every moment, I shall soon be in a
position of complete independence. Alas! I have been fearfully deceived. I expected to have been one of the wise
counsellors of Ansbac ; but no ! my talk is to oppose the arms
of the French. Never have I seen so many false moves, so
much stupidity, so many orders and counter-orders given at the
same time. This adds to my annoyance, and I see no remedy
but in a complete turn of affairs. Being a person of so much
consequence, it is not to be wondered at that I have very little
leisure. Therefore I must tell you everything in three words.
A buon intenditore poche parole. We have been negotiating
with the noble city of Nuremberg, which was a troublesome
business. Then we wished to negotiate with the Ordre teu-
tonique ; *l'homme de bois* was not at Mergentheim. Next we
wanted to negotiate with the pope of Eichstädt : popes will not
hear reason. But it is absolutely necessary that we should
negotiate with somebody. Very well ! the holy father brings
the French. These are the very people. Our happiness is complete. We will negotiate with them. The proceeding is perfectly natural. By all means follow the game until we find it,
and if there should not be any we can but make some. The
thing now to be done is to send some *très-habile* person to
General Moreau to insist upon the neutrality of the Franconian
provinces and the territory of Hohenlohe being respected. This

très-habile person may have the brain of a hundred, but I fancy it would be more to the point if he had the arms of 25,000. But this is a foolish notion I have brought with me from Bayreuth! Herr von Hardenberg wished at first to send Sieur W., because he is a man of energy and of high character. As the French are fond of pictures, he might perhaps have presented them with some of his or (if you will) with the portrait of Madame de H. But, *sic Dii non voluere*. Prince Hohenlohe conceived the unfortunate notion of objecting to Sieur W., and of proposing me. Thus I was caught. Without consulting me, they placed the matter before the king ; and how on earth was I to refuse ?' . . .

A fortnight later, while still at Ingelfinden, he writes to Freiesleben on August 2 :—

'I have returned from the French army to the head-quarters of the Prince of Hohenlohe. I have been engaged in so much that is in direct opposition to both my nature and habit of thought, that it is quite as much a necessity as a pleasure to write to you, dear Karl. I have been for twelve days marching about with a detachment of hussars, transacting some negotiations in Suabia. The successful conclusion of this business, and its importance for the welfare of so many who will not have any occasion now to leave their homes, has often excited within me a feeling of self-gratulation. On the other hand it is a melancholy sight to watch the Germans creeping away from the French into the interior of the country, and to hear all Germany discussing what they call a treaty of peace ; it is enough to make one's heart ache.

'I hope in a few weeks to be set at liberty, so that I may return to Berlin. The troops of Saxony are greatly respected by the French. General Moreau frequently said to me at his head-quarters at Schorndorf: "The Saxons are very brave soldiers ; they are no enemies of ours, their only crime is that of adopting the Imperial policy. Had I ever taken any of them prisoners, we should certainly have set them at liberty. I hope the Elector will soon make peace." The Saxons are much pleased with Hardenberg's behaviour, since he has provided them with bread, forage, and money. The march through, and the little that we were able to do, gratified me exceedingly.

You know how dearly I love your country, to which I owe so much. I will however not weary your patience, dear Karl, with politics,yet there are two observations which I am anxious to make.

'I have recently had an opportunity of admiring the high cultivation of feeling and mode of expression exhibited by the French, to which no other European nation has ever attained. I was speaking to one of the common sentries allotted to us as a guard of honour upon the cruelty of the Imperialists, who often killed their prisoners. I remarked, "But you will grant, citizen, that, nevertheless, they are good soldiers?" "Soldiers!" replied the youth, who was scarcely twenty years of age and insufferably dirty,—"No, citizen; to be a soldier one must be a man. Those people have no notion of humanity!" Might not this be a quotation from Racine, and where would you hear such an expression from a German soldier?[1] The other remark I wished to make was that these men really seem to possess the spirit of the old republicans. They despise the present constitution *en détail,* but the general idea of a republic is never alluded to without calling forth visible enthusiasm. The army on the Rhine rejoices at the progress of Bonaparte in Italy, "because a republic ought to adopt a comprehensive plan." These people are of the same mind with Wilhelm Meister, who never saw a puppet show without reflecting upon the elevation of mind in the human race.

'General Desaix is one of the most remarkable and, I may add, one of the most amiable men I ever met. He has a head like Cromwell, but possesses a larger share of good nature. His exploits testify to his military talent. There is a touch of gentleness and melancholy about him which renders him attractive notwithstanding his rough exterior. He is well acquainted with the antiphlogistic chemistry, and had a vague notion of my safety-lamp, "as a German invention communicated to the National Institute," probably by Dolomieu, who lately read before the Institute a paper of mine on "Expérience sur l'influence du gaz acide mur oxygène sur la fibre animale."

---

[1] Since Humboldt wrote these words, what a change has been effected in the relative state of culture of the French and German soldiers, especially in regard to the feelings of humanity displayed by both nations!

I exhibited the drawing of my new lamp to the general officers, and explained its use in military service. This and Desaix's taste for chemistry drew us much together, and I found my intimacy with him of essential advantage in the execution of my mission.

'At Stuttgard I witnessed some balloon practice. General Reynier made the ascent, and invited me to accompany him, but unfortunately I had not long enough notice, and I could not keep the hussars I had with me waiting. The sight of the ascent is quite enough to remove all trace of fear, and I shall regret as long as I live that I missed this opportunity.'[1]

After his return to Bayreuth in the autumn of 1796, he was again occupied in various scientific labours, while he at the same time prosecuted his investigations on subterranean meteorology. In connection with this subject he endeavoured to enlist national co-operation in the erection of a series of eudometric stations—a plan he was unable to accomplish, from the unsettled state of Europe at that time. After his discovery of the magnetic serpentine rocks at Gefrees, he devoted himself with great zeal to the study of terrestrial magnetism, and published several small treatises on the subject in various periodicals. He was anxious by this means to incite others to make investigations of a more extended character, and thought 'it would be well to throw such a shell into the world of science as should arouse to action.'

In the meantime his resolution of leaving all official employment in order to gratify his increasing desire for extensive travel continued to gain strength. He had already determined to go to Italy the following May, 'even if my mother be still living.' The minister had indulged the hope of being able to bind him to engage in the service at some future time, by offering him the continuance of his salary, but Humboldt addressed Baron von Schuckmann on the subject in the following terms :—

'I cannot consent to the minister's proposal that I should retain my salary. I would willingly follow the counsel of my

[1] General St.-Cyr had reconnoitred the enemy for the space of a month with a Conte Balloon (Ballon captif), while Moreau was accustomed to say : 'Moi, je préfère le chemin des ânes.'

friends, for I feel that I am not rich enough to forego readily
even a small additional income, and I have besides the vanity
to think that princes might do something for men of my stamp;
but I prefer to occupy the independent position in which I am
placed. The more we seek to influence the moral conduct of
others, the more strictly must we ourselves obey the laws of
morality. The merit of not having abused the friendship of a
minister is the only merit I shall leave behind me in this
country. The funds, too, at the disposal of the Government
here are quite inadequate to such an expense, for I think that
country must be called poor in which the head-master of an
important school—one of the most valuable institutions in a
State—is allowed to starve with five or six children upon a
miserable salary of from seventy to ninety florins. Kr. and H.
will divide the money between them. If others show want of
principle, that is no reason why I should too. But why
should I urge this upon you, my dear friend, who out of love
to me have alone been induced to deviate from those principles
in which we both agree ? '

In the midst of these engrossing occupations and extensive
projects, Humboldt received the intelligence of the death of
his mother, which took place at Berlin on November 19, 1796.
'I had long been prepared for this event. It has not taken me
by surprise ; rather have I felt comforted that at the last she
suffered so little. She was only one day worse than usual, only
for one day were her sufferings more than ordinarily severe.
She expired without a struggle. You know, my dear friend,
that this is not an event by which my heart will be very deeply
wounded, for we have always been strangers more or less to one
another ; but who could have remained unmoved at the sight
of her unremitting sufferings ! '[1] Humboldt also wrote to
Willdenow :—'A happy release has at length been granted to
my poor mother. On the mere ground of humanity, her de-
cease was to be desired.'

The intelligence could not have been unexpected either at
Jena, where William von Humboldt at that time resided, for
Schiller as early as June wrote to Goethe that Humboldt's

---

[1] Alexander von Humboldt to Freiesleben, under date, Bayreuth, Nov.
25, 1796.

mother was dying, and that was probably the cause of his continued detention at Berlin.

As a noble trait in the character of Frau von Humboldt, it may be mentioned that by her will she set apart 500 thalers to be unredeemably secured upon the estate of Falkenberg, the interest of which at four per cent. was to be applied in perpetuity for the preservation of the church tower and family grave at Falkenberg, with the provision that the excess of interest was to be applied to raise this capital to 1,000 thalers, the interest from which, after deducting the necessary amount for repairs, was again to accumulate to the formation of an additional capital of 500 thalers. The interest of this third capital of 500 thalers was to be applied to increasing the salary of the school-master at Falkenberg, while the accumulated savings of the 1,000 thalers above mentioned were to be expended on the improvement of the school, in suitable alterations in the building, and in the purchase of necessary books, for all of which an exact account was to be rendered. The administration of this endowment, which is still in existence, was vested in the consistory of the province.

The death of his mother was a turning-point in the life of Humboldt. The obligations of filial duty which had hitherto formed a barrier to the accomplishment of many of his extensive plans were cancelled by this event, the ties that bound him to home were severed, and ample means were placed within his reach for gratifying his long-cherished wish for visiting the tropics. Like Bacon, he renounced the service of the State that he might devote himself exclusively to science.

One of the last letters written by Humboldt from Bayreuth was addressed to Willdenow on December 20, 1796. It gives a valuable review of the labours he had accomplished during that year:—

'Notwithstanding the serious interruption caused by my diplomatic mission to the French general, and my detention with the army during the months of July and August, I have yet been able to accomplish some good work this summer. My great work in physical science, on muscular excitability and the chemical process of life, is nearly completed. It contains the results of some 4,000 experiments, besides much relating to the physiology of plants. . . . In the winter I intend publishing some of the chemical treatises which I have by me ready for the press: experiments upon light[1] and nitrogen; transformation of morils[2] into animal fat by the application of nitric acid; a new description of barometer based upon a recently discovered principle, an instrument with which some highly satisfactory measurements have been obtained here; experiments with phosphorus as a eudiometer; on two new kinds of gas, oxygenised carbonic acid and azoture de phosphore oxydee. . . . A work of mine in French is to be printed at Geneva, "Lettres physiques a M. Pictet;"[3] it is a collection of papers contributed by me from time to time to the National Institute, and published in the Memoirs of that society. I have made several experiments this summer on the respiration of plants. . . .

'This may convince you, my dear Willdenow, that though I have been less of a scribe than other people, I have certainly not been less industrious. Pray see that my little godson grows up quickly, that I may take him to India with me. My journey is absolutely fixed. I shall continue my preparations for another year or so, and provide myself with various instruments; my plan is to spend a year or more in Italy, for the purpose of studying volcanoes, then to visit Paris and England, where I should like to remain another year, for above all things I mean to avoid being in a hurry, and finally, when all is complete, to sail in an English ship for the West Indies.

'*Should I not live to accomplish this plan, I shall at least have commenced it with energy and made good use of the position in which fortunate circumstances have placed me.*'
. . .

----

[1] 'I consider this to be the most delicate chemical experiment I have ever made.'—*Alexander von Humboldt to Freiesleben.*

[2] [*Morchella esculenta,* a kind of edible mushroom.]

[3] It was never published.

CHAPTER IV.

WEIMAR AND JENA.

State of Society in Weimar and Jena—Goethe as a Natural Philosopher— Early Recognition of Humboldt's Genius—Opposition to his Views and subsequent Recantation—Humboldt's Opinion of Goethe—Contrasts and Harmonies afforded by the two Characters—Schiller's Medical Studies—Humboldt a Contributor to the 'Horen'—The Genius of Rhodes—Schiller's Harsh Judgment—Körner's Mediation—Schiller's Idealism—Humboldt's Empiricism and Love of Formula—Friendly Counsel and Avowal—Reconciliation—Humboldt and the New Philosophy—Humboldt and Karl August—The Duke's Love of Nature— Frequent Visits to Weimar and Jena—Inscription in an Album.

OUR narrative has already anticipated the time when Alexander von Humboldt was first introduced to the men of genius, forming the distinguished circle of literati at Weimar and Jena. If we deem his relationships with these eminent men to be of sufficient importance to devote a chapter to the subject, we shall scarcely feel that an apology is due to our readers—at least not to those who claim him as a fellow-countryman.

A new phase had passed over the court of Weimar since the arrival of Wieland; he had been summoned to the court in 1772, by the Duchess Anna Amelia, who, early left a widow, was anxious to commit the education of her two sons, Karl August and Constantine, to the care of one who by his recently published work, the 'Goldener Spiegel,' [1] had shown himself well fitted for such a task. With Wieland rose the first star of that brilliant constellation which rendered Weimar so illustrious.

The young princes grew up to man's estate, distinguished as much for their physical as for their mental endowments, and

[1] ['The Golden Mirror, or the Kings of Scheschian.' A series of important lessons which the rulers of mankind should derive from history.]

Karl August soon fulfilled the prophetic words of Wieland to Jacobi : ' If Heaven only grant life to our young prince and to the excellent friends by whom he is surrounded, we shall in a few years have a little court here worthy of being visited from all parts of the world.' With Goethe's arrival at the court of Karl August in November 1775, began the halcyon days of Weimar's celebrity. Votaries of every branch of the arts and sciences were welcome guests at the court, and every friend of the Muses was certain of a cordial reception. ' Bethlehem of Judah is not deserted, the wise men visit it,' writes Herder to Knebel, on September 11, 1784.

While classical poetry was being cultivated at Weimar to its highest stage of development, a fresh interest in scientific inquiry was simultaneously awakened in the neighbouring town of Jena. It is undoubtedly to the genius of Schiller, who, since 1789, had occupied the chair of history at the University, that the revolution of thought in Germany, emanating from Jena, is mainly to be ascribed.

In this position of influence Weimar and Jena stood alone during the last decade of the past century. Even Berlin, when exercising so wide and important an influence in later years, never possessed the charm, nor attained the height of intellectual power and activity, which at that time animated all hearts and minds in Weimar and Jena. The eminent men of Berlin had in almost all periods of her history stood in intellectual isolation; no social intercourse had ever existed whereby to bridge the chilling separation. A striking contrast was afforded by the mode of life at Weimar. The prince and the scholar, the poet and the statesman, lived there one common existence, mutually stimulating to intellectual activity. And the result of this activity was no tender hot-house plant, fostered only in princely circles; it flourished equally in the study of the scholar, the garret of the thoughtful poet, and in the boudoir of gifted and cultivated women. Weimar and Jena constituted *one* intellectual family. When poetry, with weary wing, could no longer soar at Weimar to her accustomed heights, science at Jena would be fired by a fresh impulse ; and when the formalities of university life exercised too contracting an influence at Jena, the spirit found full freedom at Weimar under the genial auspices of poetry and the arts.

It was during the winter of 1789 that William von Humboldt made the acquaintance of Schiller at Weimar, where the two sisters Charlotte and Caroline von Lengefeld then resided, and where the poet was almost a weekly visitor. This acquaintance soon ripened into friendship, which was further cemented by Schiller's marriage with Charlotte von Lengefeld, on February 20, 1790, and the union of William von Humboldt with Caroline von Dacheröden, the intimate friend of the two sisters, in July of the following year, upon which commenced, to use the words of Frau von Wolzogen, 'those happy days of close intimacy that existed between the two newly-married couples at Jena.'

Those bright days were succeeded by years of equal happiness, when William von Humboldt, in the spring of 1794, took up his residence at Jena, having resigned his position in the service of the State, on which he had but recently entered, in order that he might devote himself to scientific pursuits. With his æsthetic criticism, he occupied a middle place between Schiller and Goethe. Mind stimulates mind, 'the souls of poets enkindle one the other.' Speculations of the deepest nature were clothed in language of the most elevated and graceful character, and daily some fresh flight was reached in the highest problems of life and art.

In consequence of the position held by his brother at Jena, Alexander von Humboldt, already Superintendent of Mines, was at once admitted into the intimate circle of his brother's friends. As Schiller had been attracted to William, so Goethe felt drawn to Alexander von Humboldt.

We shall doubtless be excused if we here revert to certain well-known phases in the development of Goethe's genius, which must have operated largely in the creation of this bond of union.

In Goethe there had been early developed an intense love of Nature, accompanied by a strong bent towards the investigation of her phenomena and the study of her laws. This characteristic had manifested itself in early youth, and often found expression in his poems in such lines as the following :—

> Boyhood's pleasures e'en were tinted
> With the longing felt within,
> To enquire of Nature's secrets,
> And her confidence to win.[1]

Even in his earliest poetic effusions there are to be found traces of his love for the contemplation of nature, and during his residence at Strasburg, in 1770, he evinced a marked preference for the study of the natural sciences. He attended lectures on chemistry and anatomy, was present at Lobenstein's clinical instruction given at the bedsides of his patients, and he even attended a course upon midwifery. Soon after his settlement at Weimar he was led by his official employment, in the cultivation of land and the management of forests, to devote himself to the science of botany, while almost at the same time his attention was directed, by the influence of Merck, to osteology and comparative anatomy. And, since the project had been started for reworking the mines at Ilmenau, the science of mineralogy had become the favourite and fashionable study at Weimar. 'Everyone,' says Böttiger,[2] 'was absorbed in mineralogy, even the court ladies found a deep meaning in stones, and arranged them in cabinets.' Goethe himself always set a very high value on his scientific studies. 'I have,' he relates in later years, 'devoted a great portion of my life to the study of science, for which I early conceived a strong passion. The results of my investigations were not attained by a mere inspiration, but through quiet, persevering, and indefatigable labour.' In his zeal for mineralogy 'no mountain was too high, no mine too deep, no passage too contracted, no labyrinth too intricate.' . . . Bohemia and Carlsbad were pre-eminent in the influence they exerted over his inquiring mind.

> To say what rapture through me thrilled,
> To paint what joy my heart then filled,

---

[1]   'Freudig war vor vielen Jahren
>   Eifrig so der Geist bestrebt,
>   Zu erforschen, zu erfahren,
>   Wie Natur im Schaffen lebt.'

[2] 'Literarische Zustände und Zeitgenossen,' vol. i. p. 22.

What thoughts within me sought expression
Would need too long to make confession.[1]

Not merely was his interest excited by the study of plants, animals, and minerals, but investigations into the nature of the atmosphere and clouds, of light and colour, possessed for him a powerful charm.

Of this prince of poets, therefore, it may well be remarked, in the language of Alexander von Humboldt,[2] that 'the great creations of poetic fancy did not withhold his penetrating glance from investigating the depths of Nature's secrets.'

As early as 1786 Goethe wrote his paper on the intermaxillary bone, which was followed in 1790 by his 'Metamorphoses of Plants,' and in 1791 and 1792 by his first treatises on optics, the precursors of his subsequent work on colour, 'Farbenlehre.'

It was about this time that Alexander von Humboldt first crossed his path. Steffens[3] relates that Goethe, who usually kept all younger poets ungraciously at a distance, eagerly welcomed to his side those youths who were engaged in any branch of natural philosophy. It was therefore his love for the natural sciences that constituted the mental elective affinity, the strong power of attraction which drew him irresistibly to Alexander von Humboldt, in whose experiments on the galvanic action of the filaments of the nerves and muscles he had taken the liveliest interest.[4] According to Eckermann, he has acknowledged in express terms how much he was indebted to the circumstance that, just as he was beginning to get tired of the world, he had been brought under the inspiring influence of the two brothers Von Humboldt, then in the height of their fresh youthful energy. 'The debt I owe to Fichte, Schelling, Hegel, the brothers Humboldt, and Schlegel,' he writes, 'may some day

---

[1]
        'Was ich dort gelebt, genossen,
        Was mir all dorther entsprossen,
        Welche Freude, welche Kenntniss,
        Wär' ein allzu lang Geständniss.'

[2] 'Rede bei der Eröffnung der Versammlung der Naturforscher und Aerzte in Berlin,' 1828.

[3] Steffens, 'Was ich erlebte,' vol. iv. p. 101.

[4] A. von Humboldt, 'Versuche über die gereizte Muskel- und Nervenfaser,' vol. i. pp. 76, 77.

be gratefully acknowledged if I should ever be able to accomplish my wish of furnishing from my own point of view a sketch, if not a history, of that epoch which proved so memorable to me—the last decade of the past century.'

In the various remarks relating to the two Humboldts to be met with in Goethe's writings, he alludes more to their scientific labours than to any of their other studies, and particularly mentions their investigations in anatomy and galvanism, in which he actively participated. The following entry occurs in the ' Jahres- und Tageshefte' for the year 1794 :—

' The long-expected arrival of Alexander von Humboldt from Bayreuth was the signal for turning our thoughts exclusively to science. His elder brother, also at Jena, who manifested a keen interest in almost every branch of knowledge, lent to these investigations his advice and practical assistance. It is worthy of remark that the Court Counsellor Loder was then lecturing upon the ligaments of the animal frame—a highly important branch of anatomy, for do not these constitute the necessary connection between the bones and the muscles? And yet from a strange infatuation this branch is quite neglected by the medical students. We three, in company with our friend Meyer, used to wade of a morning through the deepest snow to hear, in an almost empty lecture-room, this portion of physical structure lucidly explained and exemplified by the most accurate anatomical preparations.'

In 1795 he further adds :—

' By the advent of the two brothers Von Humboldt at Jena, towards the end of the year, I was entirely diverted from the creative art of poetry, and brought back to the study of nature. Just at this time they both took great interest in the pursuit of science, and in the course of conversation I could not refrain from communicating my ideas upon comparative anatomy and its systematic arrangement. As my views appeared to display consistency and some measure of completeness, I was urgently requested to record them on paper, and at once acceding to this proposition, I proceeded to dictate to Max Jacobi the ground-plan of a system of comparative osteology. In thus gratifying the wishes of my friends, I gained for myself a platform upon which to erect new theories. The influence exerted by

Alexander von Humboldt requires to be dealt with separately. His presence at Jena promoted the study of comparative anatomy; I was incited by him and his elder brother to record the outline of my scheme of anatomy. During his residence at Bayreuth my correspondence with him was of great interest.' Unfortunately, none of these letters have as yet been found.

In the appendix to the ' Osteology,' Goethe alludes almost in the same words to the stimulus he at that time derived from the two brothers :—' My time in this way was closely occupied, till in the year 1795 the brothers Humboldt, who, as Dioscuri, had often illuminated my path, came to reside for a considerable time at Jena. I brought the subject of my anatomical scheme so perseveringly and pressingly forward, that at last it produced some amount of impatience, and occasioned the request that I should set down in writing the thoughts with which my whole being—heart, soul, and mind—was imbued.'

The Humboldts are again alluded to by Goethe, when describing in 1797 the prodigality of intellectual life then at Jena :—' The Humboldt brothers were there, and almost every aspect of nature was discussed in a philosophic and scientific spirit. My osteological type of 1795 afforded a motive for a more systematic use of the Museum, as well as of my own private collection. I drew out a system upon the metamorphoses of insects which I had had in view for many years. The drawings of the Hartz Mountains by Krause gave rise to geological disquisitions. Experiments in galvanism were instituted by Humboldt.'

In a letter to Schiller of April 26, 1797, Goethe writes, after a visit from Alexander von Humboldt at Weimar:— ' During Humboldt's visit my time has been usefully and agreeably spent; his presence has had the effect of arousing from its winter sleep my taste for natural science.'

It is thus evident that the important significance of Alexander von Humboldt's character and attainments was early appreciated by Goethe, who continued to regard him with an ever increasing admiration till the close of life.

When, however, Humboldt avowed himself a convert to the theory of volcanic agency in the formation of the crust of the earth, Goethe, who still remained an adherent of the

aqueous theory, could not restrain a manifestation of anger. As
if to prove that even a genius of the highest order is not always
permitted to comprehend Nature in all her aspects, Goethe could
never be induced to abandon these views, which the researches
of modern science had rendered obsolete through the trium-
phant establishment of the theory of volcanic agency. His
indignation found vent in some expressions of Mephistopheles.
It cannot, however, be admitted for an instant, as Boas and
Saupe have both assumed, in conformity with earlier writers,
that the following epigrams bear any reference to Alexander
von Humboldt :—

161. *Creation by Fire.*

Poor columns of basalt! 'Tis said, ye were by fire thus moulded,
And yet none saw thee ever by the flames enfolded.[1]

163. *Short-lived Pleasures.*

At length beneath the floods primeval is their fate fast sealed,
And thus the strife so long enkindled is for ever healed.[2]

At that time Humboldt was an avowed advocate of the aqueous
theory, as may be seen from his 'Mineralogical Observations
on some Basalts of the Rhine.' It was only when travelling in
the New World that he became a convert to the volcanic theory.

Goethe's whole nature was so organised that he could
scarcely fail to hold Werner's views with regard to the aqueous
theory, which taught that in the long intervals separating the
primeval catastrophes everything was slowly formed by the
'moisture of life.' He says:—'I have spent several years of
my life in studying the mutual connection of the stratified
formations. The views I formed coincided with Werner's
doctrine, and I continued to hold them even after I had good
reason to believe that they left many problems unsolved.'

---

[1] 161. *Schöpfung durch Feuer.*

'Arme basaltische Säulen! Ihr solltet dem Feuer gehören,
Und doch sah euch kein Mensch je aus dem Feuer entstehn.'

[2] 163. *Kurze Freude.*

'Endlich zog man sie wieder ins alte Wasser herunter,
Und es löscht sich nun bald dieser entzündete Streit.'

In the controversy between Thales and Anaxagoras in the second part of 'Faust,' Goethe brings forward the old controversy between the Neptunists and Vulcanists, symbolising thereby the scientific movement of his own time, and gives unmistakable evidence, especially in the discourse between Faust and Mephistopheles, in the beginning of the fourth act, of the theory which the bent of his own mind was disposed to favour. Thus, speaking of Nature, Thales says in the second act :—

> She makes each form by rules that never fail,
> And 'tis not Force, even on a mighty scale.[1]

And Faust continues the strain in the fourth act :—

> When Nature in herself her being founded,
> Complete and perfect then the globe she rounded,
> Glad of the summits and the gorges deep,
> Set rock to rock, and mountain steep to steep,
> The hills with easy outlines downward moulded,
> Till gently from their feet the vales unfolded!
> They green and grow; with joy therein she ranges,
> Requiring no insane, convulsive changes.[2]

Goethe was no friend of wild commotions, and even in his own daily existence kept at a distance from him everything that could disturb his habitual equilibrium. He shrank from the idea of the mighty plutonic forces having risen from the lowest depths with demoniac power, whereby the mountains seem but the awful charnel-houses of former creations :—

> Basalt, that ebon devil's moor,
> Bursts forth from hell's remotest floor,
> And earth is riven, rocks are rent,
> Till all is topsy-turvy sent:

---

[1] ' Sie bildet regelnd jegliche Gestalt
Und selbst im Grossen ist es nicht Gewalt.'

[2] ' Als die Natur sich in sich selbst gegründet,
Da hat sie rein den Erdball abgeründet,
Der Gipfel sich, der Schluchten sich erfreut
Und Fels an Fels und Berg an Berg gereiht,
Die Hügel dann bequem hinabgebildet,
Mit sanftem Zug sie in das Thal gemildet:
Da grünt's und wächst's, und um sich zu erfreuen
Bedarf sie nicht der tollen Strudeleien.'

> Thus would geologists assert,
> The rocks were formed now so inert. [1]

In a still more diabolic manner Mephistopheles thus ridicules the volcanic theory of upheaval:—

> When God the Lord—wherefore, I also know—
> Banned us from air to darkness deep and central,
> Where round and round, in fierce, intensest glow,
> Eternal fires were whirled in Earth's hot entrail,
> We found ourselves too much illuminated,
> Yet crowded and uneasily situated.
> The Devils all set up a coughing, sneezing,
> At every vent without cessation wheezing:
> With sulphur-stench and acids Hell dilated,
> And such enormous gas was thence created,
> That very soon Earth's level, far extended,
> Thick as it was, was heaved, and split, and rended!
> The thing is plain, no theories o'ercome it:
> What formerly was bot om, now is summit.
> Hereon they base the law there's no disputing,
> To give the undermost the topmost footing. [2]

Humboldt was fully aware of the reference these verses bore

---

[1]   'Basalt, der schwarze Teufelsmoor,
>   Aus tiefster Hölle bricht hervor,
>   Zerspaltet Fels, Gestein und Erden,
>   Omega muss zum Alpha werden:
>   Und so wär' denn die liebe Welt
>   Geognostisch auf den Kopf gestellt!'

[2]   'Als Gott der Herr—ich weiss auch wohl warum—
>   Uns aus der Luft in tiefste Tiefen bannte,
>   Da wo centralisch glühend, um und um,
>   Ein ewig Feuer flammend sich durchbrannte,
>   Wir fanden uns bei allzu grosser Hellung
>   In sehr gedrängter, unbequemer Stellung.
>   Die Teufel fingen sämmtlich an zu husten,
>   Von oben und von unten auszupusten;
>   Die Hölle schwoll von Schwefelstank und Säure:
>   Das gab ein Gas! das ging ins Ungeheure,
>   Sodass gar bald der Länder flache Kruste,
>   So dick sie war, zerkrachend bersten musste.
>   Nun haben wir's an einem andern Zipfel:
>   Was ehmals Grund war, ist nun Gipfel.
>   Sie gründen auch hierauf die rechten Lehren,
>   Das Unterste ins Oberste zu kehren.'

to himself, and in a letter of January 24, 1857, to Herr von Kobell, author of the poem 'Die Urzeit,' he especially thanks him for his avowed adherence to the eruptive theory, since ' in the jubilant conclusion of the second canto of his poem, " Die Urzeit," he had ventured to avenge him (Humboldt) for the bad treatment he had received in the second part of " Faust." '

Goethe's aversion to the modern theory of upheaval almost turned him into a volcano, vomiting fire and flame. ' The case may be as it pleases, but it must be written that I curse this execrable racket and lumber-room of the new order of creation ! ' ' Able, clever, and bold thinkers dress up for them-selves such a theory out of mere probabilities ; they manage to gather around them followers and adherents, and these from sheer numbers gain a literary power ; the theory gets pushed to an extreme, and is carried forwards with a reckless impetu-osity. . . . It is then spoken of as the unanimous belief of scientific investigators.' Undoubtedly Alexander von Humboldt was included among these ' able, clever, and bold thinkers.'

And when, on the death of Werner in 1817, not only a younger generation but also the older adherents of the Neptunic theory of the school of Freiberg began to view with favour the modern volcanic theories, his ' abhorrence was increased of the forced explanations such as extensive upheavals, volcanic fires, floods, and other titanic occurrences, to which they were obliged to have recourse.' He was painfully affected by the modern revolutions in science.

> As kings are now from thrones subverted,
> So granite must be disconcerted,
> And gneiss, the child, is father held
> Till it in turn is downward felled ;
> For Pluto's trident threatens soon
> To upturn all beneath the moon.[1]

And so at last, in a spirit of proud resignation, he gave vent to the discontented feelings of an old man who sees the well-

---

[1]   ' Wie man die Könige verletzt,
      Wird der Granit auch abgesetzt ;
      Und Gneis, der Sohn, ist nun Papa !
      Auch dessen Untergang ist nah ;
      Denn Pluto's Gabel drohet schon
      Dem Urgrund Revolution.'

arranged classification of ideas to which he had been all his life
accustomed ruthlessly upset by the younger generation :—

> The noble Werner scarce his back had turned,
> When Neptune's kingdom rudely is o'erthrown.
> Though all to Hephæstos obedience have learned,
> I cannot yield him homage :
> My faith can only on conclusions rest ;
> Before me many diverse views have passed,
> Each one by me more hated than the last,
> New Deities and Idols.[1]

Doubtless Alexander von Humboldt is to be reckoned among
these new deities, for he and Leopold von Buch were the titanic
destroyers of the realm of Neptune. The chancellor Von
Müller [2] relates that Goethe criticised with some bitterness
Humboldt's lecture upon volcanoes in 1824. ' Our friend,' said
he, ' has no true genius, he is possessed only of much sound
sense and a considerable amount of zeal and perseverance. In
æsthetics everyone may think and feel as he pleases, but in
science what is untrue and absurd is quite insupportable.'
He expressed himself subsequently, in 1828, with an equal
amount of pique : [3]—' If Alexander von Humboldt and the other
Plutonists incense me too much, I shall not hesitate to expose
them to obloquy. I am already secretly preparing several
epigrams against them ; posterity shall know that there lived
in this century at least one sensible man who could see through
these absurdities.'

Time, however, somewhat modified this angry vehemence.
He wrote to Varnhagen on July 8, 1829, in the following
strain : [4]—' The sciences (geology, geography, and mineralogy)

---

[1]   ' Kaum wendet der edle Werner den Rücken,
       Zerstört man das poseidaonische Reich.
       Wenn alle sich von Hephästos bücken,
       Ich kann es nicht sogleich,
       Ich weiss nur in der Folge zu schätzen :
       Schon hab' ich manches Credo verpasst,
       Mir sind sie alle gleich verhasst,
       Neue Götter und Götzen.'

[2] Burckhardt, ' Goethe's Unterhaltungen mit dem Kanzler von Müller '
(Stuttgart, 1870), p. 56.
[3] Ibid. p. 124.
[4] Paulus, ' Geisterrevue,' p. 390.

now engaging our attention are advancing with dispropor-
tionate strides, in most cases on a sure foundation, but occasionally too hastily, and sometimes even under the influence
of fashion. We must not therefore follow too precipitately,
since we cannot afford time to wander thoughtlessly in
error.' In a still milder tone he writes to Carus : [1]—' When
I observe the advance that science has lately made, I feel like
a traveller who at early dawn is journeying eastward, gladly
watching the increasing light in earnest longing for the
appearance of the great luminary, but who, upon the first
sight of the King of Day, is obliged to turn away his eyes,
because he cannot bear to look upon the splendour that he has
so ardently desired.'

At length, notwithstanding his proud words, the poet hero
bowed before the man of science. Late in the evening of life
he writes to Zelter on October 5, 1831 : [2]

' I have received the two volumes of " Fragmens de Geologie
et de Climatologie asiatiques, par Alexandre de Humboldt ; "
in looking them over, a rather curious observation occurred to
me, which I will communicate to you. The extraordinary
talent of this extraordinary man is shown in his forcible mode
of expressing himself, and it is clearly evident that every
discourse will persuade the hearer and make him believe that
he is convinced. Few men are capable of being convinced ;
most men allow themselves to be persuaded, and thus the
treatises here brought before us are real speeches, delivered
with great facility, so that people are at last brought to
imagine that they comprehend the impossible. That the
mountains of Himalaya can have been upheaved to the height
of 25,000 feet and yet point so proudly and inflexibly towards
heaven as if nothing had happened, is quite beyond the powers
of my comprehension, and lies in the misty regions haunted by
transubstantiation ; my cerebral system would have to be entirely reorganised—which would be rather a pity—were any
space to be found for the reception of such a wonder.

' There are people, however, whose minds are so constituted
as to be able to receive such articles of faith side by side with

[1] Carus, ' Goethe zu dessen näherm Verständnisse ' (1843), p. 34.

[2] ' Briefwechsel zwischen Goethe und Zelter,' vol. vi. p. 308.

propositions of the highest reason; I cannot understand it,
though I see it take place every day. But is it necessary to
understand everything? I repeat: our mighty conqueror of
the world of science is perhaps the greatest orator. Not only
are all facts present to his mind in consequence of his prodigious
memory, but he knows how to use them with the greatest skill
and boldness. The initiated see clearly enough where weakness
has entwined itself round strength, while strength is not averse
to be decked in the garb of fragile beauty.

'And so the effect of such a paradox, when skilfully and
energetically propounded, is powerful; many of our boldest
scientific investigators are brought thereby to imagine that
they can grasp the incomprehensible. To them, on the contrary, I appear as an obstinate arch-heretic, wherein may God
graciously maintain and confirm us. Selah!'

With a yet fuller appreciation of Alexander's views, Goethe
thus wrote to William von Humboldt, on December 1, 1831 : [1]—

'I have been greatly indebted to your brother, for whom I
can find no suitable epithet, for several hours of unreserved
and friendly intercourse. For although his method of viewing
the facts of geology and his mode of reasoning upon them is
altogether opposed to my cerebral system, yet I have observed
with warm interest, not unmixed with astonishment, that what
I could not myself receive is logically apprehended by him,
and incorporated among his vast stores of knowledge, to the
value of which it will owe its preservation.'

Of an earlier date there are also many expressions testifying
to his appreciation of Humboldt's surprising gifts. In the
'Elective Affinities' [2] Ottilie writes in her Journal :—' That
investigator of nature alone is worthy of homage who is able
graphically to reproduce with all the force of local colouring,
and faithful delineation of surrounding accessories, everything
that to our minds is most foreign and unfamiliar. How
glad I should be, if only for once, to hear Humboldt recount
his travels!' In the 'Maximen und Reflexionen' occurs the
passage :—'The most remarkable men of the sixteenth and
seventeenth centuries were in themselves schools of learning,

---

[1] Schlesier, 'Leben Wilhelm von Humboldt's.'
[2] 'Die Wahlverwandtschaften,' vol. ii. chap. vii.

just as Humboldt is in our own time.' In the 'Correspondence with Knebel'[1] Humboldt is described as 'a rich cornucopia, distributing his gifts with prodigality.'

Eckermann relates[2] having on one occasion found Goethe in a state of joyous excitement, exclaiming as he advanced to meet him: 'Alexander von Humboldt has been with me for some hours this morning; what an extraordinary man he is! Though I have known him for so long, I am always struck with fresh amazement in his company. He may be said to be without a rival in extent of information and acquaintance with existing sciences. He possesses, too, a versatility of genius which I have never seen equalled. Whatever may be the subject broached, he seems quite at home in it, and showers upon us treasures in profusion from his stores of knowledge. He resembles a living fountain, whence flow many streams, yielding to all comers a quickening and refreshing draught. He will remain here a few days, and I already feel that I shall have lived through years in the time.'

It is very remarkable and altogether inexplicable—unless some clue may be discovered hereafter among the poet's papers preserved at Weimar—that Goethe never discussed with Humboldt any of his own researches in botany or optics, although, as before stated, he had been eagerly engaged upon these studies since the year 1790.

That Humboldt on his part estimated Goethe very highly as a botanist, is evident from his dedicating to him his work, 'Thoughts upon the Geographical Distribution of Plants, illustrated with a View of Nature in Tropical Regions,'[3] forming the first part of his Travels in America; the dedication page was illustrated after a design by Thorwaldsen, and represents the Genius of Poetry as Apollo crowned with laurel unveiling Isis, at whose feet lies a book inscribed 'The Metamorphoses of Plants,' thus indicating that to poetry we are also indebted

[1] 'Briefwechsel mit Knebel,' vol. i. p. 243.
[2] 'Gespräche mit Goethe in den letzten Jahren seines Lebens.' Monday, December 26, 1826.
[3] 'Ideen zu einer Geographie der Pflanzen, nebst einem Naturgemälde der Tropenländer.'

for the unveiling of nature. Goethe[1] felt highly honoured 'by the dedication in a manner so gratifying of this important work,' and remarks that 'out of early and ever increasing friendship for the noble author, and incited by this fresh token of a sympathy so flattering,' he threw himself eagerly into the study of the work, and as the illustration belonging to it was not yet completed, he himself drew, in illustration of the text, a 'conventional picture,' and dedicated this 'symbolic landscape' to the friend to whom, as he said, its existence was to be ascribed.[2]

When Humboldt sent to Goethe, on June 6, 1816, a copy of his work, 'Thoughts on the Physiognomy of Growth,'[3] the poet was in affliction from the death of his wife, and acknowledged the gift on June 12 by the following verses :—

> In sorrow's darkest hour
> Your book arrived to cheer me in my grief,
> It urged with power
> That I should seek in labour for relief;
> The world still wreathes itself in blossoms fair
> By Nature's laws, eternal in their force :
> These were to thee of joy the constant source,
> Then let such pleasures banish thy despair.[4]

Humboldt recognised in Goethe not only a distinguished botanist, but also a successful student of optics and osteology. He classed his descriptions of nature with those of Forster, Buffon, and Bernardin de St.-Pierre, as characterised by inimitable truth, and he frequently quoted his 'Aphorisms on the

---

[1] 'Jahres- und Tageshefte.'

[2] This print first appeared in 1813, with some pages of letterpress, in vol. xli. of 'Geographische Ephemeriden;' it was subsequently published in a separate form in folio, and has also been reproduced in Paris. Thorwaldsen's print has become very scarce.

[3] 'Ideen zu einer Physiognomik der Gewächse.'

[4]   'An Trauertagen
> Gelangte zu mir Dein herrlich Heft,
> Es schien zu sagen :
> Ermanne dich zu fröhlichem Geschäft ;
> Die Welt in allen Zonen grünt und blüht
> Nach ewigen beweglichen Gesetzen :
> Das wusstest du ja sonst zu schätzen,
> Erheitre so durch mich dein schwer bedrängt Gemüth !'

Physical Sciences.' His genuine admiration of the great poet culminates in the fine passage with which he concludes the chapter in 'Cosmos'[1] on the poetic description of nature. 'Where is the nation of the imaginative South who might not envy us our great master of poetic art, whose works are deeply imbued with an intense love of nature displayed with equal fervour in the "Sorrows of Werther," the "Reminiscences of Italy," the "Metamorphoses of Plants," and the "Miscellaneous Poems"? Who has so eloquently incited kindred minds "to unravel the profound mysteries of the universe," and renew the bond by which, in the primitive ages of the world, philosophy, physical science, and poetry, were united? Who has so attractively portrayed that land in which his imagination found a home, where

> Soft blows the air beneath that southern sky,
> Where myrtles bloom and laurels tower high"?'[2]

And yet the remarkable love of nature distinguishing these two men, who held each other in such high esteem, was inspired by feelings of a very opposite character. Goethe was pre-eminently characterised by an inborn love of nature and all natural phenomena, a subjective comprehension and a keen susceptibility for the impression produced by the forces of nature, while, as the enemy of all objective exact investigation, he viewed with the utmost abhorrence the 'physico-mathematical guild.' It was only in later life that he felt any impulse towards scientific investigation, and then he accomplished important results, not merely by means of isolated discoveries, but by developing a comprehensive view of nature and method of observation. He attained that insight into nature which led to the perception of the reciprocal action of cause and effect, by which he was enabled to rise from the individual to the type, from the special to the universal, from the small to the great, from the part to the whole.

The characteristic of Humboldt's mode of thought, on the contrary, was an early and overpowering impulse towards

---

[1] 'Kosmos,' vol. ii. p. 75.
[2]   'Ein sanfter Wind vom blauen Himmel weht,
            Die Myrte still und hoch der Lorbeer steht?'

objective scientific knowledge. He was one of those scrupulous empirics who observe and collate nothing but facts, and who, rejecting all bold hypotheses, confine themselves exclusively to the realm of experience. He was, according to Schiller's severe criticism, the impersonation of ' keen cold reason, which would have all nature shamelessly exposed to scrutiny,' 'with no power of imagination, no tender sympathy, no sentimental interest.' That, however, he could rise from the bare facts of observation to a highly poetic representation of nature, is evident from the ' Aspects of Nature,' from many descriptions in his ' Travels in America,' and from isolated passages in ' Cosmos.'

Both he and Goethe were familiar with all branches of natural philosophy ; they held in common the intimate connection of the sciences, and, above all, the unity of nature displayed in the constitution of the universe.

In reviewing Humboldt's intercourse with Schiller, many points of mutual interest might be looked for from the attention early bestowed by the poet to the study of medicine. There is, however, but little known of Schiller's achievements in the medical art, and the little that is known is not much to his credit, for his practice seems to have been distinguished more by boldness than by success. Although in the pursuit of medicine Schiller met with no encouragement and but little gratification of his natural tastes, yet during his early professional studies he had devoted himself by preference to the consideration of the severest problems in physiology. When only eighteen years of age, he selected as a subject for a theme ' The Philosophy of Physiology,' and in contending for his degree in 1780 he wrote and defended a thesis ' On the Connection between the Animal and Intellectual Nature of Man.' He ever retained a sympathetic interest in the study of natural philosophy and physics, and upon taking up his residence at Jena, was gratified at the prospect of living under the same roof with Göttling, professor of chemistry, and of assisting him in his experiments.

This taste for medical science must have been revived and strengthened by his personal intercourse with Alexander von Humboldt, who was often his guest for days together. In the

preface to the 'Aspects of Nature,' Humboldt remarks:—
'During my long sojourn at Jena, Schiller frequently conversed
with me upon physiological subjects while recalling the medical
studies of his youth, and the experiments upon which I was
then engaged on the effect produced on the muscles and
nerves by contact with chemically different substances often
gave a specific turn to our discourse.'

Freiesleben also refers to many delightful evenings spent
amid the intellectual circle at Schiller's house, when Goethe
and the two Humboldts engaged in interesting converse on
anatomy while discussing their zoological preparations.

Alexander von Humboldt was one of the first to be invited
by Schiller to contribute to his new periodical 'Die Horen,'
and was, indeed, the only natural philosopher whose assistance
was sought by the poet. By a fortunate chance, Humboldt's
answer to Schiller's request has been preserved; it is dated
from the head-quarters of General von Möllendorf, and is as
follows :—

'Cantonment Quarters, Flörsheim: August 6, 1794.

'How can I excuse myself, my honoured friend, for my delay
in replying to your letter ! Never has my vanity, and that of a
noble kind, been so highly flattered as by your invitation to
assist you in the dissemination of philosophical ideas. I have
hitherto pursued my literary course in a manner so isolated
and unobtrusive that I could hardly suppose I had attracted
any attention. How can I then sufficiently value the distinction
of your notice?

'The rapidity of my movements while travelling with the
minister Von Hardenberg, to whom I am at present bound both
by duty and inclination, prevented me from gratifying my wish
of visiting you at Jena. Just now, my ill-fortune has led me
hither, where I am engaged in diplomacy—to me an untried
career—and I am obliged for the most part to follow the army
under the command of the field-marshal. Pray ascribe my
long silence, as well as the want of connection in this letter, to
my present unsettled state. I hope soon to regain my freedom,
when I shall devote my energies to the great task I have set
myself, to the accomplishment of which all my efforts are
even now directed.

'Never have my expectations been raised so high by any literary undertaking as they are by yours, in which, from the intellectual powers employed, great results may be anticipated. I am delighted to find that scientific investigation is not to be excluded from your plan. *Res ardua vetustis novitatem dare, omnibus naturam et naturæ suæ omnia.*[1] So long as the present method of studying botany and natural history continues to be followed, in which attention is directed only to the varieties of form, the physiognomy of plants and animals, whereby the study of characteristic distinctions and the law of classification is confounded with the true objects of science, so long must botany, for example, fail to furnish a worthy subject of speculation to thoughtful men. But you feel with me that there is something higher yet to be attained, that there is something even to be regained; for Aristotle and Pliny, whose descriptions of nature were addressed to the æsthetic feelings, and were aimed at the cultivation of a love of art, undoubtedly possessed a wider range of view than the modern naturalist who contents himself with the mere registration of nature. The universal harmony of form, the problem as to the existence of a primeval form of plant which is now developed in a thousand forms of different gradations, the distribution of these forms over the surface of the globe; the causes of the various emotions of joy and sorrow produced upon the mind by the varied aspects of the vegetable kingdom; the contrast to be observed between the dead immovable mass of rock or the apparently inorganic stem of a tree and the living garment of vegetable life by which it is clothed in beauty, just as the flesh imparts a soft outline to the skeleton; the history and geography of plants, or the historical representation of the general distribution of vegetation over the face of the earth, a portion of the general history of the world hitherto uninvestigated; the search for the oldest forms of vegetation as imprinted on their stony sepulchres (fossils, coal, peat, &c.); transformations in the earth's condition gradually fitting it for the habitation of higher grades of life; the characteristics of plants and their migrations, both in groups and individual

---

[1] Plinii Hist. Nat., præf. sect. 15.

species; maps showing the plants that have closely followed in the wake of certain nations; the general history of agriculture, comparisons instituted between cultivated plants and domesticated animals, and the probable origin of both; causes of degeneracy, a catalogue of the plants which adhere most closely to the law of uniformity, and those which deviate from it most readily, the degeneracy of cultivated plants, such as the American and Persian plants, which are wild from the Tagus to the Obi; the general confusion produced in the geography of plants by colonisation—these appear to me to be subjects worthy of thought, though hitherto for the most part neglected. I am incessantly engaged on these subjects, but I am too much distracted by the noise just now about me to explain myself very systematically. I see that in some instances I have expressed myself awkwardly, but I trust that on the whole you will be able to grasp my meaning.[1]

'Should I be in a position at any future time to send you, my much-esteemed friend, some essays on these subjects as specimens, your approval of them would make me inexpressibly happy; but anything I can do will, I fear, contrast unfavourably with the contributions you will receive from your other coadjutors.

'In the mean time farewell, and may you be as happy as the elevation and purity of your nature is fitted to render you. Remember me to your amiable wife, and give my love to my brother William, to whom I have not written for an age.

'Yours, &c.,

'HUMBOLDT.'

Schiller, delighted at this consent, writes to Körner on September 12, 1794:—'Jacobi of Düsseldorf has at length promised to contribute to the "Horen." From Humboldt's brother (Alexander von Humboldt), Superintendent of Mines in Prussia, we may expect some excellent treatises on Nature in her philosophic aspect. He is certainly the most gifted

---

[1] It is very interesting to compare these remarks with the prospectus of the 'Geographie der Pflanzen' in Berghaus' 'Hertha,' vol. vii., in 'Geogr. Zeitung,' pp. 52-60, and in the 'Humboldt-Berghaus Briefwechsel,' vol. i. p. 63.

man in Germany in this department of science, and perhaps even excels his brother in intellectual power—gifted as he undoubtedly is.'

From a letter from Humboldt to Pfaff at Helmstädt a few months later, dated from Goldkronach, in the Fichtelgebirge, November 12, 1794, we learn that at that time he was very busily engaged on this branch of botanical science. 'I am at work,' he writes, 'upon a portion of the world's history that has been hitherto entirely neglected. The results will be published in about twenty years, under the title of " Suggestions for a future History and Geography of Plants, or an Historical Account of the gradual Spread of Vegetation over the Earth's Surface in its connection with Geology." ' From some letters that passed between Theodore Körner [1] and his father at a subsequent date (Dresden, November 22, 1811, and Vienna, January 15, 1812), it appears that Humboldt left at Dresden under the care of Körner, in the year 1797, a small box containing various manuscripts and preparations for this work, labelled ' Catalecta Phytologica,' and ' Physik der Welt.' Humboldt's contribution to the ' Horen ' was not, however, on any botanical subject, but consisted of a physiological treatise upon the chemistry of life in the form of an allegory.

What is Life—the Principle of Life ?

This perplexing question has resounded as a mysterious enigma through every succeeding century of the world's history. Scientific investigators and philosophers, materialists and spiritualists, have in all ages sought by the most varied means to explain the fact of life. From the restless desire of mankind to discover the secret conditions of life and contemplate them in imagination originated the symbolic poetry of the ancients, the dream of the alchemists in the middle ages, and the theories and philosophic problems of modern times. The youthful god of the Greeks, pointing mysteriously with his finger to his lips closed in silence, the Enormon,[2] the Archeus of Van Helmont, were similar poetic symbols, similar philosophic impersonations of a power existing independently of matter, to which it is

---

[1] Ad. Wolff, ' Th. Körner's Gesammelte Werke' (Berlin, G. Mertens), vol. iv. pp. 206, 220.

[2] [The ancient name in physiology for the aura vitalis, or vital principle.]

extraneous, but yet by which it is supposed to be governed. These phantoms succeeded each other Proteus-like, till the newer chemistry developed by the discoveries of Galvani and Volta gave a fresh direction to the enquiry and seemed to point to a solution.

In the early vigour of his genius, Humboldt was the first to take up this question from a chemist's point of view.[1] He had as early as 1793, in his 'Aphorisms from the Chemical Physiology of Plants,'[2] defined the principle of life as that 'inner' power which dissolves the bond of chemical affinity and prevents free combination taking place in organic bodies. But the mind of Humboldt in these early years was deeply imbued with the poetic imaginative power of the spirit of Plato, and stimulated by the charm· of the poetic circles of Weimar and Jena, 'who delighted to adorn the truths of science in the elegant garb of poetry,' he was incited to symbolise and personify the problems which science had still left unsolved. He was thus led to contribute to the 'Horen,' a paper entitled 'The Principle of Life, or the Genius of Rhodes.'[3]

Two pictures by unknown artists, to give an outline of the story, are supposed to form the subject of a variety of criticism in the Hall of Arts at Syracuse. In one of them a group of youths and maidens are seeking with passionate gestures to embrace, while above appears an allegorical figure, called the Genius of Rhodes—from the pictures coming originally, as it was thought, from that island—a butterfly on the shoulder, and holding on high a flaming torch, seeming by a gesture of command to warn them from a nearer approach. The other picture represents the same Genius without the butterfly, with torch extinguished and head hung down, while the youths and maidens fling themselves below into each other's arms with every expression of rapturous gratification.

While some of the critics regard the Genius as the expression of spiritual love forbidding the indulgence of sensual pleasure,

---

[1] Du Bois-Reymond, 'Untersuchungen über die thierische Elektricität,' vol. i. p. 75.

[2] 'Aphorismen aus der chemischen Physiologie der Pflanzen.'

[3] 'Die Lebenskraft, oder der rhodische Genius.' See Ule, 'Die Natur,' 1856 No. 45.

others recognise it as the symbol of Reason governing Desire. The hoary-headed philosopher Epicharmus alone sees in the pictures the image of Life and the image of Death, according to the teaching of the Pythagorean school of philosophy, and is represented by Humboldt as instructing his pupils in the following strain :—' Here, in the Genius of Rhodes, in the expression of his youthful strength, the butterfly upon his shoulder, the commanding glance of his eye, we recognise the symbol of the principle of Life as it animates every germ of organic being. The terrestrial elements at his feet strive eagerly to follow their own instincts, and mingle with each other. With authority the Genius threatens them with his uplifted blazing torch, and, unmindful of Nature's rights, constrains them to obey his law. There, in the second picture, the butterfly has disappeared, the torch is extinguished, and the head of the youth is bowed down. The principle of Life is annihilated. This is the image of Death. The youths and maidens joyfully clasp each other, the terrestrial elements in their chemical affinities assert their ancient rule. Loosed from their fetters, they follow wildly their impulse for assimilation. The day of death is to them a bridal day, they are united in chemical affinity.'

Through the philosopher Epicharmus, Humboldt gave expression to his own thoughts. The narrative was a homage from the natural philosopher to the poet in accordance with the scientific views of the time. The origin of this Orphean poem and its explanation is not to be found in the halls of art at Rhodes, but in the class-books of physiology and the lecture-rooms of the medical schools of Humboldt's own time.

But the essay with its deep symbolism was compelled to yield to the searching investigations of pure science. The germ of its own destruction was contained within itself, since it attempted the solution of a problem in science. Such problems, however, are not solved by the poet, but by the natural philosopher. Would Humboldt succeed, the mere poet must give place to the investigator of science.

This poetical effusion appeared in the ' Horen' in June 1795, and on December 14 of the same year Humboldt wrote to Freiesleben :—' I have discovered a new definition of the principle of life which is wholly irrefutable; I set great store by it, and think I shall now be able to upset the old definitions.' These new views rapidly gained in strength. On February 9, 1796, he wrote again to the same friend :—' I expect very soon to be able to cut the gordian knot of the processes of life. . . . These are the principles of my new physiology.' In December of that year he sent to Van Mons a treatise entitled ' Sur le Procédé chimique de la Vitalité,' and in 1797 he made the following statement in his celebrated work, 'Experiments upon the Excitability of the Nerves and Muscles':—'After much thought and continued study of the laws of physiology and chemistry, my former belief in any actual principle of life, as it is called, has been completely shaken. I no longer view that as an independent power which is perhaps only the joint action of individual well-known substances and their material forces. The difficulty of tracing back the phenomena of organic life to the laws of physics and chemistry lies mainly, as in the case of predicting the meteorological changes in the atmosphere, in the fact that the complication of phenomena and the multitude of forces simultaneously at work are necessary conditions of activity.'

Subsequently, however, he also renounced this view:—'I call those bodies inorganic in which the particles are commingled according to the laws of chemical affinity, and those bodies organic the particles of which, upon forcible separation, change their form of combination without any alteration in their external conditions. There is some hidden law, therefore, controlling all the particles of an organism; the law is only in force so long as all the particles are in mutual operation as a means and an end to the whole. Whether this definition may ever be made available is another question.'[1]

Thus the dream of symbolic myths and poetic allegories vanished before the efforts of science to solve the problems of nature. The province ascribed to the principle of life—to that ' common jade,' as Du Bois-Reymond terms it—became increasingly circumscribed, till in the present state of science the very expression has become distasteful.

[1] 'Briefwechsel und Gespräche Alexander von Humboldt's mit einem jungen Freunde (Berlin, 1861), p. 35.

In the mean time 'the Genius of Rhodes' remains as an eloquent memorial of the affection which Humboldt bore to Schiller. It was this feeling that induced him, when at the zenith of his fame, to republish this youthful production, 'which Schiller had regarded with favour' in the latest edition of the 'Aspects of Nature.'

On the whole, however, the allegory could not be considered as a successful contribution to the 'Horen;' the meaning remained as mystical and as susceptible of a variety of conjectures as the Syracusan pictures themselves were represented to have been. We have not met with any good critique upon the work—at least none written at the time of its appearance. In the 'Jenaische Allgemeine Literaturzeitung' for 1796, No. 6, A. W. Schlegel writes:—'The narrative contains a striking allegory upon a subject in natural science which is rarely expressed in so ingenious a manner, for instruction in morals is usually loaded with platitudes. The little essay is gracefully and pleasantly written, and its perusal is calculated to awaken tender emotions.' William von Humboldt[1] was himself even unable, or else had not sufficient interest in the subject, to satisfy the enquiries of his correspondent Charlotte by any more precise explanation of the mystery than to say:—'The object of the entire essay is the development of a physiological idea. It was much more the custom of the time in which it was written than it now is to clothe serious truths in such a semi-poetic garb.' Words of warmer praise were written by Gustav von Brinckmann on June 27, 1795, to Rahel,[2] who was then at Carlsbad :—'Should any numbers of the "Horen" cross your path in Bohemia, do not omit to read a paper in the last number entitled "The Principle of Life, or the Genius of Rhodes." It is full of deep meaning, and, as it appears to me, is very well written. And now I may tell you it is by Humboldt; but by the first!—that is to say the second; for such another man certainly does not exist. He has written a letter to Herz full of fixed air, vital force, and nervous fluid, which is quite enough

---

[1] 'Briefe an eine Freundin,' vol. ii. p. 39.

[2] This letter has never been printed; a transcript of it communicated by Varnhagen was found among Humboldt's papers.

to take away one's breath, extinguish the vital force altogether, and shatter all one's nerves—such learning ! '

Schiller's judgment of the ' Genius of Rhodes' was somewhat unfavourable, and was only incidentally expressed in a few words in a letter to Goethe[1] on the subject of the ' Epigrams,' in which he says he has been reading a review of the ' Horen' by Reichardt in his journal, ' Deutschland,' ' where he had indulged in a frightful amount of license. The essays of Fichte and Woltmann are represented as models of composition, and copious extracts from them are given. The fifth paper, the worst of all, is described as being the most interesting; Vossen's poems and the " Genius of Rhodes " by Humboldt are greatly extolled along with some other such rubbish.'[2] This expression is all the more remarkable, since Reichardt's criticism was contained in these few words :—' Humboldt's essay is a masterpiece of composition.' It may be said that Schiller only expressed himself in the same excited epigrammatic tone with which he hurled ' lances into the flesh of his colleague,' and which caused him to remark concerning this just criticism that it ' thoroughly exasperated him, since a stupidity was less to be censured.' But this humour had long passed away when he gave his opinion of Alexander von Humboldt in a tone of much greater severity.

During the visit paid by the two Humboldts to Dresden in the summer of 1797, Körner wrote to Schiller, on July 17, in the following terms :—' Alexander von Humboldt excites my admiration by the zeal with which he devotes himself to his scientific studies. As a companion William is the more agreeable, from his good nature and the greater repose of his temperament ; for there is some degree of irritability and bitterness about Alexander such as is often to be noticed in men of great activity. I have become greatly attached to William, with whom I have many points of sympathy.'

To this Schiller replied on August 6, 1797 :—' I was glad to hear that you enjoyed so much your intercourse with (William) Humboldt. He is pre-eminently formed for conversation. He

[1] ' Briefwechsel zwischen Schiller und Goethe,' vol. ii. p. 4.
[2] Reichardt, ' Deutschland ' (Berlin, 1796), vol. i. pp. 8, 9.

takes a keen, intelligent interest in the subject under discussion,
rouses every dormant idea, exacts from everyone a careful
precision of expression, guards against one-sidedness of view,
and shows his appreciation of every effort to render the meaning
clear by the remarkable aptitude with which he seizes and
weighs the thoughts expressed. Agreeable as this quality is
to those who have a rich store of thought to communicate,
to him it is almost indispensable to be set in action by something from without which shall yield him material upon which
to exercise his intellectual powers; for he has no originality,
he can only analyse and combine. He is too often deficient in
a passive and unpretending surrender of himself to the subject
under consideration; he is at once too active and too restlessly
intent on certain conclusions. You know him well, however,
and no doubt sympathise with me in this opinion. . . . As to
Alexander, I have not been able to form any satisfactory judgment of him; I fear, however, notwithstanding his great talents
and restless activity, he will never accomplish anything truly
great in science. *A trivial, restless vanity is the main-spring
of all his actions.*[1] I have not been able to discover in him a
spark of pure objective interest in outward things; and, however
strange it may appear, with all the wealth of material that he
possesses, he seems to me to show a poverty of intellect very
disastrous to the subject he handles. His is that keen, cold
reason which would have all nature, which is always incomprehensible, and should be in every point reverenced as unfathom-
able, shamelessly exposed to scrutiny; and with an effrontery I
cannot comprehend he employs his formulæ, which are frequently
only mere words and always narrow in conception, as a universal
standard. He seems to me, in short, to possess an organism far
too dense for the object he has in view, and is besides much too
circumscribed in understanding. He has no imagination, and
is therefore deficient, in my opinion, in the very faculty most
necessary for his scientific labours, for the same susceptibility
of temperament is needed for the contemplation of Nature in
her smallest phenomena as in the grandest of her laws. Alex-

---

[1] These words, as well as those in italics in the following page, are
omitted in the published correspondence of Schiller and Körner, and are
here given for the first time as they stand in the original manuscript.

ander imposes upon many people and gains much in comparison
with his brother William ; he *has the gift of the gab, and* [er
ein Maul hat] can assert his own value. As to absolute worth,
I cannot compare the two, for I consider William so much
the more deserving of esteem.' [1]

In his next letter, dated August 25, Körner, with characteristic
delicacy of feeling endeavours to justify his friend and moderate Schiller's unfavourable estimate of his character :—' Your
opinion of Alexander von Humboldt appears to me to be almost
too severe. I have not, indeed, read his work upon the nerves,
and know it only by report; but granted that he be deficient in
imagination, and so deprived of entering into the fullest sympathy with nature, yet it seems to me he may be able to do
much for science. His endeavour to measure and analyse
everything arises from his acute powers of observation, and no
useful material can be accumulated by the investigator of
nature without this faculty. As a mathematician he is not to
be blamed for applying number and proportion to everything
that comes within the sphere of his operations. By so doing
he seeks to form the various materials scattered through nature
into a harmonious whole, he values the hypotheses which enlarge
his range of view, and is led by them to seek the solution of
fresh problems in nature. That the susceptibility of his character is not equal to his energy, I am ready to admit. Men
of this stamp are always too busily occupied in their own
sphere of action to take much notice of what goes on in the
outer world around them. This gives them the appearance of
harshness and heartlessness.' [2]

It may appear superfluous after such a justification for us to

---

[1] Seldom has such an erroneous and harsh judgment been pronounced by
one distinguished man upon another equally distinguished, though in a
different sphere. Fichte alone may have experienced something similar
from Goethe when obliged to resign his professorship at Jena in 1799, on
account of his supposed atheistic principles. At that time Goethe wrote
to Schlosser :—' Fichte's foolish presumption has thrown him out of a mode
of life which he will not be able to find again all the world over. I fear
he is lost to himself and the world.' And yet the most important and
valuable portion of Fichte's life proved to be the years he subsequently
spent at Berlin !

[2] In later years Humboldt thought it necessary to defend himself from

enter upon any defence of Humboldt, more particularly as nearly every word of censure from the poet is in fact an acknowledgment of merit in the scientific investigator. But one is involuntarily led to seek a better grounded explanation of this severity, and it seems to us to be found in the dissimilarity of their mental constitution and mode of intellectual development. The qualities regarded by Schiller as essential to the scientific investigator—the faculty of imagination and a sympathetic feeling for nature—were in fact the principles upon which the new natural philosophy was founded, which shortly after took its rise in Jena, and was zealously advocated by Schelling, Hegel, and Steffens. These philosophers were characterised by intellectual power, depth of feeling, susceptibility of temperament, and power of imagination. And by this very school of natural philosophy has it been convincingly demonstrated that imagination is out of place in the investigation of nature, that nowhere is intuitive perception, and the indulgence of fancy more dangerous and mischievous than in the province of natural science, where the laws of nature in all their precision and purity must be comprehended and interpreted without a taint of subjective feeling.

In Schiller's criticism we see the reflection of his own pecu-

accusations of a similar nature. He thus wrote to Pictet on January 3, 1806 :—

' I have been reproached in a matter in which I think you will be able to justify me. I am often accused of engaging in too many studies at once—botany, astronomy, comparative anatomy, &c. My reply is, how can a man be prohibited from desiring to know and comprehend everything that surrounds him? It is impossible to be writing at the same time on chemistry and astronomy, but it is quite possible to carry on at the same time accurate observations of lunar distances and experiments on the absorption of gas. To a traveller, a variety of knowledge is indispensable. Let the small treatises which I have written on various subjects be appealed to in proof whether I have not shown myself well acquainted with those subjects, and whether (as for instance in my memoir with Gay-Lussac, and my work upon the nerves—the result of four years' experiments) I have not had perseverance in following the same object. And, in order to obtain comprehensive views and recognise the bond uniting the various phenomena —a bond to which we give the name of *Nature*—it is first necessary to become acquainted with the individual parts before we can unite them organically under the same point of view. My constant travels have greatly contributed to expand my interest in a variety of subjects.' (' Le Globe, Journ. géogr. de la Soc. de Genève,' 1868, vol. vii. pp. 8, 177.)

liar feelings and of the mode of thought characteristic of a
poet, concerning whom he sings in dithyrambic inspiration :—

> His soul by the gods was so gifted with light,
> That the world in its clearness lies imaged;
> He beholds in this mirror whate'er has transpired,
> And what in the future lies buried;
> He sat with the gods when in council convened,
> And witnessed the secrets of Nature unscreened.[1]

But easy as it may be to the poet to fathom the mysteries of
nature, it is not by any means so easy to the scientific investigator. These are by no means the only verses in Schiller's
poems in which the characteristic tone of his mind found
expression whence we may found an explanation of his judgment of Humboldt. His poetic idealism is more readily
affected by the imaginative contemplation of nature than by the
study of severe science. By the former process, nature seems
to be animated by creative beings, full of grace and beauty,
while science on the contrary scares away these heavenly creations, disenchants the world, and reduces everything to number
and law, establishes restrictions, and removes all superfluities.
The most pointed expression to this feeling is given in the
poem entitled ' Götter Griechenlands.'

And yet he failed himself to realise complete satisfaction in
the idealistic views of nature held by the Greeks. At the
close of his treatise ' Ueber naive und sentimentale Dichtung,'
he remarks :—' It cannot fail to excite our wonder that so little
trace is to be found among the Greeks of the sentimental
interest with which we moderns are accustomed to invest
natural scenes and the features of nature. The ancient Greek is
certainly in the highest degree accurate, truthful, and circumstantial in his description of nature, but evinces no more heartfelt emotion than would be called forth by the description of
a garment, a shield, or a piece of armour. Nature seems more

---

[1]   ' Ihm gaben die Götter das reine Gemüth,
>   Wo die Welt sich, die ewige, spiegelt;
>   Er hat alles gesehn, was auf Erden geschieht,
>   Und was uns die Zukunft versiegelt;
>   Er sass in der Götter urältestem Rath
>   Und behorchte der Dinge geheimste Saat.'

to interest his understanding than his moral sense ; she fails to
awaken in him the same fervour and sweet sadness with which
we moderns are inspired.'

It is thus apparent that even when adopting the abstract
form of thought, characteristic of philosophy, in questions relating to the rudiments of beauty, or the laws of morality,
everything beautiful or great still remains to the poet an object
principally for the affections.   He is not satisfied merely to see
with the eyes, to hear with the ears, and to think with the
mind ; his heart must overflow with feelings of the purest and
noblest character.

The enthusiastic character of Schiller's love of nature is
apparent in the following lines :—

> As with a lover's fervency
> Pygmalion once a statue clasped,
> Till o'er the marble's icy coldness
> The warmth of feeling spread a glow ;
> So in the heat of youthful ardour
> Nature was held in my embrace,
> Till a responsive look of feeling
> Was seen reflected in her face.[1]

This feeling dictated also his exhortation :—' If thou wilt step
out of thine artificial circle to enjoy the contemplation of
Nature, she will appear before thee in her profound peacefulness,
her childlike beauty, innocence, and simplicity ; linger, then,
before that image, cherish the feelings she inspires, for they
are worthy of the most exalted humanity.   Receive them within thyself, and strive to combine her infinite perfections with
thine own prerogative of immortality, that from the union of
the two divinity may spring.'   This passage shows that proportion and number must appear to the poetic mind as the
uninviting skeleton of every creature and every work of art.

---

[1]   ' Wie einst mit flehendem Verlangen
> Pygmalion den Stein umschloss,
> Bis in des Marmors kalte Wangen
> Empfindung glühend sich ergoss ;
> So schlang ich mich mit Liebesarmen
> Um die Natur, mit Jugendlust,
> Bis sie zu athmen, zu erwarmen
> Begann an meiner Dichterbrust.'

Mathematics are uncongenial to a poetic mind. As the bee constructs its cell without knowledge of number or proportion, and leaves it to the mathematician to demonstrate that the form it has selected is that most consonant with reason, and best adapted to the purpose it has in view, so the poet dictates in lofty inspiration the smoothly-flowing verse, and leaves it to the philologist to discover the laws of measure and to form by rule and number the theory of the metre which has often flowed from him unconsciously.

Even Goethe, who, according to his own admission, valued mathematics more highly than any other branch of study, because it could accomplish that which came not within the province of his art, inveighed against 'the whole physico-mathematical guild' in these terms :—

> Your sin is not a modern one, forsooth,
> To deem that theory may pass for truth ;
> And since exactness is the soul of science,
> To all who differ you present defiance.[1]

And in another place in ' Faust ' :—

> By that, I know the learned lord you are !
> What you don't touch is lying leagues afar ;
> What you don't grasp is wholly lost to you ;
> What you don't reckon, think you, can't be true ;
> What you don't weigh, it has no weight, alas !
> What you don't coin, you're sure it will not pass.[2]

In Schiller there existed a boundless subjectivity, an ideal world, in which the facts of experience were thrown aside as ballast in order that he might, with the wings of a cherub, fly

---

[1] ' *Das* ist eine von den alten Sünden,
  Sie meinen, Rechnen, das sei Erfinden,
  Und weil ihre Wissenschaft exact,
  So sei keiner von ihnen vertrackt.'

[2] ' Daran erkenn' ich den gelehrten Herrn !
  Was ihr nicht tastet, steht euch meilenfern ;
  Was ihr nicht fasst, das fehlt euch ganz und gar ;
  Was ihr nicht rechnet, glaubt ihr, sei nicht wahr ;
  Was ihr nicht wägt, hat für euch kein Gewicht ;
  Was ihr nicht münzt, das meint ihr, gelte nicht.'

towards that light in which the actual was merged in the ideal. Thus it was that, amid the realities by which he was surrounded, he never met with anything that accorded completely with his ideal.

> Imperious truth impedes free thought,
> Deprives the mind of liberty,
> Destroys whate'er's by fancy wrought,
> And rends the veil of poetry.[1]

This subjective disposition of mind, greatly intensified by the influence of Kant's philosophy, regarded everything—nature, history, and physical enjoyment—only as mental impressions. ' Nature only charms and delights us by that with which we have ourselves invested her. The grace in which she clothes herself is but the reflection of the inner grace in the soul of the beholder, and in our magnanimity we kiss the mirror in which we have been surprised by the sight of our own image.'

If in the constitution of Schiller's mind we have found the clue to his censure of Humboldt's method of scientific investigation, there appears to have been also a special and immediate cause for this strange severity. This lay in the fact that a few weeks previously Humboldt had been working at Jena on his ' Experiments upon the Excitability of the Fibres of the Nerves and Muscles,' in which he completely set aside the explanation of vital force given in the ' Genius of Rhodes,' thus considerably weakening the confidence hitherto reposed in his scientific investigations (see above pp. 184, 185).

About this time, on April 18, 1797, Humboldt thus wrote to Freiesleben during a flying visit to Goethe at Weimar:—' I have been living since the 1st of March at Jena, entirely among my books, and occupied with chemical experiments and anatomy. I have actually returned to my old student-life, for my sphere is limited and exclusively restricted to my own pursuits. As I am industriously preparing myself for a voyage to the West Indies, and intend to devote myself there prin-

---

[1]   ' Die Wirklichkeit mit ihren Schranken
         Umlagert den gebundenen Geist,
         Sie stürzt die Schöpfung der Gedanken,
         Der Dichtung schöner Flor zerreisst.'

cipally to living organisms, I am now mainly directing my attention to anatomy. I am receiving a course of private instruction from Loder, and devote two hours a day to making anatomical preparations, so that I spend daily from six to seven hours upon this subject. The remainder of my leisure I employ on a great physiological work on the ' Excitability of the Muscles,' of which the first volume, consisting of thirty-two sheets, is to appear at Easter, while the second volume is already in the press. I am gratified by finding many at Jena engaged in successfully prosecuting my experiments upon exciting vital energy by chemical means, and on increasing or diminishing the powers of susceptibility. The conviction is spreading that these experiments may some time lay the foundation of a practical art of healing, and that I may thus become the originator of a new science—that of vital chemistry.'

In the new work above alluded to, Humboldt, by the employment of letters after the manner of algebraic formulæ, expressed the various combinations produced by different metallic conductors and the interposition of liquids. He laid great stress upon this use of signs or formulæ, and remarks : [1]— ' It would be impossible, either by the most careful perusal of my work or the attentive examination of the tables, to gain a comprehensive view of so complete an array of facts. It seemed to me, therefore, important to devise a method by which this want might be remedied. The convenience afforded in mathematics of being able to represent a variety of propositions by a few analytical signs, led me to try by similar means to express the changes in the galvanic apparatus where the substances usually fall in a chainlike arrangement.' He accordingly denoted all metallic and carbonaceous substances possessing power to set up a phlogistic process by the letter P, and homogeneous metals—two bars of gold, for instance—by PP, heterogeneous metals, such as gold and zinc, by Pp, and moist conductors by Hh (humida). The formula

$$Pp\ P$$

denotes, therefore, that a heterogeneous metal or piece of carbon is in combination with two homogeneous metals ;

[1] ' Versuche über die gereizte Muskel- und Nervenfaser,' vol. i. p. 90.

Pp Pp

shows that four metallic or carbonaceous substances form the alternate links of an endless chain;

Nerv. PH Pp HP

indicates that two points of a nerve are placed in connection by a chain in which moist substances alternate with various metals, only one of which is heterogeneous.

To these formulæ he adds the signs + and −, by which to express the commencement or discontinuance of the muscular action, and, further, the signs → and ←, to distinguish between the positive and negative phenomena.

It is exceedingly probable that it was these very ' formulæ' which proved so offensive to the 'sentimental interest,' the 'heartfelt emotion,' the 'moral sense,' the 'fervour,' the 'sweet sadness,' characterising the mind of Schiller, and which came before him just at the time when he had been confessing concerning himself: 'I am willing to admit that I judge too hastily,' while on the other hand Humboldt might well have been impatient of such 'sloppiness of feeling.' [1]

There could hardly have been devised a more fortunate method for arranging facts and impressing them upon the memory than such a system of formulæ, and in geology Humboldt also employed the symbolic language of algebra side by side with the pictorial configurations that were of universal comprehension. And, need any further remark be made in favour of their practical utility, an additional instance may be adduced in the boundary lines of Flora and Fauna, and the graphic representation of the thermometric, barometric, and magnetic conditions now of universal use even in school-books.

Upon the value and necessity of studying science in a severely empiric manner, especially from the stand-point he then occupied, Humboldt often expressed himself in the most

---

[1] 'Schiller's judgment of Humboldt,' remarks Palleske, in his 'Life of Schiller,' 'was mainly grounded upon his work on the muscles'—adding the startling conclusion: 'in which I have been assured by competent judges not a spark of the great mind is visible which has rendered immortal the compiler of " Cosmos." '

decided manner. He followed Bacon's precept, which he often called to mind during the undertakings he was at that time engaged upon—that nature should first be observed, and as many of her phenomena as possible collated. His method consisted in 'assembling mere *facts*, and never admitting anything which lay beyond the boundary of actual experience.'

'Facts,' he writes to Blumenbach, in the year 1795, 'facts remain ever the same, when the hastily-erected edifice of theory has long since fallen in ruins. I have always kept my facts distinct from my conjectures. This method of dealing with the phenomena of nature appears to me to be the one best grounded, and the most likely to succeed.' He gives expression to the same sentiments in a letter to Pictet of Geneva, dated from Bayreuth, January 24, 1796 :—' I have been drawing up a scheme for a universal science ; but the more I feel its need, the more I perceive how slight the foundations yet are for so vast an edifice. . . . . I shall confine myself, however, to giving you the facts that have hitherto escaped the notice of men of science. For in every branch of physical knowledge there is nothing stable and certain but facts. Theories are as variable as the opinions that give them birth. They are the meteors of the moral world, rarely productive of good, and more often hurtful to the intellectual progress of mankind.' [1]

It is remarkable that, at the very time that Schiller was pronouncing his severe judgment upon Humboldt, Fourcroy, the physicist of Paris, was giving expression to a not less severe criticism upon opposite grounds, namely, that he experimented too little and built too much upon the experiments he made. Almost at the same time that Schiller wrote his criticism, Fourcroy, in alluding to the epistolary treatise ' Sur le Procédé chimique de la Vitalité,' which Humboldt had addressed to Van Mons at Brussels in December 1796, remarked :—' I think Herr Humboldt is a little rash in his conclusions ; it seems to me he will be obliged to abandon some of his views. I fear that he admits too many hypotheses, and does not repeat an experiment sufficiently often before basing a theory upon it.'

[1] Millin, 'Magaz. encyclop.,' vol. vi. p. 462; reprinted in De la Roquette, 'Humboldt, Correspondance, etc.,' vol. i. p. 4.

If we turn now from Schiller's judgment of Humboldt's scientific investigations to his remarks upon his personal character, we shall find that on this subject his views are more in accordance with general testimony.

We have already seen in p. 47 that 'vanity and a love of approbation' had been early pointed out by William von Humboldt as the chief failings in his brother's character. Freiesleben expressed himself in a still more pointed manner in a letter to Humboldt from Marienburg on December 23, 1796 :—' Now, my dear friend, I venture in the strictest confidence, and in dependence on your generosity of feeling, to write a few words which you must destroy as soon as you have read them, and which I shall forget that I have ever penned as soon as I have committed them to paper ;—in speaking of your discoveries, let it be with that cautious reserve and quiet modest seriousness for which you used to be distinguished. I know I shall appear to be wanting in delicacy in presuming to make you this request, yet I feel it to be a duty ; since it has come to my knowledge that in some of your letters, as well as in conversation in certain scientific circles, upon the subject of your physiological discoveries—in which you may perhaps have expressed yourself with some vivacity while entering enthusiastically into the defence of acute but paradoxical hypotheses— you have given occasion to some erroneous judgments, to which you are the more exposed from the envy you have excited among the learned—a feeling that is increasing from day to day. Pray do not seek any further particulars as to the facts which have prompted me to make this communication, which may perhaps seem to you rather abrupt, but which is dictated by true-hearted interest in yourself; the knowledge could do no good, and might lead to bitterness of feeling. In you these remarks can produce no irritation ; for blame of this character, which is directed only against the acuteness of your penetration, would be to others a subject of envy.'

Humboldt's reply is dated from Bayreuth, February 26, 1797: —' For your beautiful and accurate experiments I thank you much, but for your brotherly counsel concerning myself and the impression I produce upon others, I would repay you, my dear Karl, with the tenderest regard of my thankful heart. You are quite right, and your counsel shall not be lost.'

Humboldt, in fact, was quite aware himself that he was not free from vanity. He begins his letter to Schiller with the expression that his ' vanity was highly flattered.' He repeatedly acknowledges his 'vanity as an author' to Von Schuckmann, afterwards minister, his coadjutor in Franconia, and in a letter dated Jena, May 14, 1797, he speaks of his industry in a manner interesting to notice from its accordance with Schiller's criticism written about the same time.

' You know,' he writes to this friend, ' my busy idleness, this perpetual activity which leads me to begin so much without bringing anything to a conclusion. I have never been so pressed with work, so industrious, so variously occupied, as I am here. . . . I am really quite busy with acquiring information, and with arranging what I have already acquired. I must work with immense assiduity in order to carry out the scheme that I.have planned for myself; therefore do not be surprised, if you are always hearing of my commencing some new work. The fact is I cannot exist without making experiments, but this is not the sole object of my labours.'

There is also abundant evidence that Humboldt, when occasion served, ' could assert his own value.' To Fourcroy's censure, already alluded to, he replied :—' My early youth was devoted to the study of botany and geology. I have always been occupied with the direct contemplation of nature. Everyone about me knows that I am unceasingly occupied with chemical experiments. I have lately been experimenting on mephitic exhalations, in which my health has been subjected to some risk. Surely this is not the mode of life of a man whose only object is to increase the number of brilliant hypotheses.'

He always took pleasure, however, in giving an account of his labours, for, as he wrote to Wattenbach (see p. 114), 'to blow one's own trumpet is part of an author's trade.' But in all such expressions there lies the most graceful irony. ' Nos poma natamus!' was often his concluding word when informing friends of the importance of his labours. He called the prospectus of his American travels a 'carte de restaurateur,' and jokingly remarked to Pictet:—'I think, therefore, the charlatanry of literature will thus be combined with utility.'[1] Even

---

[1] ' Le Globe,' vol. vii. p. 162.

in his youth Humboldt was too really great to be vain in the ordinary sense of the word. Where he appears vain, and where he himself confesses to vanity, he employs the word only as a means to indicate the importance of the subject upon which he is engaged.

In later years Schiller referred to many of his early expressions of opinion with almost a feeling of horror. Among them must, without doubt, be classed his censure of Alexander von Humboldt. He lived to see the universal homage paid to Humboldt on his return from America, on August 3, 1804. The friendship existing between them to the death of Schiller, the high esteem which Humboldt during his long life ever cherished for the man of genius so early separated from him, the friendly interest he manifested towards Schiller's family, were never interrupted by the slightest veil of misunderstanding, not even when Schiller's severe judgment was brought under his notice at the time that the correspondence between Schiller and Körner was preparing for the press. He used to term hasty judgments of this nature, into which he was himself frequently betrayed through the excitement of feeling, 'momentary ebullitions.' The same feeling that led Körner, upon the rumour of Humboldt's appointment as President of the Berlin Academy in September 1804, to at once anticipate that such a step would be productive of advantage to Schiller, likewise instigated Schiller's sister-in-law, Caroline von Wolzogen, to cherish till her latest hour the valued letters of Alexander von Humboldt as one of her greatest treasures.

Let us now consider the attitude assumed by Humboldt, the scientific investigator, towards the modern school of philosophy emanating from Jena.

Humboldt had been educated in the liberal school of philosophy rendered so popular by Mendelssohn and Engel, side by side with the severe rules of thought and perception inculcated by Kant. The fundamental separation now recognised between philosophy and science was not then acknowledged. The object of Kant's philosophy was not to increase knowledge by the exercise of pure reason, for its chief proposition was that all knowledge of truth must be learnt by experience; its aim was only to test the accuracy of our knowledge, and investigate the sources whence it was obtained.

Fichte, who occupied a professor's chair at Jena from 1794 until the controversy concerning his atheistical views compelled his retirement in 1798, much as he was opposed to the intuitive method of contemplating nature, stood in no direct opposition to science; on the contrary, his representation of the process of thought has been found to agree precisely with the conclusions since arrived at by physiology through the study of the brain and the facts of experience.

It was in 1798 that Schelling became professor at the University of Jena.[1] In his ' System of Transcendental Idealism ' he gave an outline of the principles of his new scheme of natural philosophy :—'The necessary tendency of science is to rise from the study of Nature to that of Mind. To this alone is to be ascribed the endeavour to bring theory to bear upon the phenomena of nature. The most complete theory of nature would be that which had power to resolve all nature into one intelligence. The inanimate and unconscious products of nature are only her unsuccessful efforts to reflect herself, but nature, though called inanimate, is in reality an immature intelligence, therefore, yet unconscious in her phenomena, through which there nevertheless shines a character of intelligence. Nature has only fulfilled her highest aim of reproducing herself in her last and most perfect reflection— Man, or in other terms Reason, in which Nature for the first time reverts to her primeval form, and thus reveals herself to have been originally identical with that which is within us— intelligence or consciousness. If the aim of all philosophy must either be to show that an intelligence is derived from nature, or that nature is derived from an intelligence, then transcendental philosophy, which undertakes the latter task, must necessarily assume the first proposition as the ground of its philosophy.'

---

[1] His ' Ideen zu einer Philosophie der Natur ' was published in 1797. It was succeeded in 1798 by ' Die Weltseele, eine Hypothese der höhern Physik;' in 1799 there followed the ' Erster Entwurf eines Systems der Naturphilosophie,' and in 1800 the ' System des transcendentalen Idealismus.'

The transcendental idealism of this new natural philosophy did much towards restricting the rough empiricism of that age, and training the scientific investigator to close thought. The dissensions between theory and experiment seemed even advantageous to progress, and called forth Schiller's invocation to the

Scientific Investigators and Transcendental Philosophers.

Union is premature. For if truth you would secure,
Every searcher must divide, seeking truth on every side.[1]

Even after Humboldt's return from America, philosophy and science were still in a position to expect mutual assistance; thus, too, Schelling wrote to Humboldt from Würzburg in January 1805:[2]—

. . . . 'I venture to address you on the subject of natural philosophy, since I have been assured that this new school of philosophy, which has again laid hold upon her ancient possession, nature, has already excited your attention. Great exception has been taken to it in Germany, where there is always so much opposition to everything that is new. It has first been misunderstood, then misrepresented, and the strongest prejudice against it entertained. Natural philosophy has been represented as despising experiment, and rejecting its employment at the very time when individual investigators were conducting their experiments under the guidance of philosophical ideas. None of the scientific investigators of Germany have as yet fully grasped this philosophy before giving forth their judgment upon it. At most have they raised doubts against certain points, perhaps on just grounds; but this could not affect the theory as a whole, since it lies upon a deeper foundation.

'If a man of your genius, early imbued with the spirit of the ancients, and possessed of a depth and variety of infor-

---

[1]   '*Naturforscher und Transcendentalphilosophen.*

'Feindschaft sei zwischen euch ! Noch kommt das Bündniss zu frühe.
Wenn ihr im Suchen euch trennt, wird erst die Wahrheit erkannt.'

[2] 'Aus Schelling's Leben. In Briefen' (Leipzig, 1870), vol. ii. pp. 47–50.

mation including, were such a thing possible, the whole
range of modern science, whose knowledge is not confined
merely to the present generation and its immediate predecessor,
but extends also to the past century, so important in the
world's history—if a genius of such universality would put this
new theory to the proof, how soon might its fate be decided, and how greatly would this further the development of
thought!

'Reason and experience can never be more than apparently
opposed, and I have therefore a firm conviction that you
will not fail to notice the most surprising agreement between
theory and experiment in many points of the new philosophy.
Your mind has in an empirical age so daringly overstepped
the boundaries prescribed to physics, that you must be already
acquainted with the bold views of the present theory. If,
true to your character of an empirical scientific investigator, you have, with the exercise of a prudential reserve,
admitted only such ideas into your works as are confirmed by
experience, you will not on this account deem them of less
value because they have received through philosophy the sanction of reason.' . . .

To this Humboldt replied in a letter dated Paris, February
1, 1805:—

. . . . 'You will doubtless have heard from Herr W. how
desirous I am to adopt all that is great and beautiful in the
new system of philosophy which you have been propounding
during the last few years. What, in truth, could have more
completely roused my attention than such a revolution occurring in the study of those sciences to the pursuit of which my
whole life is devoted? After being absent from Europe for six
years, without books, and closely occupied with nature, my
mind has been kept more free from prejudice than was possible
to most physicists, who, through the deleterious effects produced
by literary disputes, have become more attached to their former
interpretations of nature than to the object of their study—
nature herself. No! I regard the revolution which you have
produced in science as one of the happiest events of these
impetuous times.

'Wavering between the theory of chemical action and that of violent eruption, I have always suspected that something higher and better was to be attained, to which the origin of everything could be traced, and for this higher primeval cause we are indebted to your discoveries.

'Do not however let it distress you that these discoveries, like everything else meant for the well-being of the world, should act with some people like a poison. Philosophy can never prove a hindrance to the advance of empirical science. On the contrary, she traces every new discovery back to fundamental principles, and thus lays the foundation for fresh discoveries. Should there arise a class of men who regard it as more convenient to work out chemistry in their heads, rather than soil their hands in its pursuit, this cannot be considered as your fault, and certainly not that of the scheme of your philosophy. Ought we to decry mathematical analysis, because our millers are able to construct more efficient machinery than any that a mathematician could devise? For this mathematics are not to blame, but rather their hasty unphilosophical application where the necessary link is wanting. I have thus given you, my excellent friend, a candid explanation. Though habitually contemplating nature in her external aspect, there is no one possessed of greater admiration than myself for the creations educed from the depth and fulness of human thought.' . . .

In the year 1807 Humboldt thus gave public expression to his sentiments: [1]—'Not wholly unacquainted with Schelling's system, I am far from believing that the pure study of natural philosophy can be injurious to empirical investigation, or that the investigator and the philosopher should necessarily be opposed. Few physicists have more loudly complained than I have against the unsatisfactory nature of the theories hitherto advanced and against the terms employed; few have so distinctly expressed their disbelief in the specific differences of the so-called elements.[2] Who has more reason, therefore, to advocate a system which, while undermining the atomic theory on the one hand, and on the other far removed from the one-

[1] Preface to the 'Ideen einer Geographie der Pflanzen,' p. 5.
[2] 'Versuche über die gereizte Muskel- und Nervenfaser,' vol. i. pp. 367, 422; vol. ii. pp. 34, 40.

sided system formerly adopted by me, in which all differences in matter were ascribed merely to the different arrangement of particles, endeavours to throw light upon the inexplicable phenomena of organic life, heat, magnetism, and electricity ? '

When on the death of Fichte, in 1814, Schelling in Southern Germany and Hegel in Northern Germany assumed the lead in the world of science, it was expected that philosophy would be able by means of pure reason and abstract ideas to attain those results which formerly had only been reached by aid of experiment. She gave to abstract thought a position she denied to the sober empirical labours of the scientific investigator. According to her rules, nothing was to be left unexplained, ignorance was never to be acknowledged ; and often a complete want of comprehension was veiled by a superfluity of words, though these might be entirely irrelevant to the subject. In this way there arose a system of chemistry in which no hands need be soiled, and a system of astronomy in which neither measurements nor calculations were required. Even men of distinguished merit and acute powers of observation, such as Nees of Esenbeck, Oken, Döllinger, Walther, Schubert, Carus, &c., were carried away by this delusion ; while the men of true science, Blumenbach, Sömmering, Meckel, Treviranus, Pfaff, and Erman, were left to pursue their investigations in solitude. It was not within the power of Humboldt to stem this desolating torrent. This deplorable epoch was termed by him a 'mad saturnalia,' the 'bal en masque of natural philosophy run mad.' [1]

Humboldt was on terms of intimate friendship with the Grand Duke Karl August. The prince had been an industrious student of the various sciences of chemistry, botany, mineralogy, zoology, and meteorology, while of anatomy, according to the somewhat too flattering testimony of Walther the anatomist, ' he had attained a deeper insight than his instructor, Professor Loder.' [2] ' Science,' writes Karl August to Knebel, [3] ' is so human, so true, that I feel inclined to congratulate anyone who has become even partially engaged in its pursuit. It has been

[1] A. von Humboldt, 'Briefe an Varnhagen,' p. 90.
[2] Wagner, 'Leben Sömmering's,' vol. ii. p. 46.
[3] Knebel's 'Literarischer Nachlass,' vol. i. p. 143.

rendered so accessible, that even indolent persons may now find inducement to become its votaries. By its means truth may be so easily attained that the love of the unreal is subdued. The teachings and demonstrations of science are so conclusive that by their light the most mysterious and incomprehensible phenomena of nature are seen to take place in the most regular order, and entirely without the influence of the supernatural; this must at length effect a cure in those poor ignorant mortals who thirst after the mystery of extraordinary manifestations, by showing them that the extraordinary is after all so ordinary, the inexplicable so easy of explanation. I pray daily to my good genius to be preserved from every other form of doctrine or method of observation, and to be always kept in that unobtrusive path that leads to truth.'

The love of scientific investigation animating alike Karl August and Humboldt originated in a similar temperament and constitution of mind. One instance of their sympathy on these subjects will suffice.

In the momentous year of 1808 Humboldt dedicated, as is well known, his 'Aspects of Nature' to 'those oppressed spirits who, glad to escape from the stormy waves of life,' are willing to follow him through the dense glades of primeval forests, across immeasurable plains and over the rugged heights of the Andes.

In equal need of such a refuge, Karl August sought and found this peace of mind in the study of botany. When Röhr, the court chaplain, once expressed to the prince his astonishment at the extensive knowledge he displayed on the subject, the prince replied:—' I can easily give you the explanation, my dear Röhr. When in the year 1806 our country was overwhelmed with misfortune, and I saw around me so much falsehood, treason, and deception, my faith in man was destroyed. In my despair the only thing that sustained me was my inextinguishable love of nature. Disgusted with mankind, I turned to the study of plants, and made flowers my companions, and the flowers have never deceived me!'

Humboldt was, in fact, early honoured by the prince's confidence, and indeed to such a degree that upon his recommendation in 1797, Scherer was summoned to Weimar as Counsellor of Mines, and afterwards sent to England to acquire a more extensive knowledge of technical chemistry.

At Weimar Humboldt was always the welcome guest of the prince, and his chosen companion during the prince's visits to the court of Berlin. It was therefore a source of mutual gratification that Karl August passed his last days at the court of Berlin in almost constant companionship with Humboldt. 'Even here in Berlin,' writes Humboldt to the chancellor Müller,[1] ' he wishes to have me constantly with him. I never saw this great and benevolent prince more animated, with a clearer intellect or occupied more tenderly and earnestly on the improvement of his people, than during the last few days of his life that he spent here. I often remarked to my friends with anxiety and foreboding that this activity of mind, this mysterious clearness of perception, accompanied with such complete physical prostration, was to my mind a very alarming symptom. As for himself, he seemed to vibrate between the hope of recovery and the anticipation of his end. The day before his death, I sat with him on the sofa for several hours alone at Potsdam. He drank and slept alternately, got up to write to his consort, and then fell again into a doze. He was cheerful, but very exhausted. In the intervals he plied me with questions on many of the unsolved problems in physics, astronomy, meteorology, and geology—upon the transparency of a comet's nucleus, the existence of a lunar atmosphere, the colours of the double stars, the influence of the solar spots upon the temperature, the evidence of organic life having existed in the primeval world, and the internal heat of the earth. He then touched in a desultory manner on religious subjects. He bemoaned the spread of pietism, and the connection between this fanaticism and the political tendency to absolutism and the suppression of all free thought. "Then arise false fellows," he exclaimed, " who fancy that they will thereby gain the favour of princes, and so win for themselves decorations and places of honour. They have insinuated themselves into favour by a professed predilection for the poetry of the middle ages." He dozed repeatedly during our conversation, and was often very restless, apologising for his apparent inattention by remarking, in a kind and friendly tone : " You see, Humboldt, it is nearly over with me." '

On the following day the near approach of death was manifest. The prince died on his journey home at Graditz, near Torgau, in the buildings connected with the establishment for rearing horses, on June 14, 1828, in the seventy-first year of his age.

Jena and Weimar were ever associated by Humboldt, during the whole of his long life, with the most grateful and elevating reminiscences. In 1836 he read before the scientific society of Jena one of the early chapters of ' Cosmos,' ' On the Variety of Enjoyment afforded by the Contemplation of Nature and the Study of her Laws,' and ' On two Ascents of Chimborazo.' Even in his eighty-ninth year he thus expresses himself :—-' Jena, which I visited at the height of her intellectual glory, for the purpose of preparing myself by the study of practical anatomy for my proposed expedition to America, and which under the rule of beneficent princes continues to play an important part in the free thought of Germany, has ever remained as a bright spot in my memory amid the crowded recollections of a life extended beyond the ordinary limits.' He regretted that, owing to the rapid increase of bodily weakness, he was prevented from attending the celebration, in 1858, of the Jubilee of the University ' to which he felt attracted by the most agreeable and soul-stirring remembrances and the most heartfelt feelings of gratitude.'

The autograph letter of the Grand Duke Karl Alexander to Humboldt [1] on August 7, 1857, inviting him to the ceremony of laying the foundation stone of the monument to his grandfather, Karl August, and the unveiling of the statues of Goethe, Schiller, and Wieland, contains the following passage :—' *You are so inseparably connected with all that is great and noble in our country, so intimately associated with the period of which these names speak so proudly, that I cannot think of engaging in this festival without you.*'

This will suffice to show the living ties that bound Alexander von Humboldt to the circle of intellectual heroes by whom the classic literature of Germany was founded.

---

[1] Among the papers left by Alexander von Humboldt.

In one of the apartments of the palace at Weimar, called the Poet's Chamber, is preserved a magnificent album, consecrated in 1849 by the Princess Augusta, now Empress of Germany and Queen of Prussia, to the memory of that brilliant epoch. Alexander von Humboldt there recorded his impressions in the following words :—

'As in nature life passes by turns through the phases of luxuriant growth and restricted development, so the intellectual life of man passes through similar alternations. At times the intellectual giants upon whom will be directed the admiration of posterity stand as it were in isolation, at other times they appear grouped together, mutually heightening the beneficent effect of their influence.

'The cause of this unequal distribution of intellectual power and of this simultaneous outgrowth of intelligence, appears to lie almost beyond the reach of investigation. By the careless multitude it is viewed as a matter of chance, but may it not rather be regarded as closely allied to that phenomenon visible in the vault of heaven, where the stars of brightest lustre are seen to shine either in isolated glory scattered like the isles of Sporades in an immeasurable sea, or else in beautiful constellations, whence the thoughtful mind is led in contemplation to deduce the eternal plan on which the universe is framed, though as yet the laws by which it is controlled lie undiscovered?

'Though the simultaneous appearance of master minds cannot be ascribed to any earthly power, its cause may yet be found in the union and combined action of hidden forces. A noble spectacle is afforded when a race of princes illustrious for generations have been imbued with the elevating thought of securing by such an assemblage of men of genius, not merely the glory of their court or the pleasures of social life, but, through the inspiring influence of mutual association, of stimulating to higher flights of genius each of the distinguished minds thus gathered in concert.

'To the memory of such an influence directed towards the elevation and extension of the free range of thought, the expression of tender sentiment, and the enlargement of the powers of language (one of the products of intelligence upon which is imprinted the character of a people, the wants of an age, and the colouring of individual feeling), these pages are thoughtfully dedicated. They commemorate, in conformity with the artistic decorations of the surrounding walls, a brilliant epoch in the intellectual life of the German people.'

And in this brilliant epoch the name of Humboldt also shines in double glory.

CHAPTER V.

THWARTED PLANS, AND THEIR ULTIMATE ACCOMPLISHMENT.

State of Prussia in 1797—Visit to Jena—Dresden—Vienna—Salzburg—Paris—Journey to Marseilles—Marseilles and Toulon—Wanderings in Spain—Madrid—Corunna.

FOR some years prior to the death of Frau von Humboldt, the evils resulting from the misgovernment of the Prussian States had been gradually approaching a climax. In the political coalition against France, the country had been sold for the sake of the subsidy alternately to England and Austria, and without keeping faith with either, the money thus acquired had been lavishly spent. By the treaty of Basle the limits of France had been extended as far as the Rhine, while the partition of Poland had brought the boundaries of Russia to the banks of the Vistula. Hedged in between the two, Prussia rushed on without restraint towards the catastrophe of Jena.

While the perpetual need of money produced severity, injustice, and selfishness in the home government, the rich possessions acquired by the secularisation of church property in Westphalia, the annexation of Franconia, and the confiscations in the recently annexed Polish provinces, were wasted on the most unworthy creatures of the Government. Notwithstanding the severity of discipline, the State officials had become contaminated by the evils of foreign administration, while a succession of inglorious wars had completely demoralised the army. Under theological despotism, the poisonous weeds of official piety and hypocrisy grew rampant, and the heads of orthodoxy, by means of lawsuits and the censorship of the press, checked the expression of opinion and hunted down every appearance of heresy. The compilers of the great legislative work —the common law of Prussia—Klein, Carmer, Cocceji, stood isolated beside the administration of such men as Goerne, Hoym, Struensee ; the theory of law and justice was powerless against the corrupt practices of the officers of the crown. Notwithstanding the principles of justice recognised by the State, the burdens of feudalism, the privileges of the nobility, and the distinctions of rank remained undisturbed.

Such a condition of things produced in manners and in literature a frivolous scepticism, which reached its climax in Schlegel's ' Lucinde,' a mere glorification of passion, in which love and marriage are regarded as synonymous.

The better part of the community viewed these proceedings with grief and horror. To Humboldt the society in Berlin of every grade had long been in the highest degree distasteful. As early as 1795 he—the young mining official, the court-bred son of a royal chamberlain—gave open expression to the sentiment,[1] ' that court life robs even the most intellectual of their genius and their freedom.' Even at that time the Berlin Royal Academy of Science well deserved the appellation he gave it somewhat later of ' a lazar-house,' ' a hospital in which the sick sleep more soundly than those in health.'[2] At the time when he was investigating the processes of life and the principles of a practical art of healing with untiring energy and self-sacrifice, quacks, charlatans, adepts, and magnetisers were carrying on their deceptive quackeries in the royal sick-chamber in the marble palace at Potsdam by the aid of magnetic passes from the hands of women, by kittens, and the entrails of unborn calves.

What was there, then, in Berlin to detain Humboldt in such a home after the death of his mother ?

After a short stay at Berlin, whither he had been summoned by family affairs, Humboldt returned to Bayreuth to wind up his official engagements in Franconia, and by March 1 we find him at Jena with his brother, who had likewise been seized with so strong a propensity for roving, that he had confided

---

[1] In the ' Genius of Rhodes.'

[2] De la Roquette, ' Humboldt, Correspondance, etc.,' vol. i. p. 184. (' Le Globe, Journ. géogr. etc.' p. 179.)

his intention to Schiller of ' never having a fixed residence, but only residing at Jena in the intervals of travel.' By the arrival of Haften at Jena the circle of intimate friends was rendered complete.

Of the unceasing occupation of Humboldt at this time, ample details have been given in the foregoing pages. It was during this period that the brothers were thrown in close personal intercourse with Goethe, who joined the friendly circle at Jena towards the end of February, and remained there till the beginning of April, occupied in the completion of his epic poem ' Hermann und Dorothea.' On his return home Goethe was accompanied by William von Humboldt, who delivered at Weimar ' an exhaustive criticism on the prosody ' of the new poem, and availed himself of the opportunity afforded him by this visit to select suitable works in preparation for his journey to Italy.

It was during this visit to Jena, on May 14, 1797, that Alexander von Humboldt wrote the letter to Schuckmann from which copious extracts have already been given ; the letter is of extraordinary length—' a letter like a newspaper '—in which he narrates his discoveries and discusses his plans for the future. He then proceeds :—' After the 1st of June I intend to pass some weeks at Dresden, where I am anxious, under the superintendence of Köhler, to acquire some proficiency in the use of my large sextant of fourteen inches ; thence I hope to visit Freiberg, to gain some information from Werner " about the origin of volcanoes." I expect to be in Venice early in September, and shall probably pass the winter at Naples.' . . .

He writes further, that at Jena there is great intellectual languor among the professors, but much activity of mind among the students, with whom, therefore, he lived almost entirely. Though Loder was not possessed of much mental power, he yet taught the principles of mechanics well. The day previously a labouring man and his wife had been killed by a flash of lightning. He had himself dissected the body of the man, and learnt from this experiment what powerful conductors of electricity the bones are. The back of the skull was perforated by the lightning as by shot, and in less than twelve hours after death decomposition had set in. At the close of his letter he remarks:—

'Goethe is almost always here; he has just completed his grand epic poem "Hermann und Dorothea." It is one of the most beautiful poems he has written, and proves him to be still in the freshness of youth. The entire composition of this masterpiece occupied only six weeks. He is already engaged upon another work. You will be surprised to see in "Hermann" how a simple story of common life may be succesfully treated in a truly Homeric spirit. Schiller is still busily occupied on his tragedy "Wallenstein." My brother William has finished several of the choruses from the tragic poets, besides the whole of the "Agamemnon" of Æschylus: the latter will soon be published. Thus, my dear friend, you see that everything here is prospering. I shall leave Jena with regret: where shall I again meet with such a circle of friends?'

Humboldt's next plan was to accompany the whole family and Haften to Italy, travelling by way of Dresden and Vienna. Alexander von Humboldt was exceedingly anxious to study while in Italy the phenomena of volcanoes, and he intended after a sojourn there to proceed alone through Egypt to Asia. Probably the travellers would already have left Jena but for the protracted illness of the wife of William von Humboldt, whose health had never been quite re-established since the birth of her second son, Theodore, in January, added to which, William von Humboldt and the children were also prostrated by an attack of ague, so that almost the entire household were laid aside. 'And yet,' writes Schiller to Goethe on April 14, 'they are still always talking of the approaching great journey.'

In concluding the history of this period we may again allude to Humboldt's epistolary treatise 'Sur le Procédé chimique de la Vitalité,' addressed to Van Mons, which, as previously stated in p. 197, called forth from the physicist Fourcroy a severe criticism, though of an entirely opposite character to the one expressed by Schiller—a criticism however which, after a long correspondence not entirely free from asperity, was considerably softened by the following explanation offered by Fourcroy:—

'Your discoveries in galvanism are the result of researches

instituted with too much accuracy, and give promise of exerting too great an influence upon animal physics for me to have given expression to such an opinion. It could not therefore have been your labours, which I hold in the highest esteem, and which I contemplate with renewed pleasure every day, to which I alluded in my letter to citizen Van Mons. Pray rest assured that I value the indefatigable investigators of nature too highly as the true interpreters of her mysteries, as, in a word, the true physicists, among whom your name is already so distinguished, ever to have thought of calumniating your efforts, damping your ardour, or confounding you with the dangerous inventors of hypotheses.'

In the beginning of June we find the entire Humboldt family established at Dresden, in company with Von Haften and Fischer; the latter of whom had, in the meanwhile, taken his degree as Doctor of Medicine. Here William von Humboldt, through his intercourse with Körner, the friend of Schiller, was introduced into the congenial society of Count von Kessler, the Prussian ambassador, and of Adelung, the librarian and distinguished philologist, while Alexander, who had recently become possessed of a sextant by Hadley, was familiarising himself, with the assistance of Köhler, curator of the Philosophical Museum, in the use of astronomical and meteorological instruments, and those employed in geodesy and hypsometry. The fifth volume of his 'Journal' contains whole pages of figures, the results of the observations which he carried on in and around Dresden, Pillnitz, Königstein, Töplitz, and Prague, and in letters from Salzburg he mentions Köhler in grateful terms as his teacher and friend.[1]

This sojourn at Dresden proved of the highest value to Humboldt in preparation for his subsequent travels, since here was preserved the extensive collection of Spanish and American minerals in the possession of Baron von Rackwitz, which he studied with much interest. It was probably through the acquaintances he made while at Dresden that he was afterwards able to secure the interest of Baron von Forell, ambassador from Saxony at the Spanish court, in furtherance of his expedition to America.

[1] 'Allgemeine geographische Ephemeriden,' vol. ii. p. 267.

In a fragment of a letter written by Humboldt from Dresden
to his friend Freiesleben, announcing an intended visit to
Freiberg, we see the complete reflection of his tone of mind at
this period, which is all the more interesting from the picture
it affords of his inner life, in the disclosure of the various plans,
thoughts, and hopes, with which his mind was then occupied.
' All the inanimate objects surrounding Freiberg,' he writes to
his friend, ' without excepting even the stage over the " Ascen-
sion," possess a deep interest for me; yet in the thought of
seeing them all again I have a painful feeling that I shall look
upon them from another point of view, and thus be deprived
of those bright images that now stand associated with those
happy days spent among the mines. Five years ago I was met
at every turn by joyous, friendly faces, and it was a pleasant
feeling to be everywhere so greeted; but now it will be altogether a new world, while to the old one I have become so
changed, so absorbed by other thoughts and habits, that I am
now quite a stranger.' In commenting upon this letter Freiesleben remarks that notwithstanding these gloomy anticipations
the emotion experienced by Humboldt on the occasion of his
visit was characterised by a tone of joyous excitement.

Humboldt greatly enjoyed the refreshment of the social
intercourse afforded him at the house of Körner, the friend of
Schiller, and amid the family circle of Neumann, Secretary at
War; in both houses he was a constant and ever welcome guest.
At the court, too, both William and Alexander von Humboldt
were received with marks of distinguished favour.

An important business occupying their attention during
this residence at Dresden was the division of the property
they inherited from their mother—a transaction conducted
for them with characteristic care by their faithful friend
Kunth. It may be well, perhaps, to take this opportunity
of correcting the misapprehension hitherto prevalent with
regard to the estate of Ringenwalde, which has been represented as one of the possessions of the family of Colomb,
sold by Alexander von Humboldt, while in America in 1802, to
defray the expenses of his travels. It has been proved, however,
by documentary evidence, that in the sixteenth century, Ringenwalde was in the possession of the family of Schönebeck,

and was sold by the last representative of that family, in 1763, to Captain von Hollwede, through whose widow, Elizabeth von Colomb, it passed into the hands of her second husband, Alexander George von Humboldt, the father of the two brothers. Major von Humboldt died in 1779, and was interred at Ringenwalde, but his remains were afterwards removed to Falkenberg; the estate was sold by his heirs as early as 1793, for 72,000 thalers, of which sum 45,000 thalers remained as a secret mortgage upon the estate until 1803. On the division of the property after the death of their mother in 1796, all claim upon this mortgage was relinquished by his brother William and his step-brother Captain von Hollwede in favour of Alexander von Humboldt, who shortly after endeavoured, without success, to realise this property in preparation for his extensive travels.

In addition to the 45,000 thalers from the estate at Ringenwalde, Alexander von Humboldt inherited a mortgage upon the property at Tegel of 8,000 thalers, besides various investments and ready money, which brought up his inheritance to the sum of 91,475 thalers 4 groschen. After deducting the sum of 6,100 thalers in payment of some debts, he remarks in his journal, while still at Dresden:—'The whole of my property this June 16, 1797, amounts to the sum of 85,375 thalers 4 groschen, from which the yearly income will average 3,476 thalers.'[1]

This sojourn in Dresden, prolonged on account of the increased illness of his sister-in-law through a relapse of fever, occasioned a postponement of all their plans. 'It is likely to prove an agreeable journey,' writes Schiller to Goethe on June 30, 'when before starting they must have already exceeded their time!'

At length they were able to leave Dresden towards the end of July, 1797. 'The Humboldts are gone,' Schiller writes to Goethe on July 30, 'and have left you their heartiest greeting.' It was at this period that the correspondence between Körner and Schiller, quoted in pp. 187-189, took place relative to the characters of William and Alexander von Humboldt.

The sympathy of their friends followed the travellers in their

[1] [In English money, reckoning the thaler at 2s. 11d., this amounted to 12,450l., and his income to 506l.]

journey through Prague and Vienna. In the capital of Austria they found much to amuse and interest them during the first part of their stay, for while William spent his days with Bast, a young philologist, in poring over the manuscript treasures of the Imperial Library, Alexander became engrossed in the study of botany, availing himself of the valuable herbarium and collection of rare plants in the Imperial Gardens at Schönbrunn, placed at his disposal through the kindness of Jacquin and Van der Schott.

Some letters written by Alexander at this time contain some humorous descriptions of the intellectual society of Vienna.

To Freiesleben he writes, in a letter which bears no date, but is evidently the first addressed from the Austrian capital :—
' Though in the midst of Vienna, I am living in complete solitude, for I am working very hard and not wholly without success. The second part of my book (" Experiments in Galvanism ") is now nearly completed, and my other works will soon be ready. The journey here from Prague was somewhat wearisome. The southern part of Bohemia is just as flat and uninteresting as those northern districts which I visited with you are beautiful and romantic. Or did the journey here seem only dull to me because you were not by my side ? Buch writes me word that he " is going to Italy to cast off his outer garments, and clothe himself in ether." (He may manage thus to look quite handsome.) He neither says *when* he goes, nor where I should be likely to meet him.

' Not only is my new book much read here, but also my former works. At Schönbrunn they make use with advantage of oxygenised muriatic acid. I have seen trees grown from seed twenty-four years old. Otherwise everything here is. . . . They know nothing about the magnet, and young Jacquin, to whom I was speaking of it, had not even the curiosity to wish to see a specimen.

' My " Subterranean Gases " reached me here at Vienna, my dear Karl, at my present address—first floor of No. 1224, Kärntnerstrasse.[1] I shall certainly remain here till October 4.

---

[1] Reference is here made to the manuscript of his work ' On Subterranean Gases,' which had been sent to Freiesleben to revise. Upon receiving the

Whether I go hence to Italy is still uncertain, partly because there is now a prohibition against anyone leaving Vienna, partly because Italy itself is in a state of commotion. The winter, Haften's children—all create some amount of anxiety. It is most probable that my brother will not visit Rome either, but will go direct to Paris. He intends to remain eighteen months in France before visiting Italy. I shall most likely pass the autumn and winter in Switzerland, either at Zurich or Geneva, and proceed to Italy in April by way of the Tyrol. I shall thus gain sufficient leisure to complete several of my new undertakings, and I shall hope during the autumn and winter (when I must, without fail, revisit the St. Gothard) to gain valuable material for my book upon the atmosphere.

'Young Böthlingk has arrived here and is quite determined now to go with me to the West Indies. We think of commencing our journey by Spain, and thence to Teneriffe. He has an income of 40,000 roubles.'

To Professor Loder,[1] at Jena, Humboldt writes more in detail :—' I spent a glorious time at Vienna. I passed several weeks at Schönbrunn, and though the claims of social life and the friendship accorded me by Count Sauzau (?) (a sort of

work again into his possession, Humboldt writes :—' I could scarcely recognise my work. You have taken more trouble with it than the thing is worth, not only in arranging the old materials, but in adding new matter. It will now be an easy task for me to make a book out of it, and I should certainly dedicate it to you if your assistance in its preparation were not so widely known, and if I had not determined to dedicate to you my " Geology," a much more important work.'

[1] Loder had sent Humboldt a pecuniary acknowledgment for an article ' On the Employment of Galvanism in Medical Practice,' which he had contributed to the 'Journal für Chirurgie, Geburtshülfe und gerichtliche Arzneikunde,' vol. i. pp. 441-71, and this was apparently the occasion of Humboldt's letter. The letter begins :—' You send me a pecuniary acknowledgment for a few pages, to which you have accorded an honourable place in your journal. Your liberality is equal to that of Herr Cotta when one dances with one of his Hours ; only your Hours can carry on the dance much longer than his, for they frequently show signs of fatigue. Fifteen thalers fifteen groschen for those few pages! Well, this is the first money I have ever received from a periodical; and you, my dear friend, have with this unprecedented generosity raised for yourself an imperishable memorial in my gratitude which enables me to esteem it a great pleasure to receive this gift from you.'

prime minister, and only thirty-four years of age!) drew me into the wide circle of ordinary society, I have yet had leisure to enjoy familiar intercourse with Franck and Jacquin, and to avail myself of the public institutions. For many weeks I have visited the clinical hospital, merely for the purpose of becoming better acquainted with the elder Franck (Johann Peter), and I freely confess that I have seldom been more deeply impressed by anyone. What clearness of thought, what powers of perception, what solidity side by side with the most vivid flashes of genius! Of the unlimited use of sthenical[1] treatment at this hospital, which is raising such an outcry in Vienna, I have seen nothing. Franck has been attending my sister-in-law, and has treated her case, even though there is unmistakeable evidence of debility (from too great a flow of milk), according to the old system—a course which would certainly have been approved by you, my dear friend, and by Hufeland. What has particularly charmed me in Franck is that with all the peculiarities of a man of eminence he is yet so simple in his domestic habits. Up till eleven o'clock he is engaged at the hospital, from eleven till two he drives about to his patients, from three till five, while reclining on his couch, (after the manner of Tissot), he receives visits from patients, princes as well as ladies, and in the evening from five till ten he again visits professionally. He conducts his correspondence at night and of a morning before seven o'clock. There is certainly no need for me to vaunt such activity to you, who already set such an example of indefatigable industry. It is therefore only possible to enjoy social intercourse with Franck during the intervals between his numerous engagements; but in these leisure moments he is always collected in mind and ready for conversation on physiology or natural science. But this will suffice for a description of one with whom perhaps you are personally acquainted.

'There is still another name I must mention to you, that of Professor Porth, who is without exception the most genial man in Vienna. As he resides near the Botanic Gardens, I was able to visit him very frequently, in his *one* room, in which he lives

---

[1] [Sthenic, a term employed in the Brunonian system of medicine, which was then exciting much attention in Germany.]

surrounded by a heterogeneous collection of foreign plants and animals, statues, dried specimens, and coins. Indolence and wealth have alone prevented him from attaining great celebrity, for no one has ever so nearly approached Lieberkühn in his successful method of making anatomical preparations by injections. What a collection he has of them !—all hidden away in dusty cases, so protected in order that the delicate preparations intended for the microscope, and the drawings, which are executed in a masterly style, should be kept free from injury. " Je ne pense plus à ces balourdises," remarked the old hospitaller ; " ne les louez pas, cela n'en vaut pas la peine." It is a great misfortune that the old man is so rich (he possesses between 200,000 and 300,000 florins, for with all my efforts I could not induce him to part with even a shred of prepared cuticle. There is no one in Vienna so well acquainted as he is with modern chemistry and the latest discoveries in physiology. Everything around him bears the impress of his peculiarities. Thus he wears a waistcoat provided with sleeves and terminating in breeches and stockings, and looks as if in a sentry-box. He eats only once a day, and that at ten o'clock at night, in order, as he says, not to carry food about in the body, which is very fatiguing and burdensome. Almost the only society he has now is that of my young friend Van der Schott, Director of the Botanic Gardens of Vienna. One of his treasures is an antique statue of one of the sons of Niobe, for which he paid 15,000 florins. It stands in the same corner of the room in which he carries on his chemical experiments, and where chickens are being hatched. He is now contriving a hat for his own wear, which upon pulling a string will expand into an umbrella three feet in diameter. In short, it is impossible to find united in one person more genius, learning, and practical ingenuity than he exhibits, combined with which is an amount of eccentricity bordering upon madness. He has been compared with Beireis, but unjustly so, since he is wholly devoid of charlatanry, is a great lover of truth, and is very modest. Among the poor he often operates for cataract, and that without assistance.'

William von Humboldt and his family, accompanied by Burgsdorf and the sculptor Dyk, left Vienna on October 11, and, passing through Munich, Schaffhausen, Zurich, and Basle, were to reach Paris by November 26. The project entertained by Alexander of going to Switzerland with Haften to await the restoration of tranquillity in Italy was somewhat interfered with by the arrival of Leopold von Buch at Vienna.

The description given by Humboldt in a letter to Freiesleben of the appearance and manners of this distinguished geologist is quite a gem in its way, and remained a true picture of his peculiarities even in later years. ' I was heartily glad to see him,' he writes ; ' he is an excellent man and most genial, and has observed a great deal and with extreme care : in appearance, however, he looks as if he had come from the moon. It seems to me as if his having been alone on the journey had been still further prejudicial to him. I took him round with me to see some friends, but the visits were not generally successful. After the first visit he would, as a rule, put on his spectacles and become entirely absorbed in the examination of some cracks in the glazed stove at the farthest corner of the room, or else he would glide round the room like a hedgehog, examining the shelves against the walls. He is notwithstanding extremely interesting and amiable, a perfect treasury of knowledge, from whom I shall gain much that is valuable. He stays here a fortnight before proceeding to Ischl and Salzburg, where he intends to spend a few weeks with me, and is to start for Italy in the winter by way of the Tyrol.'

In the same letter Humboldt makes the following communication :—' I have great hopes (but this is a secret), indeed almost amounting to certainty, of being able to secure a most advantageous travelling companion in the person of Van der Schott, a truly excellent young man, of most estimable character, and possessing great botanical knowledge. He is director of the Botanic Gardens here, and has received permission from the emperor to travel : I intend to attach myself to the expedition, so pray congratulate me on my good fortune. We must, however, first catch our hare.'

Since Bonaparte's campaign in Italy had, for the time, completely closed every prospect of accomplishing a scientific journey through that country, Humboldt made arrangements with Buch to pass the winter at Salzburg, in order to institute with him some meteorological observations both there and at Berchtesgaden. They set out by way of Steiermark early in October 1797.

These remarkable men were alike distinguished for their talent and zeal for science; they equally enjoyed a position of independence, and were free to follow the bent of their genius; and both had devoted themselves to the same studies in which even their earliest efforts had produced works of intrinsic value.

Buch published the results of his own labours during this expedition in a complete form in his ' Geognostische Beobachtungen auf Reisen,' but Humboldt, pressed with preparations for more important enterprises, merely recorded his results in the course of his widely extended correspondence. Some results of Humboldt's labours are also to be found in Buch's work, which could hardly fail to be the case in an undertaking carried on conjointly; a great number of measurements for determining the height of various places between Salzburg and Aussee constituted a part of these labours, besides some general meteorological observations, and various experiments with the eudiometer for testing the purity of the air. The value of these labours was enhanced by the circumstance ' that it would not again be easy to combine a place so favourable for observation with observers so eminently accurate and so great a variety of experiments.'

Some letters written by Humboldt during his stay at Salzburg to Von Zach, director of the Observatory at Seeberg, near Gotha, were published in the ' Geographische Ephemeriden.' In January, 1798, he writes:[1]—

' You wish that I should pay some attention to the geographical determination of certain places, and you urge it upon me with so much warmth and energy that I am compelled to yield to the force of your eloquence. . . . I have started on my journey and have reached Salzburg, where I am awaiting a change in the affairs of Italy. In the meantime I am occupying myself in determining the latitude of this town. I employ a 12-inch sextant by Wright, which is unfortunately

[1] ' Allgemeine geographische Ephemeriden,' vol. i. p. 357.

extremely heavy for use; it does good work, but observing with it is exceedingly laborious.

'I remain here till the beginning of April. The proximity of the Alps, among which I frequently make excursions notwithstanding the winter, the complete solitude I can command for the prosecution of my studies, and the free use of the excellent library of Baron von Moll, all contribute to render my stay here very enjoyable. I intend shortly to publish my " Investigations on the Atmosphere during the Winter of 1798," which I think have been carried on with a chemical accuracy only to be attained by one who has, like myself, been living out of doors both day and night, and could test the air at all hours.'

He relates, further, that he and Buch had determined the height of the Geisberg to be 2,972 feet 6 inches above the level of their apartments, and expressed an intention of instituting upon that mountain some observations upon refraction. Unfavourable weather had prevented the observation of an eclipse of the moon on December 4, as well as the occultation of a star on February 28; he had however succeeded in determining the latitude and the variation of the magnetic needle, and had instituted some trigonometrical observations for the construction of maps of greater accuracy; he was also engaged in investigating the delicate chemical changes occurring in the atmosphere for the purpose of noting the phenomena of refraction, and to this end undertook a daily register of the density, temperature, and humidity of the air, as well as of the amount of electricity, oxygen, and carbonic acid present in the atmosphere. In a letter dated Berchtesgaden, April 17,[1] he thus refers among other subjects to these labours:—'When you reflect how far removed this work is from the ordinary range of my chemical and physiological observations, I shall hope to find in you a lenient judge. Do not suppose, however, that, trusting to this forbearance on your part, I shall send you any hasty or uncertain observations. No; I am seeking to determine only a few points, but these I am anxious to ascertain with all the accuracy of which I with my heavy 12-inch sextant am capable. There is no one spot in the whole of the southern

---

[1] 'Allgemeine geographische Ephemeriden,' vol. ii. p. 165.

part of Bavaria, the place of which has been determined by astronomical observation ; therefore the maps are in error in every direction to the amount of from 5′ to 6′. By sextant observations of polar stars, I have satisfactorily obtained the latitude of Salzburg, Berchtesgaden, and Reichenhall.' . . .

In November 1797, Humboldt had received a proposal from an English nobleman to accompany him in an expedition to Upper Egypt. This was no other than Lord Bristol, Bishop of Derry, who, notwithstanding his high position in the Church, was a bold free-thinker and a votary of pleasure; in the enjoyment of an income of 60,000*l.* a year, he held a conspicuous place in the fashionable world, and was an enthusiastic patron of the fine arts.[1] He had visited Greece and the coast of Illyria, and had spent many years in Italy, where at Rome he had made the acquaintance of the archæologist Hirt, subsequently court counsellor at Berlin. Strange to say, in the party selected for this expedition to Egypt, which was equipped with every appliance for securing an amount of comfort truly princelike, besides Humboldt, Hirt, and Savary, the well-known traveller who had spent eight years in Egypt, two ladies were to be included, the Countess Dennis and the Countess Lichtenau.[2]

---

[1] This eccentric bishop called forth some severe criticisms from Goethe. See ' Sämmtliche Werke ' (8vo.), vol. xxi. p. 367.

[2] In the letter of invitation addressed to Hirt, under date Triest, there occurs the following passage :—' We shall have two large spronari with both oars and sails. La Dennis and M. le Professeur Hirt are to accompany the dear Countess in her boat. M. Savary, the author of the charming letters upon Egypt, will be with me in mine. I intend to take with me two or three artists, not only for the ruins and the grand points of view, but also for the costumes, so that nothing shall be wanting to render the journey agreeable.

' Dear Hirt ! will not this be an expedition worthy of your profound knowledge and your indefatigable industry ? What splendid drawings may we not expect from our artists !—what a magnificent work will not our united efforts furnish for publication ! '

In a similar strain Lord Bristol raved in the letters he addressed to his chère amie et adorable Comtesse de Lichtenau :—' Jamais un voyage ne sera plus complet tant pour l'âme que pour le corps.' Indeed his gallantry carried him so far as to lead him to remark with a play upon the words: ' Quant aux femmes, il faut que vous passiez pour la mienne, et que pour n'être pas *violée*, vous soyez *voilée*, et alors votre personne est plus sacrée que la mienne.'

It must however be expressly stated here, that the invitations to Hirt and to the Countess Lichtenau to join the expedition had been given as early as March 1797, and that in the mean time, in consequence of the death of King Frederick William II., on November 16, 1797, the circumstances of the countess had become so altered that it was no longer possible for her to join the party. Humboldt seems to have been fully aware of the peculiarities of Lord Bristol's character. He often spoke of him as ' the mad old lord,' and wrote of him to Pictet in the following terms :—' You might possibly think the society of the noble lord objectionable ; he is eccentric in the highest degree. I have only once seen him, and that was during one of the expeditions he used to make on horseback between Pyrmont and Naples. I was aware that it is not easy to live at peace with him. But as I travel at my own expense I preserve my independence, and do not risk anything ; I can leave him at any time if he should oppose me too much. Besides, he is a man of genius, and it would have been a pity to have lost so excellent an opportunity. I might do something in meteorology. However, I must beg of you not to mention the expedition to anyone.' [1]

Although Humboldt's most cherished hopes had ever been directed towards a scientific exploration of regions within the tropics, he was yet willing, since Vesuvius and Etna were no longer to be reached, to embrace the opportunity now afforded him of visiting a country which had played so leading a part in the history of the world's civilisation. He accepted the proposal under the express condition that he should be at liberty on returning to Alexandria to pursue the journey alone through Syria and Palestine.

In prospect of this expedition, he devoted himself with diligence to those studies which would best prepare him for the full appreciation of such a tour—studies which afterwards proved of great value to him in the comparisons he was led to institute between the monuments of antiquity of the Old World in Egypt and those of the New World to be found in Peru and Mexico.

[1] Le Globe, Journ. géogr.' vol. vii. pp. 153, 185.

The preparations for the journey were soon so far completed
that on April 22, 1798, Humboldt left Berchtesgaden by way of
Strasburg for Paris, in order that he might provide himself
with good instruments and take leave of his brother and his
family.

A few days before leaving Salzburg, he thus writes to
Freiesleben :—

' . . . . Lord Bristol, an old English nobleman, with an
income of 60,000l. a year, (half mad and half a genius), who
travelled in Greece with Fortis, sent me an invitation to meet
him at Naples and join him in an expedition up the Nile.   He
was to start in August and sail in his own yacht with an armed
crew, with artists, sculptors, &c., and with a kitchen and well-
provided cellar. He was to go as far as Syene in Upper Egypt.
I was to be free of expense throughout.   In the spring of 1799,
we were to return by way of Constantinople and Vienna.   Such
a proposition was not to be declined.   I promised to go with
him, and had arranged to leave Paris by the end of June, in
order to meet him at Naples by the 1st of August.   It is
now, however, rumoured everywhere that the French are going
to take possession of Egypt.   If so, Bristol as an Englishman
will not be allowed to enter the country, and whether it will be
safe for me to go or not I can only learn in Paris, whither I set
out to-morrow.   Thus our most cherished plans are scattered
to the winds !   I am nevertheless not without hope of accomplishing somehow this visit to Egypt.   If peace be maintained
between France and Turkey, I shall go alone, setting out from
Marseilles.   I count much upon the trip, as I have cherished
the hope for so long, and it will be such an excellent way of
employing the time I now have at my disposal.'

Of the various labours he was then engaged upon he gives
the following account :—' Here, where I intended to remain
scarcely two months, I have now spent five in the deep retirement of a hermit's solitude, engaged upon various experiments with more devoted industry and with more successful
results than ever.   In Zach's " Ephemeriden " you will see
some notice of my astronomical work, such as the determination
of latitudes, trigonometrical measures of the Alpine ranges, &c.
The second part of my labours was completed in February, and

comprised a geological treatise for Moll's " Jahrbücher," on the hardening of rocks, which I am having printed off separately, an introduction to Ingenhous's treatise on manures, besides various papers on chemical experiments. The paper on subterranean gases will be finished while I am at Paris; I have been able to make several additions to it during my sojourn here. I have remained in this neighbourhood longer than I intended, for the purpose of completing a series of eudiometric observations at Berchtesgaden, Aussee, and Salzburg. The chemical part of these investigations is beginning to assume an entirely new aspect. I have collected a great mass of new facts, and I am now working uninterruptedly in arranging them; for in visiting the mines in this district my love for practical mining has been quite reawakened.'

The last letter from Salzburg is dated April 19, 1798, and is addressed to Eichstädt, court counsellor at Jena, the editor of the 'Jenaer allgemeine Literaturzeitung.' It runs as follows:—

' On the eve of starting for Paris to join my brother, allow me to recall myself once more to your remembrance. I have spent nearly five months here in a most busy solitude, though my plans have been so unsettled that sometimes twice in a week I have been on the point of starting for Italy. Political affairs, however, have now assumed such an aspect that it has become impossible to cross the Alps. I am therefore thinking of spending part of the summer in Paris; and since this unfortunate war has rendered the seas too unsafe for me to prosecute my intended voyage to the West Indies, I propose to pass the winter in the East. No sooner are my preparations for this eastern expedition completed, than rumours reach me on all sides of a campaign in Egypt, which will have the effect either of greatly facilitating or of entirely defeating my object. I am willing, however, to believe that the events now transpiring will ultimately be of service to science, but for myself, I am so hampered in all my projects that I daily feel inclined to wish I had lived either forty years earlier or forty years later. A dull uniformity, so detrimental to moral improvement, will, I fear, spread itself over the whole earth, and nations whose physical and moral position call for very different forms of government will all be compelled to submit to the one model—a Directory and two Councils; but insurrection will be just as rife under a republican despotism as under ecclesiastical rule. There is only one advantage to be gathered from the present state of things, and that is the extermination of the feudal system and of all the aristocratic privileges which have so long pressed upon the poorer and more intellectual classes of mankind—an advantage which I am happy to think will still remain, should monarchical institutions again return to be as prevalent as republicanism seems likely to be now. Amid the various emotions, mostly of a melancholy character, which have been excited within me by the events of the closing century, I have, I believe, continued faithfully to prosecute the aims I have had in view. I have never laboured with such persistent industry, nor been more successful in my experiments. For the space of five months I have daily observed the state of the atmosphere, and I hope to work out the results of these very troublesome investigations in Paris before I start on my voyage. You may perhaps have seen in Zach's journal that I have been able to determine astronomically the position of several places in this neighbourhood, which I accomplished by means of the sextant with which in former days you used to observe in your gay and beautiful garden.

'Pray do me the favour to insert the enclosed notice in the "Intelligenzblatt."[1] I hope it may prove the means of relieving me from much troublesome correspondence. I cannot possibly travel all over Germany to teach every unskilful pair of hands how to perform my experiments.' . . .

William von Humboldt, who, as we have seen, had preceded his brother to Paris, had soon learned to regard it as a home, and was on terms of friendly intercourse with various distinguished votaries of science and art. A new social existence had sprung up with the new phase in political life, and scientific circles had gradually resumed that position in society which

---

[1] Several physicists had complained to him that they could not succeed with his experiments. 'It seems a strange requirement to expect to be able to produce in a few days, or it may be hours, all the phenomena which another has only succeeded in accomplishing after five years' hard work and the observation of several hundred cases.' The 'Notice' is printed in the 'Jen. allg. Literaturztg., Intelligenzblatt,' 1798, No. 79, col. 670.

was their due. Humboldt's house, presided over by his amiable
and talented wife, became, as she described it, a *point de
ralliement* for all the Germans of any note then in Paris,
among whom may be enumerated that eccentric nobleman
Count Schlabrendorf, Gustav von Brinckmann and William von
Burgsdorf, friends they had known at Jena and Berlin, the
youthful poet Ludwig Tieck, and Schick. To this circle were
also gathered the following distinguished Frenchmen :—Villoisin, Corai, St.-Croix, Du Theil, Chardon de la Rochette, and
the artists David and Forestier.[1] In addition to this brilliant
home circle Humboldt enjoyed the *entrée* of the most distinguished houses in Paris, where he was ever received as a
welcome guest. The society he frequented with most pleasure
was that assembling at the house of Millin, editor of the
' Magazin encyclopédique,' at whose receptions,[2] held every
septidi, he met a circle of eminent and distinguished men,
among whom the arrival of Alexander von Humboldt had been
eagerly anticipated.

Before, however, Alexander could reach Paris, Bonaparte's
expedition to Egypt, the preparations for which had been conducted with unusual secresy, had been publicly announced. In
a letter to Von Zach,[3] Lalande writes, towards the end of April,
1798 :—' 160 persons have been appointed to this great scientific
expedition. I have myself selected three astronomers, Nouet,
Quenet, and Méchain, notwithstanding his youth, and if needful
I can find yet other three. Burckhardt has also received a
proposal to join the expedition, but he remains with us, as he
cannot well be spared.'

With greater detail, but in equal uncertainty as to the object
and destination of the undertaking, Burckhardt[4] writes from
Paris on April 30, 1798 :—' The expedition about which I lately
wrote is still a profound secret. It is known that the following
appointments have been made :—Berthollet, Dolomieu, Saix,
Conti, Samuel Renard, Reignaud, Costas, Geoffroy, Le Blond,

---

[1] Varnhagen, ' Galerie von Bildnissen aus Rahel's Umgang,' vol. i.
p. 143.

[2] Bertuch, ' Allgemeine geographische Ephemeriden,' vol. i. p. 686.

[3] Ibid. vol. i. p. 680.

[4] Ibid. vol. i. p. 687.

Quenet, Desgenettes, Thouin, together with Dubois as surgeon, Delille as botanist, and Nouet as astronomer, with whom Méchain goes as assistant. Prony has furnished twelve survey-ing engineers and six cadets from the École polytechnique. Berthollet, Bonaparte's physician, is director in chief of the scientific department of the expedition, in which are included the following antiquarians :—Denon, Jomard, Pouqueville, and Rozier. Prony has surrendered all his instruments, and any others that were required have been procured either by purchase without limit as to price, or by official demand.' On the 30th Floréal (May 19, 1798) the fleet weighed anchor from Toulon, and Bonaparte's campaign in Egypt was openly declared.

Soon after Humboldt's arrival in Paris, news reached him of the arrest of Lord Bristol in Milan, on suspicion that his pro-posed expedition up the Nile was for the purpose of creating an agitation in favour of England against the French. Humboldt was compelled therefore, keenly as he felt the disappointment, to renounce his journey to Egypt. This was the commencement of a long series of disappointed hopes and frustrated plans.

By a fortunate coincidence, Humboldt had arrived at Paris just as Delambre was completing the measurements between Melun and Lieursaint for determining the base for the meridian line. Delambre was on the point of concluding the northern portion of the world-famous French meridian line—a line which extended from Dunkirk on the coast of the North Sea to Barcelona on the shores of the Mediterranean, a distance of 250 leagues as measured on the earth's surface, but including a celestial meridian arc of 9° 3'. Humboldt was present at the concluding operations, and described them with great delight to Von Zach :[1]—'At about twelve o'clock this morning, the 15th Prairial (June 3, 1798), the measurements were com-pleted between Melun and Lieursaint for the base of the great line, and I hasten to communicate to you this event, so im-portant in the history of astronomical geography. In company with Lalande and our excellent friend Burckhardt, I spent two memorable days with Delambre. The weather, which during three decades had been uninterruptedly favourable for the

[1] 'Allgemeine geographische Ephemeriden,' vol. ii. p. 174.

measurement of the base line has not been less beautiful during
the last few days. At Lieursaint we met with Prony and
Bougainville, the celebrated traveller, who at the age of sixty is
still eagerly contemplating a second voyage, in which his son, a
youth of fifteen, is to accompany him. . . . In the course of a
fortnight, Delambre will leave with his assistants for Perpignan,
where Méchain will by that time have completed the last five
or six triangles, and where the southern base will be measured
twice in succession before the winter. As I am intending to
embark from Marseilles in the autumn, I shall gladly avail
myself of Delambre's invitation to stop at Perpignan on my
way and assist in some of the observations. I intend to provide
myself for this purpose with a repeating circle by Lenoir.'

Even a hasty glance at the state of science in Paris at this
time will show how many events were transpiring which were
calculated to excite the interest and stimulate the zeal of
Humboldt.

Notwithstanding the violent overthrow of every moral principle and the bloody scenes of the reign of terror, when even the
Academy was powerless to protect the lives of some of its most
noted members, and Bochart von Saron, Lavoisier, La Roche-
foucauld, Malesherbes, Bailly, and Condorcet fell victims to the
popular fury, notwithstanding the sardonic expression of the
frantic judge, 'Nous n'avons pas besoin de savans,' Paris
was yet at the close of the century the metropolis of the exact
sciences. Lalande, in writing to Von Zach[1] on January 26,
1798, remarks:—'The love of mathematics is daily on the
increase, not only with us but in the army. The result of this
was unmistakably apparent in our last campaigns. Bonaparte
himself has a mathematical head, and though all who study
this science may not become geometricians like Laplace and
Lagrange, or heroes like Bonaparte, there is yet left an
influence upon the mind which enables them to accomplish
more than they could possibly have achieved without this
training. Our mathematical schools are good, and successfully
accomplish their main object in the diffusion of mathematical
knowledge. Bonaparte attends with great regularity the sit-

---

[1] 'Allgemeine geographische Ephemeriden,' vol. i. p. 346.

tings of the National Institute, of which he is a member.'
'Bonaparte,' he writes further, on April 20, 1798,[1] 'always calls
me his grandpapa, because he is a pupil of D'Agelet, who again
was a pupil of mine. I have begged him to use his influence
with the Directory to obtain the removal of the Opera House,
which in case of fire is dangerously near the Library. I also
suggested the purchase of Paulmy's admirable library, consisting of 100,000 volumes, and recommended that some new
instruments and an increase of salary be granted to Thulis at
Marseilles—all of which has been accomplished.' Burckhardt
also writes about this time:[2]—'There is something very interesting in witnessing the modest and unaffected demeanour
preserved by Bonaparte in the midst of the universal applause
with which he is greeted. I often enjoy this pleasure at the
National Institute.'

At this period there was gathered in Paris a remarkable
assemblage of distinguished men; among mathematicians the
great Lagrange, the amiable and powerful writer of the 'Analytical Mechanics' and the 'Theory of Analytical Functions;'
Montucla, author of the 'History of Mathematics;' and
Delambre, compiler of the 'History of Astronomy;' while in
other fields of science laboured Borda, Monge, Fourier, Berthollet, Geoffroy de St.-Hilaire, Larrey, Lalande, the mineralogists Haüy and Brongniard, and Cuvier, who, born at Mont-
béliard, at that time belonging to Würtemberg, in the same
year as Humboldt, was fellow-student with Schiller at the Academia Carolina in Stuttgart.

The mere record of these names is of itself a testimony to the
advancement of science; but should further proof be needed, we
have but to pass in review the giant strides made at this epoch in
its various branches. In theoretic astronomy, the theory of the
moon's motion and of the perturbations of the planets had been
established by D'Alembert and Clairaut; the cause of the
precession of the equinoxes had been ascertained; the figure of
the earth had been more accurately determined through the
measurement of the meridian line; nutation and the aberration
of light had been discovered and elucidated by Bradley;

[1] 'Allgemeine geographische Ephemeriden,' vol. i. p. 679.
[2] Ibid. vol. i. p. 352; also p. 227.

Dollond had brought to perfection the astronomical tele-
scope, while Laplace had given to the world his admirable
'Mécanique céleste.' Mechanics and the laws of motion had
kept pace with the advance of astronomy, while physical science
had received a surprising impetus from the application of
mathematics and from the invention and improvement of every
kind of apparatus. The various phenomena of magnetism, electricity, and galvanism underwent the searching scrutiny of the
most careful observers, and became the object of manifold investigations. In botany and zoology Jussieu and Cuvier had proved
worthy successors to Linnæus and Buffon. With Lavoisier
a brilliant career had been opened to the science of chemistry.
In short, a more exact process of investigation, a more scientific method of treatment, had been introduced into every department of natural science.

Full of joyous excitement, Humboldt wrote to Pictet[1] on
June 22 as follows:—' I do not tell you about Paris, nor of my
mode of life here; you know my tastes and my love of occupation. I live in the midst of science, I work with Vauquelin in
his laboratory, and I have given some lectures at the National
Institute; I feel I have just claim to the reception that has
been accorded me.'[2]

Not less eager was he to give help to others in their labours.
' Hallé,' writes Lalande,[3] ' has laid before the National Institute
an important paper upon galvanism; Von Humboldt has rendered essential service to our commissioners, who have been
instituting a variety of experiments, and have done much
valuable work.'

---

[1] ' Le Globe,' vol. vii. p. 155.

[2] In a letter to Delambre, dated Lima, November 25, 1802, Humboldt
writes:—' In the desert plains of the Apure, in the dense forests of the
Cassiquiari and of the Orinoco, everywhere has your name and the names of
your associates been present before me; and in reviewing the various
passages of my wandering life, I often linger with pleasure at the recollection of years VI. and VII., which I spent in the midst of you all, when I
was loaded with so much kindness by Laplace, Fourcroy, Vauquelin,
Guyton, Chaptal, Jussieu, Desfontaines, Hallé, Lalande, Prony, and especially by you, my kind and generous friend, in the plains of Lieursaint.'
(' Annales du Mus. d'Hist. nat., an XII.' vol. ii. p. 170.)

[3] ' Allgemeine geographische Ephemeriden,' vol. ii. p. 172.

The lectures [1] delivered by Humboldt were on the nature of nitrous gas and the possibility of a more exact analysis of the atmosphere : they were works founded upon the observations he had made at Salzburg, and they were subsequently adduced to disprove the statements of Lavoisier concerning the saturation of nitrous gas with oxygen—statements which at that time were everywhere received with confidence. It was these lectures also which afterwards gave occasion to some severe criticism on the part of Gay-Lussac, then just entering on his career. Ehrenberg is in error [2] when he asserts that it was in these labours that Humboldt was assisted by 'his intimate friend Gay-Lussac.' It was only upon his return from America that Humboldt made acquaintance with the young chemist, and from that time first dated their joint labours. [3] These treatises, with sundry others on kindred subjects, were published in a separate volume in 1799. [4]

In one of his latest letters to Willdenow, Humboldt thus speaks of his reception at Paris and the alternate hopes and disappointments to which he had been subject:—'I was received at Paris in a manner I could never have ventured to anticipate. The venerable Bougainville was projecting another voyage round the world, with the hope of reaching the South Pole. He urged me to accompany him, and as I was just then occupied with magnetic investigations, it occurred to me that an expedition to the South Pole might prove more useful than a journey to Egypt. I became absorbed in plans for this extensive project, when all at once the Directory arrived at the heroic

[1] 'Allgemeine geographische Ephemeriden,' vol. ii. p. 176.
[2] 'Gedächtnissrede auf Alexander von Humboldt,' p. 18.
[3] Arago's 'Sämmtliche Werke,' vol. iii. p. 16 ; 'Gedächtnissrede auf Gay-Lussac.'
[4] 'Versuche über die chemische Zerlegung des Luftkreises und über einige andere Gegenstände der Naturlehre.' Brunswick, 1799. The treatises 'On the Disengagement of Heat regarded as a Geognostic Phenomenon,' and 'On the Influence of Chlorine and Oxygenated Muriatic Acid on the Germination of Plants, together with a Description of the Phenomena,' excited at that time considerable attention. During his absence from Europe, a collection of his treatises, some purely scientific, some on practical mining, was published under the title, 'Ueber die unterirdischen Gasarten und die Mittel ihren Nachtheil zu vermindern, &c.' (Brunswick, 1799), to which a preface was affixed by William von Humboldt.

determination of superseding the septuagenarian Bougainville by appointing Captain Baudin to the command of the expedition. I had scarcely heard of this decision before I received an invitation from the Government to embark on board the "Vulcain," one of the three corvettes forming the expedition. I was empowered to equip myself with the necessary instruments from the national collections, which were all placed at my disposal. My advice was asked in the selection of scientific men and in all matters connected with the preparations for the expedition. Many of my friends were opposed to the idea of my undertaking a voyage of five years' duration; but my determination remained unshaken, and I should have despised myself had I declined such an opportunity for accomplishing useful work.[1] The ships were ready for sea, and I had already acceded to the request of Bougainville that I would take charge of his son, a youth of fifteen, whom he was anxious should become early accustomed to the dangers of the sea. Our companions were well chosen, men of considerable attainments, in the prime of life, and capable of great physical exertion. How earnestly we looked at each other on our first meeting! Strangers till that moment, and then to live for several years in close intercourse! We were to spend the first year in Paraguay and Patagonia, the second year in Peru, Chili, Mexico, and California, the third year in navigating the Southern Ocean, the fourth in Madagascar, and the fifth on the coast of Africa, ... What an intolerable grief that in one fortnight all these hopes have been shattered! A paltry 300,000 francs and the apprehension of a speedy declaration of war are said to be the reasons. My personal influence with François de Neufchateau, who is well disposed towards me, and every other spring of action that could be set in motion, were all in vain. In Paris, where the expedition had excited so much interest, we

---

[1] Humboldt had indeed little confidence in the personal character of Captain Baudin, who had given great dissatisfaction at the court of Vienna on the occasion of his being commissioned to convey the young botanist Van der Schott to Brazil. As, however, he could never hope to undertake with his own means so extensive a voyage, and see so large a portion of the globe, he resolved to risk his fortunes with him, and hoped to meet with success.

were supposed to have actually set sail. By a second decree
the Directory postponed the voyage for a year. . . .

'It was impossible not to feel my position and the severity
of this disappointment; but it is the part of a man to work, and
not to yield to unavailing regrets. I immediately formed the
determination of joining the army in Egypt by the land route,
accompanying the caravan that travels from Tripoli to Cairo
through the desert of Selimar. I selected as my companion
one of the young men who had been appointed to the former
expedition, Bonpland, a very good botanist and the favourite
pupil of Jussieu and Desfontaines. He has served in the navy,
is very robust, courageous, and good humoured, and is well
versed in comparative anatomy. We hastened to Marseilles,
there to embark for Algiers with the Swedish consul, Skjölde-
brand, in the frigate " Jaramas," which was conveying presents
to the Dey of Algiers. I wished to spend the winter in Algiers
and on the Atlas, where, in the province of Constantine, there
are to be found, according to Desfontaines, above 400 new
species of plants. Thence I hoped, by joining a caravan bound
for Mecca, to reach Bonaparte by way of Sufetula, Tunis, and
Tripoli.'

The departure from Paris was delayed till October 20, 1798.[1]
On the 12th of that month Humboldt read before the National
Institute a paper upon Agriculture, on the conclusion of which
Jussieu addressed him, in few words, a graceful farewell. The
most affecting parting was that with Baudin, who termed the
separation a dissolution of marriage. In other respects Humboldt retained his courage, and continued during the journey
in such good spirits as to write humorous descriptions in his
note-book of the profane company in which he was thrown,
both in the diligence and at the table-d'hôte. Lyons was

[1] William von Humboldt wrote from Paris to F. A. Wolf, on October 22,
1798 :—' My brother, I grieve to say, left here on the 20th—the day before
yesterday. I have felt his departure exceedingly. For the last few months
we have been living under the same roof, have dined together daily, have
usually associated with the same society, have, in short, lived together in
the most perfect sympathy; and now, after having so fully enjoyed the
pleasure of our unbroken intercourse, there must needs be this separation,
which, however, in all probability, will not be of long duration.' (Wilhelm von Humboldt's ' Gesammelte Werke,' vol. v. p. 206.)

reached on October 24, whence, proceeding down the Rhone, the travellers arrived at Marseilles on the evening of the 27th. The whole of the following morning was occupied in the examination of passports,[1] a troublesome business, in which it was evident that the only knowledge of official routine possessed by the Prussian consul, Herr Sauvages of Prenzlau, was restricted to the names of such persons as were entitled to 'Your Excellency'—a list he had most faithfully committed to memory from the calendar. The Swedish frigate had not yet arrived, but was hourly expected. The intervening time was spent by the travellers, notwithstanding the extreme heat of the weather, in making an extensive collection of the plants to be found along the coast, especially of several varieties of sea-weed, as well as in the collection of specimens of crabs and mussels, of which they made drawings; they also undertook various magnetic, meteorological, and astronomical observations, notwithstanding the serious injury some of the instruments had received, and in this labour were assisted by the director of the Observatory, M. Thulis, formerly a merchant at Cairo, who notwithstanding many unpleasing peculiarities was very affable and obliging.

On November 10, an excursion to Toulon was undertaken, which lasted three days. The town, thronged with its 5,000 prisoners, did not impress them agreeably; they visited all the objects of interest, the magnificent Botanic Garden, the Arsenal, the collection of models, the great harbour, the inner and the outer roadsteads, where, next to the ship of the line 'La Hardie' of 74 guns, the interest of the travellers was most keenly excited by the frigate 'La Boudeuse,' as the vessel in which Bougainville had made his celebrated expedition round the world. 'She is now preparing to set sail,' he remarks in his note-book, 'to convoy a fleet of merchant-men to Marseilles, which they hope to reach in five hours. The whole ship was astir, and all hands were on deck working the sails. My heart swelled with joyous emotion as I witnessed the preparations around me, but when I descended to the cabin, a spacious apartment, the thought of Baudin's voyage completely over-

---

[1] See Appendix.

came me. I lay for full ten minutes in the window, contemplating the bright vision. At length they came to seek me; I could almost have shed tears as I thought of my shattered prospects.' After a short visit to the islands of the Hyères, 'where the golden apples were hanging by hundreds on the dwarf trees,' Humboldt and Bonpland returned to Marseilles on November 13.

Here they were detained for two whole months, from October 27 till the end of December, watching in vain for an opportunity of sailing to Africa. Everything was kept ready packed, and the horizon was daily searched in quest of the long-expected 'Jaramus.' At length the news arrived that the 'Jaramus' had been wrecked off the coast of Portugal, with the loss of everyone on board. 'Nothing daunted,' writes Humboldt, 'by all these disappointments, I hired a vessel from Ragusa to take us direct to Tunis. The municipal authorities of Marseilles, however, apparently warned of the troubles which were soon to beset the French in Barbary, refused to grant us passports; and scarcely had the Ragusan bark quitted the harbour when there arose a fearful tempest, which, lasting nearly a week, strewed the shore between Cette and Agele with the fragments of many shipwrecked vessels. Almost at the same time the news arrived that the Dey of Algiers had interdicted the departure of the caravan for Mecca, as Egypt, through which it had to pass, was polluted by the presence of the Christians. Thus all hope vanished of either reaching Egypt by the Levant or of joining the French expedition by the land route to Cairo.'

Were, then, the best months of the spring, January, February, and March, to be spent in weary waiting at Marseilles, among a society the leading members of which, whether consuls, diplomatists, officials, or even men of science, had thrown aside the mask, and revealed a tone both immoral in feeling and politically dangerous? Corsica and Sardinia, rich as they might be in botanical treasures, offered no facilities for any comprehensive plan. The only feasible arrangement seemed to be to proceed to Spain, and thence to embark in the spring for Smyrna. In pursuance of this plan, Humboldt, on December 15, gave a commission for the purchase of some Spanish bills of exchange to Ellenberg, a merchant with whom

Fould of Paris had placed 40,000 francs to his credit. In the afternoon of the same day he was informed that instructions had been received from Fould ten days previously not to advance him the smallest sum, and that it had only been from a feeling of delicacy that he had not been earlier made acquainted with this unpleasant information. (See p. 245.)

This, however, was not allowed to interfere with the projected journey to Spain, and towards the end of December, 1798, Humboldt and Bonpland left Marseilles. During the six weeks spent in reaching Madrid, where they arrived in the beginning of February, they found abundant occupation in collecting plants, determining the latitude and longitude of places, and their height above the sea, besides making various observations connected with meteorology, geology, and magnetism. Humboldt makes but slight allusion to this journey in his works; and this gives an increased value to his letters, full of detail, whence the following extracts of biographical interest are taken.

'My journey hither was chiefly made on foot,' he writes to Willdenow from Aranjuez, on April 20, 1799, 'and I travelled by way of the coast of the Mediterranean through Cette, Montpellier, Narbonne, Perpignan, over the Pyrenees, and across Catalonia to Valencia and Murcia; thence I proceeded across the high table-land of La Mancha. In Montpellier I passed some most agreeable days with Chaptal, and in Barcelona with John Gill, an Englishman with whom I had lived at Hamburg, and who is now a partner in an extensive mercantile house in Spain. In the valleys of the Pyrenees the pea was in blossom, while the snow-capped summit of the Canigou reared itself just above us. In Catalonia and Valencia the country is like a succession of gardens enclosed by cacti (torch thistles) and aloes; date-palms forty and fifty feet high, laden with fruit, tower over every monastery. The fields present the appearance of a forest of locust or carob trees (*Ceratonia siliqua*), olives, and orange trees, the tops of which separate into several heads in the manner of some of our pear trees. In Valencia eight oranges may be bought for a peseta. At the mouth of the Ebro, near Balaguer, is a plain forty miles in extent covered with dwarf palms (*Chamærops*), pistachios, and numerous varieties of the Erica, rose cistus, and holly rose. The heaths were in blossom, and even in the wildest regions we gathered the jonquil and the narcissus. The common palm (*Phœnix dactylifera*) grows so luxuriantly in the neighbourhood of Cambrils, that twenty or thirty stems are frequently grouped so thickly together as to present an impenetrable barrier. As white palm-leaves are much in request for the decoration of the churches, it is common in Valencia to encase the young shoots of the date-palm in a sort of conical cap, made of the Spanish feather-grass (*Stipa tenacissima*), so that the young leaves, being excluded from the light, should come out blanched. The basin in which the town of Valencia lies is not surpassed in any part of Europe for luxuriance of vegetation. The sight of these palms, pomegranates, carobs, and mallows makes one feel as if one had never before seen trees, or foliage. In the middle of January the thermometer stood in the shade at 72°. The flowers were almost all over.

'I shall attempt no description of the ruins at Tarragona, the mountain near Murviedro, nor the remains of ancient Sagentum with its temple of Diana, enormous amphitheatre, and tower of Hercules, commanding an extensive view over the sea, Cape Cullera, and Valencia, the towers of which are seen to rise from a forest of date-palms. While you poor creatures can scarcely keep yourselves warm, I am walking with heated brow among fragrant orange groves and over fields which, irrigated by a thousand canals, yield five crops in the course of a year—namely, rice, wheat, hemp, peas, and cotton. Amid such luxuriance of vegetation, and surrounded by a race of such remarkable beauty, one willingly forgets the fatigues of travel and the discomforts of the inns, where we have never once met with bread. Even close to the coast the country is everywhere highly cultivated. In Catalonia the industry of the people is as remarkable as in Holland. In every village some occupation is going on, weaving or ship-building, no one seems to be idle. Nowhere in Europe, perhaps, has the culture of land been brought to greater perfection than in the district between Valencia and Castellon de la Plana; yet sixty miles inland the country is a perfect desert. The interior is in reality the summit of a chain of mountains which remained standing from 2,000 to 3,000 feet above the sea when all else was engulphed

in the Mediterranean. To this elevated land Spain owes her
existence, but to this also is due the severity of the climate
and the sterility of the soil everywhere excepting along the
coast line. Near Madrid the olive trees are often injured by
the cold, and oranges are rarely to be seen growing in the open
air. . . . But I am beginning to write descriptions, which I
never intend to do, as I should then be sending you a book
instead of a letter.

' If, dearest friend, I have not sent you a line since I left
Marseilles, I have not the less been occupied on your behalf
and for your gratification. I am just now despatching you a
box containing 400 plants, and when you look them through,
you will be convinced that in all my wanderings, whether amid
woods, over plains, or along the sea-shore, scarcely a day has
passed in which you have not been vividly before me. Wherever I have been I have made collections for you, and only
for you; for I do not intend to begin collecting for my own
herbarium till I have reached the other side of the ocean.'

A letter to Von Zach, from Madrid, dated the 23rd Floréal,
year VII. (May 12, 1799), is not less full of detail, though
embracing a very different class of subjects.[1]

In this letter Humboldt mentions that the Paris Board of
Longitude had lent him a compass constructed by Lenoir on
Borda's plan for registering the dip of the needle, and that a
similar instrument was in the possession of Nouet, who, it
will be remembered, had been attached to Bonaparte's expedition. Humboldt goes on to remark, that but for the unfortunate circumstances which had prevented his journey to
Africa, it would have been possible, by comparing the results of
the two instruments, to have ascertained within the space of
eight months the magnetic constants for the whole length of
the Mediterranean, from the Straits of Gibraltar to the
Isthmus of Suez. He then proceeds to give details of various
methods of observation, and magnetic results obtained at Paris,
Nismes, Montpellier, Marseilles, Perpignan, Gironne, Barcelona, Cambrils, Valencia, Madrid, and other places; he further
appends a series of meteorological observations with determina-

---

[1] ' Allgemeine geographische Ephemeriden,' vol. iv. p. 146.

tions of latitude and longitude, height above the sea level, &c., of various localities, both in France and Spain.

The letter then continues :—'Although the country through which I am travelling offers nothing very favourable for astronomical geography, yet I avail myself of every opportunity that presents itself for taking observations of the sun and stars of the first magnitude. In the kingdom of Valencia I suffered greatly from the hooting of the rabble, as at that time I had not obtained permission from Government to prosecute such labours. I have often grieved to see the sun pass the meridian while not daring to unpack my instruments. I have sometimes been obliged to wait till the middle of the night, and content myself with a star of the second magnitude, which made but a sorry figure in my artificial horizon. On the 19th Nivose (January 8, 1799) I took some observations at Barcelona [1] from the same terrace of the Golden Well from which Méchain had observed. From the 29th Nivose (January 18) to the 6th Pluviose (January 25) I was occupied in an

---

[1] We may here allude to a circumstance which, though it proved of material assistance to Humboldt in the prosecution of his journey, has yet not been generally known. Humboldt had commissioned Kunth from Barcelona to send him as early as possible a letter of credit upon a substantial house in Madrid, to be transmitted direct, and not through a third house, especially not through any Paris banker, because the credit of the Paris houses had been shaken by the state of the public funds—a position of affairs which, as before mentioned, had caused him considerable inconvenience at Marseilles. Kunth attempted to get the business transacted by a Berlin house upon the production of certain Prussian official documents and every satisfactory security. But instead of the letter of credit, came the announcement that the transaction could not be completed till the validity of the security had been proved. On this becoming known, the house of Mendelssohn and Friedländer immediately undertook, without any pledge or security whatever, to remit any sum required by Herr von Humboldt to the Marquis d'Iranda, of the firm of Simon d'Arragora, one of the first houses in Madrid. The marquis, who was past seventy, had personally retired from the business, but he loaded Humboldt with proofs of kindness and goodwill, ordered the financial arrangements of his journey in the kindest and most disinterested manner, without demanding the slightest security, and afterwards honoured his bills of exchange. Humboldt wrote from Madrid on April 4, 1799:—'The Marquis d'Iranda is one of the most distinguished men in Europe. He treats me as his son, and does and will do everything for me.' (Biester, 'Neue Berliner Monatsschrift,' vol. vi. p. 193.)

excursion to Monserrat[1] where I was able to make some observations of the sun, the moon, and Sirius, not only from the convent but also from Mattorel and Colbaton. At Mattorel I observed in the open streets, surrounded by about thirty spectators, who kept shouting to each other that I was worshipping the moon.'

The astronomical observations made during the journey from Barcelona to Valencia at isolated ventas and cabins in the wild and uninhabited district, extending over a space of from 120 to 160 square miles, proved of great value in completing the geography of Spain. These places became established points in the vast plains over which the traveller journeyed as upon the ocean; as an example, we may mention that even the position of such a town as Valencia, containing 80,000 inhabitants, varied in the best maps of that day to the amount of two minutes. This series of observations was commenced at Madrid, from the palace of the Duke of Infantado, on the 14th Ventose (March 4).

The conclusion of this very long letter to Von Zach is in the following words :—' This is all that I can send you at present. Pray receive this little with indulgence, and bear in mind that there is other work besides astronomy demanding my attention.'

A part of this work consisted in obtaining, by barometric measurements, the sectional elevation in the direction from south-east to north-west of the whole Spanish Peninsula, that is to say, from the coast of the Mediterranean at Valencia to the shores of the Atlantic in Galicia. Through these observations was first revealed the existence of an extended plateau at a considerable elevation in central Spain. ' The height of Madrid,' says Humboldt, ' had indeed been estimated as early as 1776 by Lalande, from calculations based upon the barometric observations of the celebrated traveller and mathematician Don Jorge Juan, to be 1,903 feet above the sea, but geographers had at that time no knowledge of the connection

---

[1] An incident that occurred on Monserrat, of a hermit who, by dint of his rosary, courageously rescued the mule of a poor muleteer, is described by Humboldt in a letter to his brother, by whom it has been inserted in his description of Monserrat. (Wilhelm von Humboldt's ' Gesammelte Werke,' vol. iii. p. 209.)

existing between all the high table-lands of the Spanish Peninsula by which they were united into one vast plateau.' The
leading feature of this new investigation has since become of
great importance, not only in acquiring a knowledge of the
varying configuration of the earth's surface, but in determining
the zones of elevation characterised by certain forms of vegetable and animal life. It would, however, be inappropriate to
enter here on a discussion of these labours.[1]

While at Madrid, Humboldt unexpectedly met with the
fulfilment of his long-cherished plans. Herr von Forell, a
distinguished patron of science, already mentioned in p. 215,
as the ambassador from Saxony to the court of Madrid, offered
with the greatest readiness to obtain permission from the king,
through the minister Don Mariano Luis de Urquijo, a man of
liberal views, for Humboldt, at his own expense, to visit the
Spanish colonies in America.

In consequence of this application he was accorded an interview with the king, and the promise of the desired permit. His
reception by the sovereign is thus described by Humboldt:—'I
was presented at the court of Aranjuez in March 1799. The
king honoured me with a very gracious reception. I laid
before him my reasons for wishing to visit the islands of the
Philippines and to travel in the New World, and presented to
the Secretary of State a treatise in which I had expressed my
views fully on the subject.[2] My petition was supported by the
Chevalier d'Urquijo, who was successful in removing every
difficulty. The kind and courteous treatment shown me by

[1] The first publication of the results obtained from these hypsometric
measures and the sectional elevation deduced from them appeared in
Cavanilles' 'Annales de Histor. natural,' vol. i. p. 86; they were afterwards
published in Laborde's 'Itinéraire descriptif de l'Espagne, 1808,' vol. i. p.
cxiv., and incorporated into Donnet and Malo's large map of the Spanish
Peninsula, and in Humboldt's 'Atlas géogr. et phys. du Nouv. Cont.' Pl.
III.; finally enlarged in Humboldt's treatise 'Ueber die Gestalt und das
Klima des Hochlandes in der Iberischen Halbinsel, ein Sendschreiben an
Prof. Berghaus' (Hertha, 1825, vol. iv.; and reprinted in the 'Briefwechsel
Alexander von Humboldt's mit Heinrich Berghaus,' vol. i. pp. 18–48).

[2] Unfortunately, this memoir has not since been recovered; Bergenroth
had entertained hopes of finding it among the archives of Simancas, but
his early death, while investigating the records of Spain, put an end to his
labours.

this minister was all the more magnanimous since my relationships with him were by no means of a personal character.[1]
Never before has a permission so unlimited been granted to
any traveller, and never before has a foreigner been honoured
by such marks of confidence from the Spanish Government.'

Madrid itself afforded many opportunities for the acquire-
ment of useful information. Cavanilles, Director of the Botanic
Gardens, a correspondent of Willdenow; Née, who had with
Hänke accompanied the expedition of Malaspina in the capacity
of botanist, and had returned with the most extensive herbarium
that had at that time been brought to Europe; Don Casimir
Ortega, Proust, Hergen, the Abbé Pourret, the learned editor
of the ' Flora of Peru,' Ruiz and Pavon—all threw open to the
travellers, in the most obliging manner, their valuable collections. Humboldt and Bonpland inspected with the greatest
interest the products from America, including some of the
Mexican plants discovered by Sesse, Mociño, and Cervantes,
drawings of which had been preserved in the Museum of Natural
History at Madrid. But however instructive a longer sojourn
might have been, ' we felt,' writes Humboldt, ' too impatient
to avail ourselves of the permission granted by the court to
delay our journey any longer. During the past year I had
encountered so many difficulties that I could scarcely even
now believe in the final realisation of all my wishes.'

The travellers left Madrid in the middle of May. Still continuing their hypsometric measurements, they took their
journey through Old Castile, Leon, and Galicia, by way of
Villalpando, Astorga, and Lugo, to the harbour of Corunna,
where they were to take passage in a mail boat for Cuba. The
captain of the port, Don Raphael Clavijo, recommended them
to embark in the corvette ' Pizarro,' the next ship that was to
sail, which though not a fast sailer had the reputation of being
a fortunate vessel. He enjoined the captain to stay sufficiently
long at Teneriffe for the travellers to visit Orotava and ascend
the Peak, and preparations were at once commenced for their
reception on board, and for the safe conveyance of their

---

[1] No trace can be discovered of any expression of interest in Humboldt
and his undertakings on the part of the chamberlain, Count von Rhode,
Prussian Envoy Extraordinary at the Court of Madrid.

numerous instruments. Ten days afterwards, on June 4, the horizon was covered by a thick mist, a presage of favourable winds and the signal for departure.

The following letters were penned by Humboldt as farewell greetings to his friends Von Moll and Willdenow :—

To Von Moll.

'Corunna: June 5, 1799.

'In a few hours we shall be sailing round Cape Finisterre. . . . . I shall make collections of plants and fossils, and I shall hope to make some astronomical observations with the excellent instruments I have with me. I intend to institute a chemical analysis of the atmosphere. . . . But all this does not form the main object of my journey. My attention will ever be directed to observing the harmony among the forces of nature, to remarking the influence exerted by inanimate creation upon the animal and vegetable kingdoms. . . .

'A. HUMBOLDT.'

To Willdenow.

'Corunna : June 5, 1799.

'On the eve of my departure in the frigate 'Pizarro,' allow me once more, my dear friend, to recall myself to your remembrance. In five days we shall be at the Canaries, whence we sail with letters to the coast of Caracas, and thence to La Trinidad in Cuba. Give my affectionate regards to your wife and children and Hermes, and my kind greetings to Zöllner, the two Klaproths, Hermbstedt, and whoever else may remember me. I hope we may meet again in health and peace. I shall bear you ever in remembrance.

'*Man must ever strive after all that is good and great!* '

RETROSPECT.

The most marked features in the characters of eminent men may always be traced to two predominant elements—their native genius and the outward circumstances by which that genius has either been fostered or repressed. From the combined action of these two elements is determined the kind and force of impress which a great man will produce upon his century.

Alexander von Humboldt, even in the first thirty years of his life, affords an instance in which the highest natural endowments were united to outward circumstances that comprised all that could render life attractive. But of yet higher value than any of these gifts appears the moral force, the indefatigable industry, and the persevering, enthusiastic energy by which he was led to devote these gifts to the benefit of mankind and ennoble them by their consecration to the service of all that could further the extension of knowledge and the spread of philanthropy.

Humboldt's endeavours were neither directed to the attainment of office, nor the acquisition of honours, still less were they aimed at either riches or pleasure. The motive by which he was influenced in the pursuit of his studies was not so much a preference for any particular branch of scientific investigation, as a powerful impulse early experienced and dominant throughout life towards the investigation of the laws regulating organic and inorganic nature, and the discovery of 'the bond by which the unity of nature is maintained, the parts woven into a whole, and a mutual dependence observed throughout.'

The brilliant results he early achieved are not mainly to be ascribed to the completeness of his labours and the accuracy of his observations, nor to the method of their prosecution, whether mathematical, analytic, or systematic, but rather to the grasp of thought by which he was able to combine all these methods and assimilate all former theories with those of his own time.

It was these noble endowments that inspired that sense of superiority which developed in him so early a maturity, and enabled him when still a student to associate as an equal with the most distinguished men in science or art, in politics or social life.

At the close of life, Humboldt has been likened to a fountain giving forth streams of universal knowledge : he may still more be said in the days of his early manhood to have resembled a luminous centre, diffusing in every direction a flood of intellectual light, and exciting everywhere a reciprocal activity. It was impossible for him to pursue any new method of observation, or investigate any newly-discovered truth, without at once calling upon his friends and fellow-labourers to accompany him along the untrodden path.

Highly as he was gifted with mental power, he was not less endowed with moral excellence. He possessed a capacity for friendship amounting even to enthusiasm, in the exercise of which he shrank from no sacrifice, and drew around him the noblest and the best of his time ; he was distinguished by nobility of mind, a genuine devotion to all that is good and true, a willingness to acknowledge every service rendered to him, an unalienable kindness of heart, universal affability, a courteous manner, and a freedom from all unworthy motives.

With a restless industry and an inexhaustible capacity for work, Alexander von Humboldt united an ideal, an almost poetic enthusiasm. Hence his motto :—

' *Man must ever strive after all that is good and great !* '

II.

ALEXANDER VON HUMBOLDT:

TRAVELS IN AMERICA AND ASIA.

BY

JULIUS LÖWENBERG.

TRAVELS IN AMERICA.

'Es war, als wäre eine neue Sonne voll Licht und Wärme im Westen über de
Neuen Welt emporgestiegen, um auf die Alte Welt wohlthätig zurückzustrahlen.'
CARL RITTER (*Speech at the Humboldt Festival*, August 5, 1844).

CHAPTER I.

PRELIMINARY REMARKS.

Extent of the Spanish Colonies in America—Principles of Government—
Results of previous Explorations—New Direction given to Physical
Science—The Objective Character of Humboldt's Descriptions of Nature
—Their Biographical Value.

BEFORE we can fully enter into the significance of Alexander
von Humboldt's travels in America, we must revert in imagination from these days of easy communication—of steamers,
railways, and telegraphs—to those days of tedious, insecure, and
infrequent transport, when nations held themselves in distrust-
ful isolation.

The Spanish colonies in America, at the close of the last
century, extended without interruption from 38° North latitude
to 42° South latitude—from the most northern point of California to beyond even the most southern part of Chili—an
immense territory, measuring from North to South above
5,500 miles. These possessions included a large district of
North America, now forming the States of Florida, Louisiana,
Texas, Mexico, and California, the whole of Central America,
most of the West India Islands, and the whole of South
America, with the exception of Brazil, Patagonia, and Tierra
del Fuego. This vast extent of country, larger than had ever
before been united under one rule, was merely a colony or
dependency.

Spain, however, had not a conception of the true advantages to be derived from colonial possessions, or how, by securing their prosperity, they might be made a source of wealth to the mother country, and through her to the rest of the world. The government consisted of the most absolute military rule, for the support of which, strange to say, an army of 2,000 men sufficed. Commerce was carried on by a system of monopolies of a most oppressive character. The whole export and import trade was restricted to a limited number of markets, and conducted by a half-yearly convoy of vessels homeward and outward bound, plying between Cadiz and Seville, and two or three ports in the New World. A Spaniard discovered in any commercial relationship with foreigners was condemned by law to a confiscation of his property, if not to a forfeiture of his life; while he who was detected collecting statistical data, or publishing any information concerning the government, was subjected to perpetual imprisonment. To all foreigners the colonies were absolutely closed. During three centuries it would be scarcely possible to name more than six expeditions undertaken in the interests of science to any part of the Spanish colonies—either by the Spaniards themselves, or by foreigners under permission of the Spanish Government. These expeditions had been undertaken mainly for the purpose of determining by astronomical observation the position of certain places, as a means for the better construction of charts of the coast; at most the travellers returned with some dried specimens of new plants for the botanist, and some prepared skins of birds and animals as a contribution to the museums. Of this nature were the travels of Francisco Dominguez in 1577, of Feuillée in 1705, and of Frezier in 1712. Even the valuable trigonometric labours of the French academicians La Condamine, Bouguer, and Godin, the former of whom explored the valley of the Amazon, the results obtained in Quito and Peru by the Spanish geometricians Jorge Juan and Antonio Ulloa, and the more recent labours of Azara in the territory of La Plata, contributed but little towards a more extended knowledge of South America. If to this list be added the unfortunate expedition of Solano in 1754 to the Upper Orinoco and the Rio Meta, from which only

13 individuals out of 325 returned alive, the unproductive wanderings of Requena to the Rio Napo and the Amazon, and Löffling's botanical expedition to the coast of Cumana in 1751, we shall have enumerated nearly every effort that had been made before Humboldt's time towards the scientific exploration of South America, while the published results were represented by such works as Pater Gili's fabulous narrative 'Orenoco illustrato,' Gaulin's 'Historia corographica de la Nueva Andalusia,' and Dobritzhofer's 'Geschichte der Abipomer.'

If further proof were needed of this distrust and suspicion, we have only to call to mind that in the year in which Humboldt was born, the English astronomers, desirous of observing the transit of Venus, were refused a landing on the coast of California, and that Malaspina was condemned to perpetual imprisonment in return for the services he had rendered to his Government.

In contemplating Humboldt in the light of a traveller, it is important to notice the comprehensive view of nature that furnished in him the motive for travel, and led him to become the originator of a new epoch in scientific exploration. While earlier travellers seem merely to have been actuated by simple curiosity, and to have regarded with equal interest everything that came before them—accumulating in a desultory manner a large amount of heterogeneous information with which they interwove minute details of personal adventure—Humboldt carefully avoided all reference to matters of personal detail, and directed his attention mainly to the fundamental structure of the earth's surface, to the intimate relationship between all natural phenomena, and to the existence of a connection between regions the most widely separated, thereby proving the unity of nature. He was never satisfied describing merely in a conventional manner the countries he visited; he preferred to collect data for the more complete development of a science which before his time had been variously termed Physics, the Theory of the Earth, or Physical Geography, but of which till then the outlines had been but slightly sketched. In 1796, he wrote to Pictet :—' I have been drawing up a scheme for a universal science' (p. 197). The connection between facts already observed possessed a higher interest for him than the search after isolated new facts; the discovery of a new species interested him far less than the observation of the geographical connection between animal and vegetable life, whence a comprehensive view might be obtained not merely as to the distribution of these forms of life over the surface of the globe, but also as to the various plains of elevation within which their existence is limited.

For this purpose it was essential that no detail should be neglected, and no circumstance thought too trivial to be recorded. For ‘little service can be rendered to the advancement of science when the observer is too deeply engrossed with the grandeur of general ideas to stoop to the consideration of individual facts.’ By the collection of a variety of forms and the observation of different phenomena, to discover the laws governing their mutual relationship, and to combine the whole into one grand unity—this was the task in physical science which Humboldt set himself to accomplish. In view of this object, it seemed to him that he would best attain his purpose by undertaking a journey on land, whereby he could penetrate into the interior of the continent, than by prosecuting his researches merely in a lengthened voyage by sea, when little more than the coast of various countries could be within his reach.

In the expedition undertaken by Humboldt everything combined to render his investigations of the highest value; for not only was he led by his good genius into those portions of our globe where Nature offers to the student of comparative science her grandest productions and most impressive phenomena in almost prodigal profusion, but by a long course of preparation he himself had been equipped for such a journey as no previous traveller had been. By the most diligent application to various studies he had made himself master of all the scientific knowledge of his time, and at his own cost had furnished himself with the best nautical, astronomical, and scientific instruments, in the use of which he was well versed.

The published results of his travels in America surpassed, as is well known, the boldest anticipations; they filled seventeen folio, nine quarto, and seven octavo volumes. It is to be regretted, however, that the chronological order of the journey has not been preserved, and that in some places his own observations have not been recorded in a manner sufficiently distinctive as to enable the reader at once to recognise the investigations that are to be ascribed to Humboldt and to him alone. 'It was one of the essential qualities of his being, which his vast learning and extensive reading greatly contributed to strengthen, that there was no subject engaging his attention in which he did not seek to make himself master of all the information that could be obtained concerning it; and as everything remained present to his astounding memory, he readily discovered links of connection between kindred subjects, points of comparison between the labours of others and the results he had himself obtained, proofs and evidences in favour of his new theories, and supplementary data upon those points which he had either been unable as yet to explain, or which, from the mass of observations needing to be recorded, had been hitherto set aside as of least value, and were now found worthy of greater prominence. His love of completeness often led him to introduce the results obtained by the labours of others into his own observations, from which they were not separated with sufficient clearness for the reader always to discover for which he was alone indebted to Humboldt.'[1]

The statement here made by Encke in reference to Humboldt's astronomical determination of places is equally applicable to other branches of science; it must not, however, be deduced from these remarks that Humboldt laid claim to the labours of others. No one was in this respect more considerate and just, or more forgetful of self, no one was more emphatic in declining the honour of scientific achievements falsely attributed to him. The results of his own scientific labours would have stood in a still clearer light had they not been mixed up with the labours of others. It is true that science suffers no loss by this, but the compilation of his biography is rendered more difficult. This peculiarity of incorporating the labours of others is very conspicuous in the 'Personal Narrative,' a work which supplies

[1] Encke, 'Alexander von Humboldt's astronomische Ortbestimmungen in den Monatsberichten der Berliner Akademie der Wissenschaften,' October 1859, p. 639.

the best and most original source of information concerning his travels in America ; but a subject of greater regret is the incomplete nature of the narrative, as the work includes but a third of his wanderings in the New Continent. The omissions, unfortunately, can be but sparingly supplied from his journals; for, as he expresses himself on one occasion in his diary, ' when surrounded by the grand beauties of nature, and intently occupied with the natural phenomena displayed at every step, one is little disposed to note down in a journal merely personal adventures or the trivial occurrences of every-day life.' Though Humboldt kept a brief journal when actually travelling, either by land or water, in which to keep a record of the impressions of the moment when visiting an interesting locality, such as the summit of a volcano or other remarkable mountain, yet this journal, he remarks, was continually interrupted whenever he entered a town, or engaged in any absorbing scientific work, which rendered the keeping of a diary a matter of subordinate interest.[1]

The remark attributed to Humboldt that, ' the biography of a man of learning is to be found in his works,'[2] was no doubt an expression he made use of for the purpose of allaying the importunity with which he was frequently assailed for biographical material by industrious bookmakers.

We are fortunately able in this section, as in the preceding one, to avail ourselves of Humboldt's letters and portions of his journals, in which not only the details of his personal history and the scenes around him are portrayed with the vividness of life pictures, but the impressions of the moment are given with all the warmth and freshness of a first sensation.

It does not enter into the plan of this work to give an abstract of the ' Travels in Equatorial Regions,' or discuss the value of the scientific work accomplished during that journey ; our aim is rather to bring into greater prominence than has ever yet been done—the personal history of Humboldt.

---

[1] ' Reise in die Aequinoctialgegenden des Neuen Continents,' vol. i. p. 33 (Tübingen, 1815).

[2] Klencke, ' Alexander von Humboldt, ein biographisches Denkmal,' Introduction, p. 4.

CHAPTER II.

FROM CORUNNA TO PUERTO CABELLO.

Landing at Teneriffe—The Peak—The Dragon-tree—Multiplicity of Phenomena—Landing and first Sojourn at Cumana—First Impressions
and preliminary Arrangements—Scientific Labours—Visit to Caripe
and Caripana, to the Mission Stations, and to the Caves of Guacharo
—First Earthquake and Meteor Shower—Visit to Caracas and Puerto
Cabello.

On the morning of June 5, 1799, the 'Pizarro' weighed
anchor in the harbour of Corunna, and at two in the afternoon
she was under sail.   'Our gaze was fixed,' relates Humboldt,
'upon the castle of St. Antonio, where the unfortunate Malaspina
was then languishing as a prisoner of state.[1]   At the moment
of leaving Europe to visit lands which had been explored with
so much devotion by this distinguished traveller, I could have
wished for a theme less sad on which to occupy my thoughts.'
    The ship's course soon brought her to that expanse of ocean
which, according to the poetic conceptions of the ancients,
bathed the shores of the Islands of the Blessed, where the
voyager, canopied under the blue vault of heaven, glided
peacefully through the 'Sea of the Ladies'[2] along those equatorial currents glowing as if on fire with countless medusæ.

---

[1] Don Alexander Marchese de Malaspina, Commodore in the Spanish
navy, was appointed in 1789 to the command of a fleet fitted out expressly
for purposes of discovery. After a very careful survey of the west coast of
North America, he sailed in search of the North-west Passage, an expedition which proved unsuccessful. On his return in 1795, he was seized on
suspicion of political intrigue, and thrown into prison, where he is supposed
to have died. (Von Zach, 'Monatl. Corresp.' vol. ii. pp. 390, 564.)
[2] [El Golfo de las Damas, a name given by the Spaniards to the part of
the Atlantic Ocean between the Canaries and America—under the idea that
even ladies could muster courage to navigate, it, since the passage could
safely be effected in an open boat.]

Truly a region as deeply interesting as any explorer could desire to visit.

Humboldt's first letter, addressed to his brother, gives evidence of the agreeable impressions he had received :—

> 'Puerto Orotava, at the foot of the Peak of Teneriffe:
> 'June 20, 1799.

'I am quite in a state of ecstacy at finding myself at length on African soil, surrounded by cocoa-nut palms and bananas. By aid of a strong breeze from the north-west, and meeting fortunately with but few vessels to detain us, we reached the coast of Morocco in ten days ; on the 17th of June we landed at Graciosa, and by the 19th we were in the harbour of Santa Cruz at Teneriffe.

'The company on board was very agreeable ; I felt particularly attracted to a young man from the Canary Islands, Don Francesco Salcedo, who possessed the confiding manner and lively intelligence peculiar to the inhabitants of these favoured islands. I have made a great many observations, especially in astronomy and chemistry—on the purity of the air, the temperature of the ocean, &c. The nights were magnificent ; in this clear, tranquil atmosphere it was quite possible to read the sextant in the brilliant moonlight, and then the southern constellations, Lupus and the Centaur ! What splendid nights ! We caught a specimen of that very rare animal the dagyza, near the spot where it was first discovered by Banks ; we also fished up from a depth of 50 fathoms a new kind of plant, green and with vine-shaped leaves, but not a fucus. The sea was phosphorescent every evening. As we neared Madeira some birds made their appearance, and accompanied us throughout the day.

'At Graciosa we landed to ascertain if there were any English frigates cruising off Teneriffe, and on being assured there were none, we continued our voyage, fortunately arriving without even sighting a man of war. How we contrived to do so is quite incomprehensible, for an hour after our arrival no fewer than six English frigates appeared off the harbour. We have no occasion to apprehend anything further from them until we reach the West Indies. My health is excellent, and I get on capitally with Bonpland.   We have already experienced at Teneriffe something of the hospitable feeling that reigns in colonial settlements. We are invited everywhere, with or without introduction, merely for the sake of hearing the latest news from Europe ; and the royal passport works wonders.   At Santa Cruz we stayed with General Armiaga ; and here at Puerto Orotava we are the guests of an English merchant, John Collegan, at whose house Cook, Banks, and Lord Macartney had formerly received hospitable entertainment.   I was greatly surprised by the cultivated tastes and ease of manner displayed by the ladies of these households.'

He further proceeds :—

'June 23: Evening.

' I returned last night from an excursion up the Peak. What an amazing scene ! what a gratification !   We descended some way into the crater, perhaps farther than any previous scientific traveller.   No one except Borda and Mason has been even beyond the last cone.   There is little danger in the ascent, only fatigue from the trying effects of heat and cold ; for the sulphurous vapour in the crater burnt holes in our clothes while our hands were numb in a temperature of 36°.

' What a remarkable spectacle was presented to us at this height of 11,500 feet !   The dark blue vault of heaven overhead ; former streams of lava at our feet ; on either side this scene of devastation ; three square miles of pumice-stone, bordered by groves of laurel, beyond which vineyards interspersed with bananas stretched down to the sea ; pretty villages dotted along the coast, the ocean with all the seven islands, among which Palma and Grand Canary are distinguished by lofty volcanoes spread out before us like a map.

' The crater into which we descended emits only sulphurous vapour ; the temperature of the ground is 190°.   The lava streams break out at the sides of the mountain, where small craters are formed similar to those by which two years ago the whole island was illuminated.   On that occasion, a subterraneous noise was heard for two months like the firing of cannon, and stones of the size of a house were hurled into the air to the height of 4,000 feet.   I have made some important mineralogical observations here.   The Peak is composed of basalt, upon which lie beds of porphyry and obsidian porphyry. Fire and water rage below the surface; I noticed steam escaping from every point. Almost all the lava streams appeared to be melted basalt. The pumice-stone is evidently produced from obsidian porphyry; I have specimens where the two conditions are united in one piece.

'At the foot of a lava stream in front of the crater we spent the night in the open air, among the group of stones known by the name of " La Estancia de los Ingleses," at a height of 7,875 feet above the sea. At two in the morning we were already on our way towards the last cone. The heavens were bright with stars, and the moon shone with a gentle radiance; but this calm was soon to be disturbed. The storm raged violently round the summit, we were obliged to cling fast to the edge of the crater. The wind rushed through the rifts with a noise like thunder, while a veil of cloud separated us from the world below. We climbed up the cone as it stood out above the mist, isolated as a ship upon the sea. The sudden change from the beauty of a bright moonlight night to the darkness and desolation of the storm and cloud produced a very impressive effect.

'Postscript.—In the district of Orotava there is a dragon tree (*Dracœna draco*) measuring forty-five feet in circumference. Four centuries ago, in the days of the Guanches,[1] the girth was as great as it is now.[2] I could almost weep at the prospect of leaving this place; I should be quite happy to settle here, and yet I am scarcely out of Europe. Could you but see these luxuriant fields, these forests of laurel, the growth of a thousand years, these vines and these roses! They actually fatten the pigs here upon apricots. The roads are lined with camelias.

'We are to sail on the 25th (June).'

---

[1] [The name of the aboriginal inhabitants of the Canary Islands.]

[2] This dragon-tree, which is fully described by Humboldt in the 'Aspects of Nature,' lost half the crown during a hurricane in the year 1819. The lower part of the hollow stem was then supported by masonry, over which bignonias and other creeping plants were trained, but the upper portion became by degrees so rotten that it was finally broken off in the storm of January 2, 1868.

Much additional detail is contained in a letter to Delamé-therie dated 30th Messidor, VII. (July 18, 1799), three days after landing at Cumana, from which we find that Humboldt had already commenced his researches upon the currents, temperature, and phosphorescence of the ocean, together with other kindred investigations, which he prosecuted with the view of aiding in the solution of some of the grand problems in science.

After a voyage of nineteen days Humboldt came in sight, on July 13, of the lofty coasts of Tobago and Trinidad, and on the 16th inst. the ship was safely anchored in the harbour of Cumana. The voyage throughout had indeed been a prosperous one to him, for not only had he entirely escaped seasickness, but he had been preserved from a contagious fever of a typhoid character which had broken out on board. In consequence of this sickness among the crew, it was thought advisable to change the original direction of the ship's course, and instead of making for Cuba or the coast of Mexico, the vessel was run into the nearest harbour on the north coast of South America and anchored at Cumana.

Humboldt thus wrote to his brother the day he landed : [1]—

'Cumana,[2] South America: July 16, 1799.

'The same good fortune, my dear brother, which enabled us to run into Teneriffe in the face of the English, has continued with us to the end of our voyage. I have worked hard all the way, especially with astronomical observations. We intend remaining some months in the Caracas, where we are truly in a wonder-land of fertility and luxuriance ; we have on all sides of us extraordinary plants, electric eels, tigers, armadillos, apes, and parrots, besides numbers of genuine Indians—half wild— a very handsome and interesting race.

'Cumana, on account of its proximity to the Snowy Mountains, is one of the coolest and most healthy places in South

---

[1] Humboldt's letters are mostly written in the style of a journal: the date merely denotes when the letter was begun or finished.

[2] Cumana, first colonised in 1521, was very early an important place of trade. The name, however, is not down in the large chart of the Gulf of Mexico by Martin Suares ; it is also wanting in Arrowsmith's Atlas, published in 1804, in which La Guayra and Caracas are also omitted.

America, with a climate like that of Mexico ; and although it has been visited by Jacquin, the interior of the country is still one of the least known portions of the globe. Besides the charm of being surrounded by nature in a new aspect (for since yesterday we have not met with a single specimen of a plant or animal common to Europe), we have been mainly influenced in our determination to remain at Cumana, two days' sail from Caracas, by the news that just now English ships of war are cruising in the neighbourhood. It is a voyage of eight or ten days to Havana, and as all European convoys touch here, to say nothing of private merchantmen, we shall not lack opportunities of visiting Cuba. We hear too, that the heat there is at its worst just in September and October. We shall spend these months, therefore, in the cooler and healthier atmosphere of this place ; it is quite possible, even at this temperature, to sleep out at nights in the open air.

'An old commissary of the navy, who has lived many years in Paris, St. Domingo, and the Philippines, is keeping house here, with help of two negroes and a negress. We have hired a very nice new house for twenty piastres a month, and obtained the services of two negresses, one as a cook. There is no lack of food here, but unfortunately there is nothing to be had in the shape of flour, bread, or biscuit. The town is still half buried in ruins ; for the great earthquake of 1797, by which Quito was destroyed, overthrew a considerable part of Cumana. This town lies in a bay as beautiful as that of Toulon, and is situated at the foot of a range of thickly wooded mountains, that rise in the form of an amphitheatre to a height of from 5,000 to 8,000 feet. All the houses are built of white cinchona and satin-wood. Along the banks of the small river (Rio de Cumana), reminding one of the Saale at Jena, stand seven monasteries surrounded by plantations which have all the appearance of English gardens. Outside the town live the copper-coloured Indians, of whom the men nearly all go naked; their huts are made of bamboo cane covered with the leaves of the cocoa-nut palm. On entering one of them, I found the mother and children employing as seats the stems of coral that had been washed up on the shore ; they had each a cocoa-nut shell before them, to serve the purpose of a plate, from which

they were eating fish.   The plantations are all open for everyone to go in and out as they please ; the inhabitants here are
so well disposed that in most of the houses the doors are left
open through the night.   There are more Indians here than
negroes.

'What magnificent vegetation!   Cocoa-nut palms from fifty
to sixty feet in height ; *Poinciana pulcherrima*, with pyramidal
bunches of flowers a foot high, and of a splendid bright red
colour ; bananas, and a host of trees with enormous leaves and
sweet-smelling flowers as large as one's hand, all of which are
entirely new.   You may imagine how completely unexplored
this country is, when I tell you that a new genus, first
described by Mutis (see "Cavanilles Icones," vol. iv.) only two
years ago, is growing here as a wide-spreading tree, sixty feet
high.   We had the good fortune to meet with a specimen of
this magnificent plant yesterday ; the stamens are an inch long.
What a vast number of smaller plants, therefore, must have as
yet escaped observation!   How brilliant the plumage of the
birds and the colours of the fishes!—even the crabs are sky-blue
and gold!

'Hitherto we have been running about like a couple of fools;
for the first three days we could settle to nothing, as we were
always leaving one object to lay hold of another.   Bonpland
declares he should lose his senses if this state of ecstacy were to
continue.   But far more thrilling than the contemplation of
these marvellous objects individually is the overpowering sight
of the whole mass of such magnificent vegetation, the elegant
luxuriance of which is so exhilarating and, at the same time, so
soothing.   I feel sure that I shall be very happy here, and that
the agreeable impressions I am now receiving will often cheer
me by their inspiring influence.

'I do not yet know how long I shall remain here, but I think
that I shall probably spend three months between here and
Caracas—possibly, however, a still longer time.   We must
make the best use we can of what lies nearest to us.   Next
month, when the winter here ceases, and the hottest and laziest
part of the year sets in, I shall probably undertake a journey to
the mouth of the Orinoco, known as the Bocca del Drago
(Dragon's Mouth) : there is a good and safe road all the way.

We sailed past the mouth of this river, and were amazed at the spectacle presented by these mighty waters. On the night of the 4th of July I saw the Southern Cross for the first time in all its splendour.

'Postscript.—You need not feel alarmed for me on account of the heat of these regions. I have nearly been a month now within the tropics, and I feel no inconvenience from the temperature. The thermometer usually stands at about 78° or 82°—not higher; it was as low as 66° of an evening off the coast of Cayenne, when I felt quite frozen. Thus, you see, the heat is by no means extreme in this neighbourhood.

'I hope you will follow my voyage upon the map. On the 5th of June we sailed from Corunna, on the 17th we touched at Graciosa, from the 19th to the 25th we spent at Teneriffe: on leaving there we experienced strong gales from the east, with rain, and on the 5th and 6th of July were sailing off the coast of Brazil; on the 14th we passed between Tobago and Grenada; on the 15th we sailed through the channel separating Margarita from the mainland of South America, and on the morning of the 16th we anchored in the harbour of Cumana.'

Humboldt's first visit to Cumana lasted from July 16 to November 28, 1799. His letters written during this period are full of detail on a variety of subjects, and we have therefore thought it desirable to introduce here short extracts from some of them, commencing with the following letter to Von Zach:—

'Cumana: September 1, 1799.

'A Spanish brigantine from Cadiz, which has anchored here this morning, gives me the welcome opportunity of sending you a token of my existence, as well as some account of my scientific labours. I am compelled to write in some haste, as I am on the eve of setting off on an expedition to the interior, and intend starting to-morrow for the mountains of Caripe and Carapana, where only four days ago there were eleven very severe shocks of earthquake. Thence I shall proceed to the interior of Paria, among the mission stations of the Capuchins, a district abounding with objects of the highest possible interest to the scientific investigator—plants, mountains, rocks, and above all a new order of human kind, the friendly Indians and the tribes of the Caribs.

'I have now been for a couple of months in this new quarter of the globe—the Terra Firma of South America, and together with my travelling companion, Bonpland, who has proved himself an indefatigable worker, I am in the enjoyment of the best possible health. We have already collected a great number of plants, insects, and shell-fish : I have made several drawings, and been much occupied with experiments for ascertaining the composition of the atmosphere.

'Just now I am trying to investigate the problem why the refraction of light should be less in this zone of high temperature than with us. This effect cannot be ascribed solely to the heat. Much, no doubt, is due to hygrometric conditions, for I am of opinion that the greater damp of this region of the globe diminishes refraction.

'During the voyage I was much occupied in registering the temperature of the ocean and in ascertaining its specific gravity by means of one of Dollond's excellent balances. The idea suggested by Franklin and Jonathan Williams of taking soundings with a thermometer attached to the lead is as ingenious as it is happy in conception, and will eventually become of great importance to navigation. I undertook several experiments on board ship with Hadley's mirror sextant. I have an eight-inch one by Ramsden with a silvered arc divided to 20 seconds. I have besides a two-inch sextant by Troughton, which I call my snuff-box sextant : it is really incredible what may be accomplished with this little instrument. A single set of observations taken by it of the altitude of the sun when passing the prime vertical will give the correct time within two or three seconds. If this accuracy be purely accidental, it must at least be admitted that the chance happens very frequently.

'I systematically kept an astronomical journal, and whenever the weather and the calmness of the ocean permitted I made observations of the latitude and longitude of the ship or of the places where we landed ; I also observed the inclination

of the magnetic needle, which, by means of Borda's new
instrument, may be found to within 20 minutes. I enclose you
the results I obtained with it during the voyage. . . .

'The chronometer I have with me is one of Louis Berthoud's,
No. 27 ; it has been frequently made use of by travellers, and
its excellence is well known to Borda; it has preserved with
me its usual uniform rate.

' I must confess it needs a remarkable stock of patience to
make astronomical observations with sufficient accuracy in
this extreme heat, and to do one's work *con amore* ! You will
however notice that this heat, however oppressive it may be,
does not seem to diminish my activity. . . .

' I wish I could convey to you an idea of the exquisite beauty
of the nights here, or give you any notion of the transparency
of the air ; by the light of Venus I have frequently been able
with a lens to read the vernier of my small sextant ! Venus
really plays the part of a moon here. She presents a large and
luminous surface two degrees in diameter, and displays the
most beautiful colours of the rainbow, even when the air is
quite clear and the sky is perfectly blue. I am persuaded that
the splendour of the heavens is nowhere so remarkable as in
these latitudes, for south of the equator the magnificent
northern constellations soon cease to be visible. Not but what
the constellations of the southern hemisphere have their own
peculiar charm. Sagittarius, the Southern Crown, the Southern Cross, the Southern Triangle, the Altar, are all conspicuous for the brilliancy of their stars, while Centaurus almost
rivals in splendour our own Orion, which I have here seen
at such an altitude as to wring from me an expression of deep
emotion and throw me into a glow of excitement.

' Another very remarkable and surprising phenomenon which
I noticed even the second day of my arrival here is the ebb
and flow in the atmosphere.

' Remember me affectionately to our friend Blumenbach.
Oh how often I think of him as I see around me these marvellous
treasures of nature ! Tell him that the geological formation
here is of a most interesting character. Mountains of mica-
schist, basalt, gypsum, and gneiss. Sulphur and petroleum,
rising with considerable force from very small openings, throw

up, even when under water, streams of air, and are probably the occasion of the frequent earthquakes. The whole of this town is lying in ruins. The great earthquake of Cumana was the signal for the one at Quito in the year 1797, in which 16,000 persons lost their lives, upon which occasion the volcano Tunguragua threw up more hot water and mud (*terre pâteuse*) than lava—a volcano, therefore, by which nature would seem desirous of reconciling the opposing systems of the Neptunists and Vulcanists !

'We are surrounded by tigers and crocodiles (alligators), ever ready to indulge in a meal, and by no means fastidious—a white or a black man being equally regarded by them as a dainty morsel. In size they are not inferior to the African beasts of prey. And what glorious vegetation !—truly colossal organisms. A ceiba,[1] out of which four canoes can be made !

'I am particularly anxious you should tell Counsellor Blumenbach that there is a man living in this province of New Andalusia who has so much milk, that since his wife was unable to suckle her child, he has himself nursed it for the last five months. The milk does not in the slightest degree differ from woman's milk. The he-goats belonging to this old man even give milk.

'Pray receive my letter with indulgence, and regard with lenity my astronomical observations. Remember that astronomy is only a secondary object of my journey, and that I am but a beginner in the science, having only during the last two years accustomed myself to the use of astronomical instruments. It must also be borne in mind that this journey has been undertaken at my own cost, and that an expedition fitted out by a single individual, even though he be wealthy—for his own gratification and instruction—cannot for a moment compete with those extensive expeditions organised by royal command and at Government expense, which are able to secure the services of several scientific men for carrying out investigations in every branch of science. I could have wished, for the sake of accomplishing something really great in astronomy and geography, that our friend Burckhardt had joined our expedition,

---

[1] [*Bombax ceiba*, a large tree of the tropics, called the cotton-tree.]

only he would have been obliged to provide himself with larger
and better instruments than any I possess.

'Do not be surprised if my letters are full of repetitions,
for it is calculated that in this out-of-the-way region, of every
four letters despatched to Europe, three will be lost; therefore
if one wishes to apprise one's friends of any particular fact it is
necessary to repeat it frequently.'

'Cumana: November 17, 1799.

'I have re-opened my letter, for I did not venture to trust it
to the brigantine from Cadiz when I found that a Spanish
mail was expected. We have been looking for it in vain for
the last two months, but it has at length arrived, and I hasten
to give you some additional intelligence. I have just returned
from an expedition into the interior of Paria; it was a very
arduous undertaking, but one in every point of view fraught
with the highest interest. We visited the high Cordilleras of
Tumiriquiri, Cocollar, and Guanaguana, which are inhabited by
the Chaymas and Guaraunos Indians. We spent some glorious
days very happily at the Capuchin monastery at Caripe, the
central mission station. We also visited the celebrated cave of
Guacharo, which is the resort of thousands of nocturnal birds
—a new species of Caprimulgus, goat-sucker. Nothing can
equal the majestic entrance to this cave, shadowed by palms,
pothos, ipomœa, &c.

'During our stay in this province we have dried more than
1,600 plants and described about 600 new varieties, including
some unknown cryptogamia; we have also collected the most
beautiful shell-fish and insects. I have made more than 60
drawings of plants, besides illustrating the comparative anatomy
of various shell-fish. We took Berthoud's chronometer and
Ramsden's and Troughton's sextants with us, across the
Guarapichi. I determined the latitude and longitude of more
than fifteen places—observations which may be of use at some
future time in affording fixed points for the construction of a
map of the interior. By means of the barometer, I measured
the height of the Cordilleras. The loftiest peak is limestone,
and does not exceed in elevation 6,405 feet. Farther west,
towards Avila, there are mountains nearly 10,500 feet in

height, which connect these Cordilleras with those of Santa Martha and Quito.

' The oppressive and almost unbearable heat did not prevent me from observing the solar eclipse of the 28th of October. On the same day I took altitudes of the sun with Bird's quadrant; I give the results below, and I should be glad if you will kindly look them through and correct them. . . .

' In making these observations, my face was so severely burnt that I was obliged to keep my bed for two days and apply medicinal remedies. The reflection from the white limestone is distressing to the eyes and liable to injure the sight. The metal of an instrument exposed to the power of the sun's rays is heated to a temperature of 124°.

' If you have looked into my last work on "Subterranean Meteorology," you will have seen that the temperature of the interior of the earth is a problem of the highest interest. Here, under 10° of latitude, the temperature at the depth of 371 fathoms is 66°. My meteorological instruments have been compared with those of the National Observatory at Paris, and corrected according to that standard. At the sea level the thermometer in the shade, even during the hottest part of the year, does not rise above 91°; the temperature keeps very regular, rarely varying more than from 75° to 82°. Early in the afternoon, when the heat is at its maximum, a thunderstorm comes on, and is succeeded by a display of lightning, which lasts for nine hours. Truly a volcanic climate !

' On the 4th of November we experienced a severe shock of earthquake; no serious damage, fortunately, occurred. I was surprised to notice that during the earthquake the dip of the magnetic needle was reduced by the amount of 1·1°. The earthquake was followed by a succession of slighter shocks, and on the 12th of November we had a regular display of fireworks. From two o'clock till five in the morning large fire-balls passed without intermission across the sky, and kept discharging sprays of fire two degrees in diameter. The eastern portion of the province of New Andalusia is covered with numerous small volcanoes, emitting warm water, sulphur, sulphuretted hydrogen, and petroleum.

' I am to sail to-morrow for La Guayra, and intend remaining

at Caracas till January. Thence I shall penetrate into the
interior of the country, to investigate the Apure, Rio Negro,
and Orinoco, as far as Angostura, whence I shall return here to
embark for Havana.' . . .

During the earthquake above mentioned Humboldt experienced for the first time those exciting sensations invariably
produced by this appalling natural phenomenon ; he remarks,
concerning it, that by habit 'man becomes as much accustomed
to the trembling of the ground as the sailor to the staggering
of a ship amid the waves.' The phenomenon witnessed by
Humboldt on the night of November 11 and 12 was the great
shower of meteors since become so noted in the annals of
science.

On November 18, the travellers set sail from Cumana in a
coasting vessel; they separated at New Barcelona, whence
Bonpland proceeded by land, for the sake of enriching his
botanical collections, while Humboldt remained with the ship,
to keep guard over the instruments, and after a voyage of four
days he landed, on November 21, at La Guayra, the port of the
inland city of Caracas, the capital of the province and the
residence of the Governor General—a city which has since
attained a melancholy notoriety from the destructive earthquake
by which it was overwhelmed in 1812.

As the rainy season had commenced, the stay at Caracas was
prolonged for two months and a half—from November 21, 1799,
till February 7, 1800. Humboldt's journal is full of bitter complaints of the persistent bad weather. 'For twenty-seven nights
we have kept watch in hopes of being able to observe the
eclipses of Jupiter's satellites, but our labour has been all in
vain.' These watches were prosecuted even through the night
of January 1, spent at the foot of the Sina, 'notwithstanding
the prospect on the morrow of undertaking a fatiguing journey
on foot of nineteen hours ;' but their self-sacrifice on this occasion met with no better reward. This pedestrian expedition
was commenced on January 2. They climbed the Silla de
Caracas, the summit of which, rising to the height of 8,100
feet, had never before been ascended, and examined the structure
of the mountain. After passing through the fruitful valleys of
Aragua and Tui, rich with cocoa plantations, they visited the

mountains of Los Tequos, the hot springs of Mariare and Trinchera, and the northern shore of the romantic lake of Valencia, where they made the discovery of the cow-tree, so called from its yielding milk : thence they journeyed through New Valencia and over the mountain range of Higuerote, in the midst of a picturesque country abounding with glorious vegetation, to Puerto Cabello, 'a wonderful and magnificent harbour,' the most western point of their wanderings.

Shortly before leaving Caracas, Humboldt wrote, on February 3, 1800, in the following grateful strain to Baron von Forell : [1]—

' I need scarcely assure you how often I am reminded to whom I am indebted for the privilege of being here, nor to whom the thanks of the public will be due for the results, whatever may be their value, which may accrue from my expedition to the West Indies.   While traversing the vast ocean separating Europe from the continents washed by the Pacific, while exploring the wild banks of the Guarapichi, or penetrating the primeval forests of the valleys of the Tumiriquiri, the thought of my kind friend has been ever before me. Gratitude is the first law of our being, and the natural philosopher, while studying the laws of Nature, should yield them unhesitating obedience.' . . .

He then continues :—' The farther we pressed into the interior, in visiting the mission stations among the Chaymas, the more heartily we congratulated ourselves that we had not gone on to Havana. It would have been very tantalising to have been so near the coast of Paria, the wonders of the Orinoco, the immense chain of Cordilleras stretching from Quito eastward to Carupana, to say nothing of the majestic vegetation which Jacquin has so graphically portrayed in his works, and yet not to have been able to devote to the examination of these wonders of Nature more than the three days during which the mail stopped at Cumana. . . . Had we first landed at Havana or Caracas, we should have found ourselves surrounded by evidences of European civilisation ; but in the Gulf of Cariaco, where the wild Indians of the lagunes (Guaraunos del arco) are within

---

[1] De la Roquette, 'Humboldt, Correspond. scientif. et littér.' vol. i. p. 89.

fifty miles, Nature reigns with undisputed sway. Neither the tigers, nor the crocodiles, nor even the monkeys, show any fear of man; the most valuable trees, such as the guaiacum, the caoutchouc, and the logwoods of Brazil and Campechy, besides a vast number of other varieties, grow down even to the coast, and by the interlacement of their branches often present an impenetrable barrier. The air is peopled by rare birds of brilliant plumage. Every object declares the grandeur, the power, the tenderness of Nature—from the boa constrictor, which can swallow a horse, down to the tiny humming-bird balancing itself on the chalice of a flower. . . . We are already sufficiently familiar with Spanish to take part in a conversation, and my admiration has been excited by the loyalty and probity (*hombria de bien*) evinced by the inhabitants of these distant countries—qualities for which the Spanish nation has in all ages been distinguished. It is true that as a people they have not kept pace with the development of enlightened views, but on the other hand the purity of their moral character has been perhaps more perfectly maintained. A hundred miles from the coast, among the mountains of Guanaguana, the inhabitants were in such a state of ignorance that they had never even heard of Prussia; yet I should find it difficult to convey to you an adequate notion of the very cordial hospitality we received from them. After only spending four days in their society, they took leave of us as if we had been living with them for a lifetime. The longer I am here, the more am I delighted with the Spanish colonies; and if I have the good fortune to live to return to Europe I shall always recall with pleasure and interest the days I have spent here.' . . .

After a short stay at Puerto Cabello, the travellers took their departure towards the beginning of March, to pursue their important expedition to the Orinoco and the investigation of the interior of the continent.

CHAPTER III.

EXPEDITION TO THE ORINOCO.

Outline of the Expedition—Letter to William von Humboldt—Prosperity,
Leisure, and Objects of Study—Bonpland's Ability and Faithfulness—
Nocturnal Scene—Letter to Willdenow—Herbariums—Arrangements
in case of Death—Fraser's Shipwreck—Profusion of Plants; their
Preservation difficult—Hardships—'The Tropics are my Element'—
Courteous Reception and Position of Independence—Bonpland's Share
in the Work—Recollections of Berlin.

THE next subject of investigation which Humboldt had laid
down for himself was the task of tracing out the water system
of the Orinoco, in order to settle the vexed question as to the
existence of any communication between this river and the
Amazon.

To give an outline of this important expedition, the travellers
started on foot from Puerto Cabello at the close of the month
of February, 1800, and crossed New Valencia to the lake of the
same name, whence, after passing along its southern shore, they
turned southward through the Llanos de Caracas and by way of
Calabozo (from March 14 to March 24) to San Fernando de
Apure, which they reached about March 27. After a stay of
a couple of days they proceeded in an easterly direction down
the river Apure, in one of the slight native boats called *pirogues*,
so constructed as to be easily carried over land. They followed
the Apure till its junction with the Orinoco at Cabruta, whence
they ascended the Orinoco past the rapids of Atures and Maypures as far as San Fernando, the confluence of the Atabapo,
where they arrived on April 23. At this point they left the
Orinoco and followed the course of its tributary, the Atabapo,
southward, as far as the junction of the small rivers Temi and
Tuamini, arriving on May 1 at the town of San Antonio de

Javita. Before them lay Monte Pimichin, which forms the watershed between the Orinoco and the Amazon, and is notorious for the serpents with which it is infested. A land journey of three days across this portage, over which the pirogue was carried by the Indians, brought the travellers to the Rio Negro, the course of which they followed still farther south to San Carlos, on the frontier of Brazil. This, the most southern point of their travels, in 2° North latitude, was reached on May 7. Leaving the Rio Negro, a little above San Carlos, at the confluence of the Casiquiare, they followed the course of this river in a north-easterly direction until it joined the Orinoco, thus proving incontestably that a communication existed between the Orinoco and the Amazon. The furthest point to which they followed the Orinoco was Esmeralda, which they reached on May 21, a town situated opposite the mountain of Duida.

On May 23, 1800, the travellers commenced their return journey. From Esmeralda they descended the Orinoco as far as Angostura, the capital of Guiana, visiting on May 31 the Cavern of Atoruipe, the burial-place of the extinct race of the Aturs. At Angostura they remained from June 15 till July 10, when they proceeded on foot northward through the Llanos of Barcelona. Their arrival at Barcelona on July 23 brought to a close this highly interesting journey, during which they had travelled 1725 miles through a wild and uninhabited region, and, among other important results, had obtained the solution of the problem of the bifurcation of the Orinoco, the position of which was accurately determined by astronomical observation.

At Barcelona the travellers again rested for some weeks, and not till September 1 was Humboldt able to reach Cumana and find himself once more under the hospitable roof of Don Vincente Emperan, the governor of that province.

The letters written by Humboldt during this journey are naturally very inferior in matters of detail, as well as in general scientific interest, to the succinct account of his travels published in his works, but they possess the inestimable advantage of reflecting the warmth and freshness of individual feeling excited by the impressions of the moment, and vividly portray the mind of the writer. Of the many letters belonging to this period our limited space will only allow of our selecting two, from which we insert the following extracts :—

To William von Humboldt.

'Cumana, South America: October 17, 1800.

'I cannot repeat to you often enough how happy I feel in this quarter of the globe, where I have already become so completely acclimatised as almost to feel as if I had never lived in Europe.

'Throughout the world, there is perhaps no land where one could live with more enjoyment and in greater peace and security than in the Spanish colonies where I have now been travelling for the last fifteen months. The climate is very healthy ; the heat sets in about nine in the morning, and does not last beyond seven in the evening. The nights and early mornings are cooler than in Europe. The aspect of nature is extremely rich and varied, while everything is on so grand a scale as to convey an impression of majesty. The inhabitants are mild, good-natured, and conversible, and though certainly very careless and ignorant, they are simple-hearted and without pretension.

'I could not possibly have been placed in circumstances more highly favourable for study and for purposes of investigation than those which I now enjoy. I am free from the manifold distractions constantly arising in civilised life from the claims of society, while nature unceasingly offers to my contemplation the newest and most interesting objects of study. The only drawback to this solitude is the want of information as to the progress that scientific discovery is making in Europe, and the loss of all the advantages arising from the interchange of thought.

'The study of the various races of mankind would alone form ample occupation for any observer, for not only are the mixed races a subject of considerable interest, but also the Indian tribes, especially such as are as yet uninfluenced by civilisation. Of the various inhabitants of European descent I am most attracted to the colonists settled in country districts, a class of people who still preserve the simplicity of Spanish manners characteristic of the fifteenth century, and who frequently exhibit traits of humanity and principles of true philosophy which are sometimes sought for in vain among nations regarded as highly civilised.

'It is therefore with great regret that I look forward to leaving these peaceful regions for the richer and more populous districts of other colonies. I shall certainly meet with greater facilities for collecting information, but I shall oftener encounter men who, while the language of the most beautiful philosophy is upon their lips, deny its first principles in their hearts, who while lash in hand they are subjecting their slaves to cruel abuse, yet speak with enthusiasm of the glories of freedom, and at the same moment offer the children of their negroes, when scarcely more than infants, for sale like so many calves. Would not the solitude of a desert be preferable to the company of such philosophers?'

After a detailed account of the expedition to the Orinoco, he continues:—

'My friend Bonpland suffered much more severely from this excursion than I did. On our arrival at Guiana he was attacked by vomiting and fever, which caused me the greatest anxiety. Probably this was the effect of the food, to which we were both unaccustomed. Seeing he made no progress towards recovery in the town (Angostura[1]), I had him conveyed to the country-house of my friend Dr. Felix Farreras, eighteen miles from the Orinoco, in a cooler valley at a somewhat greater elevation. There is no more efficient cure in this tropical climate than change of air.

'I cannot describe to you the anxiety I endured during his illness. I could never have hoped to meet again with a

---

[1] Among the Radowitz collection of autographs in the Royal Library at Berlin is a chart of the Orinoco, sketched by Humboldt, which, though very rough, is executed with great clearness, and is accompanied by the remark:—'My first attempt to employ graphically the astronomical observations I made on the Orinoco and the Casiquiare. I made the sketch at Santo Thome del Angostura in June 1800, while Bonpland was lying dangerously ill of a nervous fever. I give you the chart, as a proof that I can occasionally write with legibility.

'Berlin: 1842. AL. HUMBOLDT.'

friend, so faithful, courageous, and active. Throughout the journey he evinced many astonishing proofs of courage and resolution, for we were necessarily surrounded during the whole time by a variety of dangers, not only from the Indians among whom we travelled, but from the nature of the country, which swarmed with tigers, crocodiles, and serpents. I shall never forget the generous attachment he showed me during a storm which overtook us on the Orinoco on April 6, 1800. Our pirogue was two-thirds full of water, and the Indians already overboard fast swimming to shore, when my noble-hearted friend, who alone stood by me in this emergency, entreated me to follow their example, and offered to swim to shore with me upon his back.

'We were not destined, however, to perish in this wild region, where, within a circle of forty miles, there was no human being to learn our fate or trace the manner of our disappearance. Our position was truly appalling : the shore was distant from us more than a mile, where a number of crocodiles could be discerned lying half out of the water. Even if we had gained the shore against the fury of the waves and the voracity of the crocodiles, we should infallibly have either perished from hunger or been torn in pieces by the tigers, for the woods upon these shores are so dense and so intertwined with lianas as to be absolutely impenetrable. The strongest man, axe in hand, could hardly make his way in twenty days for the distance of a league. The river too is so little frequented that even an Indian canoe scarcely passes oftener than once in two months. At this most momentous and perilous crisis a gust of wind filled the sails of our little vessel and effected in a marvellous manner our deliverance. We only lost a few books and a portion of our food.

'You may imagine our feelings of grateful happiness when, as night approached, we went on shore, and, assembling for our evening meal, found that none of our party were missing. The night was dark and the moon only shone at intervals through the gaps in the clouds as they were driven by the wind across the sky. The monk who formed one of the party addressed himself in prayer to St. Francis and the Holy Virgin. The rest were all lost in thought, filled with emotion, and in gloomy apprehension for the future.

'We were still two days distant to the north (?) from the great cataracts we had to pass, and we had yet nearly 700 miles to accomplish in our pirogue, which, as we had just experienced, was but a frail bark. What an overwhelming anxiety! This state of depression, however, lasted only through the night. The following day was very beautiful, and our spirits partook of the peace and cheerfulness reigning throughout nature. During the morning we met a family of Caribs who had come from the mouth of the Orinoco in quest of turtle eggs, and who had undertaken a journey fraught with so much danger more for pleasure and from a love of the chase than from any necessity. In our intercourse with them we entirely forgot our misadventures.'

To Willdenow.

'Havana: February 21, 1801.

' My dearest friend,—As I am uncertain whether these lines may not suffer the fate of so many others that I have addressed to you from these tropical regions, and never reach their destination, I will confine myself merely to the request I have to make. In contemplating the results of this extensive journey, during which I hope to circumnavigate the globe, I naturally feel exceedingly anxious as to the preservation of my manuscripts and herbariums, especially as my expedition happens at a time when a neutral passport is as little respected as a neutral ship is by the pirates with whom the ocean just now is infested. It is very unlikely, indeed quite improbable, that we should both of us—Bonpland and I—be spared to return alive from our lengthened voyage to the Philippines and round the Cape of Good Hope. In case of such a catastrophe, how grievous would it be to lose the entire results of the expedition!

'To avoid this, we have taken the precaution of making copies of our descriptions of plants; up to the present time they consist of two volumes, and contain 1,400 specimens, many of which are either new or rare species. The original manuscript we retain with us, and we are forwarding a transcript piecemeal, through the French vice-consul, to Bonpland's

brother at La Rochelle. The plants have been sorted into
three divisions, as we have two or three specimens of each variety.
One of these reduced herbariums we shall carry about with us
for purposes of comparison ; a second one, belonging to Bonpland
(for I naturally share everything with him), has been already
despatched to France, while the third I am sending off to-day
to London through Mr. John Fraser, by way of Charleston.
This last herbarium is in two cases, and contains, with grasses
and cryptogamia, 1,600 different species, most of which are from
the unexplored parts of Parime and Guiana, from the district
between the Rio Negro and Brazil which we visited last spring.
By thus dividing our treasures we shall considerably diminish
the risk of loss.

'The idea has occurred to me that as a greater variety of
subjects has come under my investigation during this journey
than could possibly be expected to interest any one reader, it
might be desirable to publish my observations in separate parts.
I would propose that the first part, for instance, should contain
only an account of the expedition from a physico-moral point
of view, touching only upon those topics of general interest
which would be sure to command the attention of every well-
educated man :—such as the characteristics of the Indian races,
their language and customs, the trade of the colonies, descriptions
of the towns, the aspect of the country, the system of agriculture, data relative to the heights of mountains, meteorological
results. Then, in separate volumes :—1. Geology and the Construction of the Earth. 2. Astronomical Observations, Determinations of Latitude and Longitude, Observations of Jupiter,
Experiments in Refraction. . . . 3. Physics and Chemistry ;
Experiments upon the Chemical Constitution of the Atmosphere; Hygrometric and Barometric Results; Observations
on Electricity, Pathology, Excitability. . . . 4. Description of
new species of Apes, Crocodiles, Birds, Insects . . .; Anatomy
of marine animals. . . . 5. The Botanical Researches in conjunction with Bonpland, to include not merely the enumera-
tion of new genera and species, but descriptions after the system of Linnæus of all those species already known, of which we
have seen more specimens than other observers—a department of
our work which will, I hope, contain between 5,000 and 6,000

varieties, for vast treasures are no doubt awaiting us at Manilla and Ceylon.   Such is the outline of my plan.[1]

' In the event of my death, I trust that Delambre will edit my astronomical observations, and that Freiesleben or Buch will kindly prepare my geological notes for the press ; my observations in Physics and Chemistry I intend to commit to Schœrer, while my zoological manuscripts will be consigned to the editorship of Blumenbach.   And now I have to request that you, my dear friend, will undertake my botanical papers, and edit them under my name coupled with that of Bonpland. My brother has been empowered to entrust the requisite manuscripts for this purpose.

' I remain true to my promise of devoting to you the whole of my share of the plants collected in this expedition.   I do not intend to retain a single specimen.   I have only to request that as I am postponing the publication of the work till my return, you will not incorporate my herbarium with your own collection until this publication has appeared, or until after my death, should I not live to see it completed.

' I was unwilling to address the two cases (1,600 specimens), which I have to-day entrusted to Mr. Fraser's care, direct to Hamburg, not only because no Spanish ship can enter a neutral port, but also because I am not sure that you may not deem it safer for the cases to remain with Fraser until peace be concluded. . . .   I have reason to think that the plants will be safe under his charge, for I have been able to render him some important services.

' You may perhaps remember that Mr. Fraser's name occurs in Walter's " Flora Carolinensis," where he is mentioned as having made four botanical journeys through Labrador and Canada, partly as botanist and partly as an agent for the collection of plants and seeds for cultivation.   In 1799 he set out upon a fifth expedition of a similar nature to the Ohio and through Kentucky and Tennessee—a region now very accessible, for goods can be sent in a month from Philadelphia to New Orleans by land as far as Pittsburg and thence by water down the Ohio and Mississippi.   Unaware of the difficulty of enter-

---

[1] This plan was not adhered to.

ing the Spanish colonies without the royal permission, Fraser embarked for Havana intending to make a collection of plants. He suffered shipwreck, and, after enduring three days of misery upon a sandbank forty miles from the coast, he was picked up by some fishermen of Matanzas, and arrived here in a state of complete destitution. His name and occupation enlisted at once my sympathy on his behalf. I received him into my house, supplied him with money and all else he could require, and, through the influence of my friends, obtained permission for him to travel through the island of Cuba, which, but for his shipwreck, he would have found great difficulty in accomplishing. I sincerely believe that both he and his son—a worthy and amiable young man—would be exceedingly pleased to do me a service. I proposed to his father that the youth should accompany me in my expedition to Mexico; but the young man is afraid to trust himself among the Spaniards, being unacquainted with their language, and is hastening back to London, to publish a description of the plants he has collected in Kentucky.

'After visiting Mexico and California, I purpose going to Acapulco, where I hope to join Captain Baudin and complete with him my proposed expedition round the world.[1]

'I think I have already mentioned to you (pray excuse my bad German, for during the last two years I have spoken nothing but French and Spanish), that on my return I intend to undertake the publication of my plants. In looking over the two cases which I have committed to Fraser, should you find any new species that seem to you particularly worthy of attention, you are, of course, at liberty to introduce specimens of them—providing you do so sparingly, so as not to include the whole of the unknown varieties—into your admirable treatise. Bonpland and I shall esteem it an honour to have our names mentioned by you in so remarkable a work. My reason for asking you not to publish all, or even many, of the new varieties, is because they cannot be so well described from dried specimens as from the drawings we have made from nature. . . .

---

[1] This plan, as is well known, was never accomplished.

' I think that, with the help of Bonpland, I have been able to make very complete diagnoses, but we do not venture to decide as yet as to how many new genera we possess. We have an immense collection of palms and grasses, of Melastomæ, Pipers, Malpighiæ and Cortex Angosturæ (bark), which is a new genus, differing from Cinchona, as well as of the Cipura Cæsalpina of Auble. . . .

' I am quite resolved to withstand every temptation to publish any of our researches during the five or six years that my journey may occupy. I feel convinced that on our return to Europe we shall find that two-thirds of our supposed new genera and species have been already described. Should this be the case, a fresh representation and description of these plants as they grow in such remote regions will not be without value in a scientific point of view.

' What an infinite store of plants are treasured up in that wonderful tract of country lying between the Orinoco and the Amazon, through which I have travelled 6,443 (*sic*) miles in a district abounding with impenetrable forests, and peopled with apes of species hitherto unknown! We were scarcely able to collect specimens of a tenth of the plants we met with. I am now perfectly convinced of a fact concerning which I was exceedingly incredulous when I was in England—although I saw reason to suspect it from the herbariums of Ruiz, Pavon, Nees, and Henken—that we are not as yet acquainted with three-fifths of the whole of the existing species of plants! The fruits are truly wonderful; on our return journey we sent off specimens to France and Madrid. What a spectacle is presented by the various orders of palms as they rear their heads among the impenetrable forests of the Rio Negro! . . .

' But alas! we grieve almost to tears when we open our cases of plants! Our herbariums meet with the same sad fate which called forth the lamentations of Sparmann, Banks, Swartz, and Jacquin. The extreme humidity of the South American climate, together with the great luxuriance of the vegetation, which renders it almost impossible to procure leaves which are wholly matured, has occasioned the destruction of more than a third of our collection. We are daily discovering new insects, destructive to paper and plants. Camphor, turpentine, tar, pitched boards, and other preservatives successful in a European climate, prove quite unavailing here; nor have we found any benefit from hanging the cases up in the open air, so that our patience has been almost exhausted. After an interval of barely four months our herbariums were scarcely to be recognised; out of eight specimens five had to be thrown away, especially of those collected in Guiana, El Dorado, and the regions near the Amazon, where we were daily exposed to torrents of rain.

'During four months of this journey we passed the night in forests, surrounded by crocodiles, boa constrictors, and tigers, which are here bold enough to attack a canoe, while for food we had nothing better than rice, ants, maniocs, bananas, and occasionally the flesh of monkeys, with only the waters of the Orinoco whereby to quench our thirst. Thus have we with difficulty toiled, our hands and faces swollen with mosquito bites, from Mondavaca to the volcano of Duida, from the limits of Quito to the frontier of Surinam—through tracts of country extending over 20,000 square miles, in which no Indian is to be met with, and where the traveller encounters only apes or serpents.

'In Guiana the mosquitoes abound in such clouds as to darken the air, and as it is absolutely necessary to keep head and hands constantly covered, no writing can be done by daylight; the intolerable pain produced by the attacks of these insects renders it impossible to hold the pen steadily. All our work had therefore to be carried on by the light of a fire, in an Indian hut, where no ray of sunlight could penetrate, and into which we had to creep on our hands and knees. Here, if we escaped the torment of the mosquitoes, we were almost choked by the smoke. At Maypures we and the Indians took refuge in the midst of the cascade, where the spray from the foaming stream kept off the insects. At Higuerote the people are accustomed at night to lie buried three or four inches deep in sand, with only the head exposed. If we had not seen it, we should have considered the account fabulous. It is remarkable that where the rivers assume a darker colour, that is to say, in the coffee-brown streams of the Atabapo, Guainia, &c., neither mosquitoes nor crocodiles are to be found.

' Abundant compensation for these annoyances is afforded by the sight of the majestic forests of palms, and by the study of so many independent races of Indians who still bear about them the evidences of ancient Peruvian civilisation. And yet these nations, while possessing considerable knowledge of agriculture and exercising the rights of hospitality, combine, like the inhabitants of Otaheite, a mild and gentle aspect with the savage custom of cannibalism. Wherever we went throughout the wild districts of South America—I refer to those parts south of the cataracts of the Orinoco, where, with the exception of some five or six Franciscan monks, no Christian had previously penetrated—we scarcely ever entered a hut without encountering the horrible remains of repasts on human flesh ! !

' My health and spirits have decidedly improved since I left Spain, notwithstanding the perpetual changes of heat, cold, and damp to which I am exposed. The tropics are my element, and I have never enjoyed such uninterrupted good health as during the last two years.

' I work very hard, and give myself but little time for sleep ; while making astronomical observations I am often exposed to the sun for four or five hours, with no covering on my head, and in places where the yellow fever was raging, as, for instance, at La Guayra and Puerto Cabello, my health was never even interrupted by a headache. I was only twice laid up, with slight attacks of fever lasting three days ; on the first occasion at Angostura, the capital of Guiana, upon my return from the Rio Negro, when, after a long fast, I ate immoderately of bread, which we had not before met with in those regions ; in the second instance at Nueva Barcelona, from being wet through in sunshine, which has here always the effect of producing fever. On the Atabapo, in a climate where even the Indians are continually suffering from a kind of putrid fever, my health kept surprisingly good.

' My reception in the Spanish colonies is as flattering as the vainest and most aristocratic of men could desire. In countries where public opinion exercises no influence, and where everything is governed by despotic rule, the favour of the court is a powerful talisman. The report that I had been personally distinguished by the King and Queen of Spain, and had received recommendations from Don Urquijo, the new and all-powerful minister, softened all hearts in my favour. Never has a natural philosopher been permitted to travel about with so much freedom. For these reasons the expedition has been far less expensive than might have been supposed, seeing that on the rivers I had twenty-four Indians for months together in my employ, while in the interior I often required fourteen mules for the transport of instruments and plants. . . .

' I value my independence more and more every day, and for this reason I have scrupulously avoided accepting the smallest pecuniary assistance from any Government : if the German newspapers should happen to translate an article which has appeared in an English paper concerning me—in which, amid much of a flattering nature, it is stated " that I am travelling by commission of the Spanish Government, and am destined to a high position in the Council for India "—you will be as greatly amused by it as I was. Should I be permitted a safe return to Europe, I shall be occupied with plans widely different from anything connected with the Consejo de Indias. Under the auspices in which I started life, I seem to be intended for activity, and should I succumb to these exertions, those who know me as well as you do are aware that it is for no mean object for which I sacrifice myself.

' We Northern Europeans have, it seems to me, a strange and, I might almost say, an extravagant prejudice against the Spanish people. I have now for two years been living on intimate terms with all classes of society, from the Capuchins with whom I spent some time at their mission stations among the Chaymas Indians, to the Viceroy, and I have become almost as familiar with the Spanish language as I am with my mother tongue : this intimacy has given me opportunity for observing that this nation, notwithstanding the tyranny of the Government and the influence of the priests, is making great strides in intellectual culture, and possesses the elements of a grand character. . . .

' I have every reason to congratulate myself upon my travelling companion, Bonpland. He is a worthy disciple of Jussieu, Desfontaines, and Richard, and is extremely active and industrious ; he possesses much tact, easily accommodating

himself to men and manners, speaks very good Spanish, and
is very courageous and intrepid—in a word, he is admirably
fitted for a scientific explorer. He has without assistance undertaken the arrangement of the plants, which, including duplicates, amount to above 12,000 specimens; but the descriptions
are joint productions, in which I have taken an equal part.
We have often each given a description of the same plant, in
order to arrive at a more trustworthy result. . . .

   ' And you, my dear friend, how is your busy life passing
amidst the peaceful happiness of domestic ties ?   How fortunate
for you that you have *not* seen the impenetrable forests of the
Rio Negro, nor the glories of these tropical palms !—life in a
pine wood would seem intolerable to you ever after.   What a
spectacle is presented by the various orders of palms as they
rear their heads among the impenetrable forests of the Rio
Negro !   It is certainly only in the country of Guiana—in the
tropical regions of South America—that the world may truly
be called green. . . .

   ' When I look back to the time when I brought you
Hordeum murinum, that you might determine the species
for me, when I remember that the study of botany inspired
me with a keener desire to visit the tropics than was ever
excited by my travels with Forster, when in imagination I
contrast the Rehberge and the Panke with the cataracts of
Atures and my house of China (*Cinchona alba*) in which I
lived for so long—it all seems like a dream.   How many difficulties have been overcome !   First disappointed in accomplishing a voyage round the world with Baudin ; then almost
on the point of visiting Egypt and Algiers; at length in South
America ! and again indulging the hope of joining Baudin and
Michaux in the Pacific Ocean.   How wonderful sometimes is
the concatenation of events !   I have now the prospect, after
visiting Mexico and California, of proceeding to Acapulco,
there to fall in with Captain Baudin and complete with him
my journey round the world.

   ' Sometimes, when I give my imagination the reins, and
allow myself to fancy that all these wanderings have been
brought to a safe and happy termination, and that I am
once more at your side, in the old familiar room at the corner

of Friedrichstrasse, enjoying the happiness of our mutual affection, the picture assumes so startling a reality that I feel almost tempted to hasten the conclusion of my journey, and to forget that in great undertakings the force of reason, and not that of inclination, should be allowed to rule. I cannot help entertaining a strong conviction that we shall see each other again.

'I have not yet received any communication either from Jacquin or Van der Schott, in whom I feel so strong an interest. When will this fearful war cease, by which so many bonds are severed? Pray remember me affectionately to your dear wife and to your mother-in-law, and kiss the children for me, and please give a special greeting to Hermes. Remember me also to our circle of valued friends, Klaproth, Karsten, Zöllner, Hermbstedt, Bode, Herz, &c. Best remembrances to Herr Kunth, whom you will doubtless seek out on the receipt of this letter. Tell my dear old friend that, consistently with my determination never to entrust more than *one* letter to each vessel sailing for Europe, I have written to him to-day by another opportunity. With a brother's affection,

'Your devoted old pupil,

'ALEXANDER HUMBOLDT.'

CHAPTER IV.

VISIT TO CUBA.

Departure from Barcelona — Landing at Havana—New Plans for the
   Journey—News of Baudin—Voyage from Batabano to Cartagena—
   Double Danger—Turbaco—Fidalgo's Commission.

ON November 24, 1800, Humboldt set sail for Cuba from the
Bay of New Barcelona in a small vessel laden with meat.   The
voyage was by no means a favourable one, the ship being alternately distressed by stormy weather and hindered by calms.   On
November 30, a sudden squall from the north-east brought the
little craft into considerable danger, which was still further increased by the breaking out of a fire on board the same evening
—an alarming accident, for had not the flames speedily been
got under, they would, amidst such a cargo of meat and fat,
have soon wrought the destruction of the vessel.

Although the Caribbean Sea was even at that time almost as
familiar to navigators as the Mediterranean, yet the travellers
were able to do good service in ascertaining with greater
accuracy the position of certain cliffs, islands, and promontories.
After a protracted voyage of twenty-five days in constant bad
weather, they at length reached the harbour of Havana on
December 19.

The dangers and discomforts experienced on the sea made the
sojourn at Havana seem all the more delightful: in the town they
were the guests of Count Orelly, while in the country they were
entertained at the houses of Count Jaruca and the Marques del
Real Socorro.   Humboldt's first efforts were directed to making
a more accurate survey of the harbour, in which he was zealously
assisted by the astronomer Robredo, Commodore Montes, and

Galiano, the faithful companion of the unfortunate Malaspina.
On the completion of this undertaking, he travelled in the early
part of the year 1801 through a portion of the island, and obtained the latitude and longitude of Rio Blanco, El Almirante,
and several other places in the interior of the country. He
returned in February to Havana,[1] and occupied himself in
collecting materials for the work he subsequently published
under the title of ' Essai politique sur l'Ile de Cuba.'

Humboldt had originally intended to proceed from Cuba
through North America to the Canadian Lakes, and return
southward by way of the Ohio and Mississippi to Louisiana,
whence he was to strike across the intervening unfrequented
district to New Biscay and Mexico.[2] On learning, however,
through the American newspapers, that Captain Baudin was on
the point of starting upon his long-projected expedition, and
purposed to sail by Cape Horn to the coasts of Chili and Peru,
he at once determined to carry out the original agreement of
joining the expedition whenever an opportunity offered, for
he had every reason to hope that a union with so many other
scientific men would considerably promote the objects of his
journey. He therefore resolved to cross the isthmus of Panama
and proceed to Guayaquil, and wrote word to Baudin that he
proposed to meet him on the shores of the Pacific. This
letter never reached its destination, and was returned to
Humboldt at Lima, for Baudin, instead of sailing round Cape
Horn, had started by way of the Cape of Good Hope ; it is as
follows :[3]—

'Cartagena, West Indies : April 12, 1801.

' Citizen,—When I bade you farewell in the Rue Helvétius at
Paris, with the prospect of sailing shortly for Africa and the East
Indies, I had but faint hopes of meeting you again and forming
one of the expedition under your command. You have heard, no
doubt, through our mutual friends, Citizens Jussieu and Desfontaines, how completely my plans have been changed. . . .

[1] Oltmann's 'Untersuchungen über die Geographie des Neuen Continents,
&c.' vol. i. p. 226 ; vol. ii. p. 1.
[2] From a letter to William von Humboldt, dated Contreras, near Ibague,
September 21, 1801.
[3] 'Briefe von Alexander von Humboldt, &c., an Varnhagen,' p. 228.

In a position of independence and at my own cost, my friend
Bonpland and I have spent the last two years in travelling
through the districts of South America watered by the Orinoco,
the Casiquiare, the Rio Negro, and the Amazon. Our health
has been preserved amid the pestilential atmosphere pervading
these rivers. While buried in these forests we have often
spoken of you, of our futile interviews with Citizen François de
Neufchateau, and of the disappointment experienced through
the shipwreck of our hopes. As we were on the point of
leaving Havana for Mexico and the Philippines, the news
reached us that your perseverance had at length been rewarded
by success, and that you had started on your voyage. On considering your probable route, we felt convinced that you would
touch the coast at Valparaiso, Lima, or Guayaquil, and therefore at once changed our plans. Undismayed by the impetuous
gales off this coast, we embarked in a small pilot-boat to seek
you in the Southern Ocean, in the hope of being able to recur
to our former projects of uniting our labours with yours, and
navigating in concert the great Pacific. . . .

' A disastrous voyage of twenty-one days from Havana to
Cartagena intimidated us from prosecuting the route to Panama
and Guayaquil, since we feared the gales would blow with still
greater vehemence in the Pacific Ocean ; we therefore intend
proceeding by land up the Rio Magdalena, and by way of Santa
Fé, Popayan, and Quito.

' We hope to arrive at Quito in June, or at the beginning of
July, where we shall await the news of your arrival at Lima,
Pray send me a few lines to the following address in Spanish :—
Al Sr. Baron de Humboldt, Quito, Casa del Sr. Gobernador Bn.
de Caroudelet. If I should hear nothing from you, I shall
probably pass the time till November in visiting Chimborazo,
Loxa, &c., and proceed to Lima with my various instruments
in December or January.

' You will perceive from these projects, that this tropical
climate has not had the effect of rendering me phlegmatic, and
that I regard nothing as a sacrifice that is needed for carrying
out plans of usefulness or of scientific enterprise. I have addressed you with frankness ; I am aware that I am asking of

you more than I can offer in return—indeed I can well imagine that you may be so situated as not to be able to receive us on board. . . . Should this be the case, this letter
may cause you embarrassment, in proportion to the feeling of
friendship that you entertain towards us. I venture to plead
that you will deal frankly with me. I shall in any case be glad
to have had the opportunity of seeing you again, and I shall
never murmur at events which lie beyond our control. Such
frankness will be to me the most valuable proof of your good-
will. I shall then continue my present journey from Lima
to Acapulco and Mexico, thence to the Philippines, Surat, Bussorah, Palestine, and Marseilles. I prefer, however, to contemplate the possibility of forming one of your party. Citizen
Bonpland desires me to convey to you his respects.

'With much esteem, believe me faithfully yours,

'ALEXANDER HUMBOLDT.'[1]

Of the events of this journey and the visits he paid to the Rio
Zenu and Cartagena, Humboldt wrote the following account to
his brother :—

'Cartagena, West Indies: April 1, 1801.

'If you have received my last letter from Havana,[2] my dear
brother, you will be already aware that I have changed my plans,
and that instead of travelling through North America, we
have returned to the southern shore of the Gulf of Mexico,
whence we intend to make our way overland to Quito and Lima.
It would occupy too much space were I to enumerate the
motives which have led me to this decision ; the considerations
by which I have been mainly influenced have been the tedious
and somewhat hazardous nature of the voyage between Acapulco and Guayaquil, together with the necessity of returning
to Acapulco to meet with an opportunity of sailing to the
Philippines.

[1] The following remark has been appended by Humboldt:—'This letter,
written to Captain Baudin on my arrival at Cartagena from Havana, was
returned to me, as Captain Baudin did not call at Lima.

'Berlin: November, 1846.   A. HUMBOLDT.'

[2] This letter never reached its destination.

'I sailed from Batabano, on the southern coast of the island of Cuba, on the 8th of March, in a very small vessel, of scarcely 40 tons' burden, and did not reach land till the 30th of March, after a voyage of twenty-five days, the time usually occupied being hardly more than a week. We were almost constantly becalmed or had only very light winds, and were carried so far west by the Gulf Stream, that, owing to the captain's incredulity as to the correctness of my chronometer, we found ourselves in the Gulf of Darien. We lost a week in working our way back along the coast, which, on account of the tempestuous east wind prevalent at this time of year, was, with so small a vessel, as difficult as it was dangerous. We cast anchor in the Rio Zenu, and spent two days in botanising along its banks, which had not been previously visited by any scientific explorer.

'The country exhibited the most wild luxuriance, and was exceedingly rich in palms, so that we were able to collect a considerable number of new plants. The mouth of the river is more than two miles wide, and is much infested with crocodiles. We met there with some of the Darien Indians; they are short, broad-shouldered, and flat in figure—in every way a contrast to the Caribs; in complexion they are tolerably fair, with more flesh on their bones, and better developed muscles, than any Indians we have hitherto met with. They lead a life of great independence, and are not hampered with any of the constraints of government. You will thus see that though our voyage was tedious and somewhat dangerous, it yet brought before our notice many objects of interest. Our greatest danger, however, was yet to come, and befell us just as we were entering the port of Cartagena.

'We were trying to force our way against the wind into the harbour. The sea was fearfully rough. Our tiny craft could not withstand the force of the waves, and was thrown suddenly upon her beam-ends. A tremendous wave broke over us and threatened to engulph the ship. The man at the helm remained undismayed at his post; all at once he called out: "No gobierna el timon!" (The rudder will not act!) We all now gave ourselves up for lost. In this, as it seemed to us, our last extremity, we cut away a sail which was flapping loosely, when the ship suddenly righted herself upon the top of another wave, and we sought refuge behind the promontory of Gigante.

'But here a new and almost a greater danger awaited me. For the better observation of an eclipse of the moon,[1] I put off to shore in a boat. I had scarcely landed with my assistants when we were startled by the clanking of chains, and a party of powerful negroes (*cimarrones*), escaped from the prisons of Cartagena, rushed out upon us from the thicket, brandishing their daggers, intent apparently on seizing our boat, as they saw we were unarmed. We fled at once to the water, and had barely time to embark and put off from shore.

'On the following day, during a calm, we quietly entered the harbour of Cartagena. It is a remarkable coincidence, that the day on which I was thus twice exposed to imminent peril was Palm Sunday (*Domingo de ramos*), and that it was on Palm Sunday in the previous year that I was placed in almost equal danger, when off the turtle banks of the Uruana, in the Orinoco, a description of which I sent you at the time.'[2]

After giving detailed instructions concerning the disposition of his collections and manuscripts, he proceeds:—

'My health continues very good, and you will now have less reason to be anxious about me, since for the future my voyages will be prosecuted in the peaceful waters of the Pacific. My plan is to go over land by Santa Fé and Popayan to Quito, where I expect to arrive in July; from Quito I hope to reach Lima, and sail thence in February, 1802, for Acapulco and Mexico; from Acapulco I expect to take ship, some time in the year 1803, for the Philippines, and in 1804 I trust we may have the pleasure of meeting each other again.

'I have been now for a long time without news from Europe. I have received only one letter from you since I left Spain; and yet I know you must have written to me frequently. No one here has received letters from Europe since March 1800.' . . .

During a sojourn of three weeks at Cartagena the travellers visited the environs of Turbaco, noted for a volcano which emits mud and water, and for trees of enormous girth. While

[1] In the night of March 29 and 30.
[2] See p. 279.

here they had the good fortune to fall in with Fidalgo, at the head of a commission for the survey of the coast, with whom they were enabled to compare observations, and effect the regulation of their instruments. 'We found a remarkable and constant agreement in the observations for longitude, and noticed that since 1798 the deviation of the magnetic needle upon these coasts has been as great to the west as it has been in Europe to the east.'

CHAPTER V.

JOURNEY TO QUITO.

Change of Route—Up the Rio Magdalena to Honda—Santa Fé de Bogota
and its Environs—Ibague—The Pass of Quindiu—Caucathale and
Popayan—The Paramos of Pasto—Arrival at Quito.

HUMBOLDT had originally intended sailing from the Rio Zenu
to Porto Bello, thence up the Rio Chagre to Panama, in order
that he might investigate the geological conformation of the
isthmus, and await at Panama an opportunity of embarking
for Guayaquil and Quito.   At a favourable season of the year,
this route is incomparably shorter than the journey from Cartagena to Quito by way of Santa Fé de Bogota, Popayan, and
Pasto, which necessitates the tedious sail up the Magdalena. At
Cartagena, however, he learnt that the trade winds of the Pacific
were over for the season, and that consequently the voyage
from Panama to Guayaquil might occupy from two to three
months : this decided him to choose the inland route up the
Magdalena.   He was further influenced toward this decision by
his great desire to cross the chain of the Andes, as well as by
his wish to visit in Santa Fé de Bogota the noted botanist Don
José Celestino Mutis, with whose collection of plants he was
anxious to compare his own.   He therefore sent his heaviest
instruments, with the books and collections he could best spare,
by sea to Quito, and after a stay of three weeks at Cartagena, he left Turbaco on the night of April 19, 1801, and,
joining the Magdalena at Barancas Nuevas, embarked with
Bonpland on April 21.

'Owing to the force of the swollen stream,' he writes to his
brother from Contreras, near Ibague, September 21, 1801, 'we
were fifty-five days in making our way up the Magdalena, pass-

ing the whole time amid forest scenery, through districts with scarcely an inhabitant. For a distance of forty leagues there is neither a house nor any other human habitation. I need not again allude to the mosquitoes, the dangers of the cataracts, or the thunder-storms, which were almost incessant, and set the heavens nightly on a blaze. Of all this I have given a circumstantial account in a number of other letters. Thus we sailed as far as Honda, situated in 5° North latitude. I have made a chart of the river in four sheets, a copy of which is retained by the viceroy; I have also taken a series of barometric levels between Cartagena and Santa Fé, and have tested the condition of the atmosphere in several localities with my eudiometer, which is still in excellent order—indeed, I am fortunately able to say that none of my valuable instruments have received any injury. The Magdalena route was also that taken by Bouguer on his return to France, only he followed the river in its downward course, and was unprovided with instruments.

'From Honda I visited the mines of Mariquita and Santa Anna, where the unfortunate D'Elhuyar met with his death. The cinnamon-plant cultivated here is a species similar to that grown at Ceylon, specimens of which I had already met with on the Guaviare and Orinoco. The well-known almond tree (*Caryocus amygdaliferus*) is also found here, and whole forests of the cinchona, as well as the otoba, which is a true myristica (nutmeg), to the cultivation of which the attention of Government is now being directed. M. Desieux, a Frenchman, who has been appointed overseer to these plantations, with a salary of 2,000 piastres (500 gold Fredericks of our money), was one of our travelling companions up the Magdalena.

'Santa Fé de Bogota is situated 8,990 feet above the level of Honda. The road is indescribably bad ; in some places it consists only of narrow steps cut between walls of rock, and being only from 18 to 20 inches wide, scarcely admits the passage of a mule.[1] On emerging from this mountain pass (la boca del

[1] Up to 1816 the road was scarcely more than a watercourse, a cleft in the rock, in many parts of which two mules could not pass, and yet this was one of the roads leading to the capital of a country containing a population of from 28,000 to 30,000 inhabitants. When the Spanish Government regained for a time possession of New Granada, the prisoners taken from the

monte) in 4° 35′ North latitude, we found ourselves suddenly upon an extensive plain embracing more than thirty-two square leagues, which, though entirely devoid of trees, was covered with crops of various kinds of European grain, and thickly dotted with Indian villages. This plain (Llanura de Bogota) has been formed by the drying up of the Lake of Funzhe, which plays so important a part in the mythology of the Muyscas Indians. The evil spirit, or the Moon—a woman—cast forth a deluge which formed the lake. But the good spirit, Bochika, or the Sun, shattered the rocks of Tequendama, where there is now the celebrated waterfall, and the waters of the Lake of Funzhe flowed away. The inhabitants, who had during the flood fled to the neighbouring mountains, now returned to the plain, and Bochika, after having given to the Indians a political constitution and laws similar to those of the Incas, retired to the Temple of Sagamun, where he resided for 25,000 years, and thence finally returned to his home in the Sun.

'Our arrival at Santa Fé resembled a triumphal procession. The archbishop sent his carriage to meet us, and with it came the persons of greatest distinction in the capital. A dinner was provided for us at some distance from the city, and we proceeded with a retinue of more than sixty persons on horseback. As the object of our coming was known to be a visit to Mutis, who, on account of his great age, high position at court, and personal character, is held in the greatest estimation by all classes here, a certain degree of ceremony was accorded to our reception, that through us the inhabitants might do him honour. As the viceroy is forbidden by etiquette to entertain any guest at his own table in the capital, he invited us to dine with him at his country house at Fucha. Mutis had prepared a house for us in his own neighbourhood, and received us with the utmost cordiality and friendship. He is an excellent old man, nearly seventy-two years of age; he has been in holy

Republican party were employed in widening and improving the road from Honda to Bogota—an undertaking rendered necessary by the requirements of military communication, and to which the Government was urged by a powerful political reaction. The road speedily assumed a new aspect; and thus was easily and rapidly accomplished, during a time of bloody civil war, a work which had never even been attempted by the Spaniards during their peaceful occupation of the country for nearly 300 years.

orders, and is possessed of considerable wealth. The king pays annually 10,000 piastres towards the expenses of botanical research. Thirty artists have been engaged during the last fifteen years in painting under the superintendence of Mutis : he has from 2,000 to 3,000 drawings in large portfolios, which are executed like miniature paintings. He possesses the largest botanical library I have ever seen, excepting that of Banks in London. Notwithstanding its proximity to the equator, the climate is decidedly cool, on account of the great elevation ; the thermometer usually stands at 46° or 48°, frequently at 32°, and never above 72°.

'I have kept perfectly well amid the river miasma and the inflammation caused by the mosquito bites, but poor Bonpland was again attacked by intermittent fever on the road from Honda to Santa Fé, and by this illness we were detained in the capital full two months, till the 8th of September, 1801. I employed my time in visiting the curiosities of the neighbourhood, and in measuring the height of some of the surrounding mountains, several of which rise to 13,000 and 16,000 feet.' . . . 'Among the sights of the neighbourhood' (as we learn from Humboldt's treatise ' On the Elevated Plain of Bogota'[1]) 'were included the following objects of interest :—the magnificent waterfall of Tequendama,[2] where, through a cleft in the rock shadowed by evergreen oaks, the water rushes down a ravine, bordered on either side by palms and ferns of most luxuriant growth ; the wide plain, Campo de Gigantes, filled with bones of the mastodon ; an extensive field of coal, and an immense bed of rock salt. The existence of these formations excites surprise from the great elevation at which they occur—a height almost as great as if the Brocken were piled upon the summit of the Schneekoppe.'

On the recovery of Bonpland, the travellers set out upon their journey in September, 1801, from Santa Fé to Quito. The road lay westward across the Rio Magdalena and through Contreras to

[1] Read before the Berlin Academy on March 19, 1838, printed in an abridged form in their monthly reports of March 1838, and published entire in the 'Deutsche Vierteljahrsschrift,' vol. v. p. 97, &c., in Poggendorff's 'Annalen,' 1838, vol. xlii. p. 570, &c., as well as in Alexander von Humboldt's ' Kleinere Schriften ' (1853), vol. i. p. 100.

[2] 'Atlas pittoresque, ou Vues des Cordillères,' Pl. VI.

Ibague, one of the oldest towns in the kingdom of New Granada, situated in the valley of Combaima, 2,305 feet above the sea, where the temperature in the day ranges between 84° and 88°, and at night between 73° and 76°. The travellers, on September 23, accurately determined by repeated astronomical observation the latitude and longitude of Ibague. They then crossed over the eastern spur of the Cordilleras by the pass of Quindiu—a route almost completely amid the snow.

'This pass presents one of the most difficult roads in the Cordilleras of the Andes.[1] It lies through a dense wood, wholly uninhabited, which cannot be traversed even at the most favourable season of the year under ten or twelve days. Neither shelter nor food can be procured, and therefore it is absolutely necessary that at all times of the year travellers should carry with them a month's provisions, since it frequently happens that, owing to the rapid swelling of the torrents through the melting of the snow, all progress is interrupted for days together. The highest point of the road is 11,494 feet[2] above the sea. The path is so narrow that it rarely exceeds 12 or 16 inches in width, and for the most part resembles an open gallery cut in the rock. The torrents, in their violent descent, have worn away ravines to the depth of from 18 to 20 feet, along which the pathway passes often through ground in a state of morass, and overhung by such thick vegetation as to be almost excluded from the light. This road is impassable to mules, and the oxen which are used instead —twelve of which were required for our luggage[3]—can with difficulty traverse these galleries, extending in some instances the distance of a mile. In the event of unfortunately meeting a string of oxen, one is obliged either to retreat along the path already trodden, or to climb up the steep side of the ravine, and hold on by the projecting roots of the trees above. In addition to other inconveniences, we suffered very much during the

[1] 'Vues des Cordillères,' Pl. V.

[2] The highest point at which they encamped was 10,800 feet.

[3] The statement therefore occurring in the Autobiography in Brockhaus's 'Conversations-Lexikon,' so often cited, is erroneous, that 'The journey over the Cordilleras from Bogota to Quito *was performed entirely upon mules* . . . . and lasted four months.' In a letter to his brother, dated Lima, November 25, 1802, Humboldt remarks :—'Oxen constituted the only means of transport for our luggage on this route.'

last few days of our descent on the western slope from incessant heavy rain. Our road lay through a swampy district covered with reeds of bamboo. The pricks on the roots of this gigantic kind of grass so completely destroyed our boots, that, as we would not allow ourselves to be carried on the backs of men (*cargueros*), we reached Cartago with bare and bleeding feet.' Humboldt describes these cargueros with some minuteness :— ' In these climates Europeans become so completely enervated, that it is customary for every director of mines to have one or two Indians in his service who are termed his horses (*caballitos*), because every morning they allow themselves to be saddled, and with their body inclined forward, and leaning on a short stick, they carry their masters on their backs. Among the caballitos and cargueros some of them are recommended to travellers as being sure-footed and possessing an easy and even pace ; it really makes one's blood boil,' adds Humboldt in a burst of generous feeling, ' to hear the qualities of a human being described in the same terms as would be employed in speaking of a horse or a mule.'

At Ibague, during a day's heavy rain, the travellers provided themselves with an impervious shelter by means of tents constructed out of the leaves of the heliconia. At length they reached Cartago ' with feet bare and bleeding, but enriched with a valuable addition to their collection of plants.'

' From Cartago,' writes Humboldt to his brother, in a letter dated Lima, November 25, 1802, ' we went to Popayan by way of Buga, and the magnificent valley of the Cauca, to the right of which rise the mountains of Choka, celebrated for their platinum mines.

' We passed the month of November, 1801, at Popayan visiting the basaltic mountain of Julusuito, the crater of the volcano of Purace, which emits with a terrific noise jets of steam impregnated with sulphuretted hydrogen, and the rocks of porphyritic granite, exhibiting the form of columns, shaped as pentagons and heptagons, similar to those described by Strange, which I remember to have seen when in Venetian Lombardy.

' The greatest difficulties of our journey lay yet before us, between Popayan and Quito. We had to cross the Paramos of Pasto, and this in the rainy season, which had already set in. Paramo is the name given in the Andes to those desert regions where, at a height of about 12,000 feet above the sea, all vegetation ceases and the cold is so intense as to penetrate to the very bones. To avoid the heat of the valley of the Patia, where malaria exists to such an extent that one night spent within its precincts may engender a fever known among the Spaniards as the " calentura de Patia," lasting from three to four months, we crossed over the peak of the Cordilleras, through a pass abounding with frightful precipices, to Almager, whence we proceeded to Pasto, situated at the foot of a terrific volcano.

'It would hardly be possible to picture a more horrible road than that by which access is obtained to this little town, where we spent Christmas (1801), and where we were welcomed by the inhabitants with a touching hospitality. Thick woods interspersed with morasses, in which the mules sank up to the girths, and narrow paths winding through such clefts in the rocks that one could almost fancy one was entering the gallery of a mine, while the road was paved with the bones of mules which had perished through cold or fatigue. The whole province of Pasto, including the environs of Guachucal and Tuqueres, consists of a frozen mountain plateau, almost above the limit of vegetation and surrounded by volcanoes and solfataras, from which wreaths of smoke continually issue. The unfortunate inhabitants of these regions live almost entirely upon potatoes, and when this crop fails, as it did last year, they are obliged to retreat to the mountains, where they live upon the achupalla (*Pourretia Pitcarnia*), a small tree of which they eat the stem. As this tree serves also for food to the bears of the Andes, it is often only by contending with these animals that they can possess themselves of this the only sustenance afforded by nature to man at this elevated region.

'In the Indian village of Voisaco, 8,990 feet above the level of the sea, situated to the north of the volcano of Pasto, I discovered some red clay and hornstone porphyry mingled with vitreous felspar, possessing all the properties of the serpentine

rock in the Fichtelgebirge. This porphyry exhibits some degree of polarity, but has no force of attraction. At length,
after being wet through by torrents of rain night and day for
two months, and nearly drowned in the town of Ibarra by the
sudden rise of the water during an earthquake, we reached Quito
on the 6th of January, 1802.'

CHAPTER VI.

QUITO.

The Town and its Inhabitants—Interest in the Ascent of Mountains—
Three Ascents of Pichincha — Ascent of Chimborazo—Letters from
Paris—Despatch of Letters to Europe—News of Baudin—Noble Self-
reliance—Friends at Lima—Humboldt's Portrait at Chillo.

THROUGH the forethought of the Marques de Selvalegre, the
travellers found on their arrival at Quito an excellent house
prepared for their reception, where they were able to repose,
after the hardships of their journey, amid all the comforts
'that could be expected either in London or Paris.'

'The town of Quito is handsome,' writes Humboldt to his
brother in the letter above cited of November 25, 1802, 'but
the sky is frequently overcast ; the mountains in the neighbourhood show no appearance of vegetation, and the cold is
considerable. The great earthquake of the 4th of February,
1797, by which the whole province was convulsed and some
40,000 persons instantaneously killed, has in every way proved
a most disastrous event. It has so completely altered the
climate, that whereas Bouguer found the thermometer usually
stood at about 67°, it now ranges from 41° to 54°, rarely rising
to 70°. Since that calamity, earthquakes are of frequent
occurrence, and occasionally the shocks are of great violence.
It seems probable that the whole of the more elevated portion
of the province is one vast volcano, and that the so-called
mountains of Cotopaxi and Pichincha are but small peaks,
the craters of which constitute the emission tubes (chimneys)
of the vast subterranean fires. The truth of this hypothesis
has unfortunately been only too clearly demonstrated by the
earthquake of 1797, for the earth then opened in all directions, casting forth sulphur, water, &c. Notwithstanding the

dangers by which the inhabitants of Quito are surrounded, and the apprehensions to which they must frequently be exposed, they are a gay, lively, and amiable people. The town breathes only an atmosphere of luxury and voluptuousness, and perhaps nowhere is there a population so entirely given up to the pursuit of pleasure. Thus can man accustom himself to sleep in peace on the brink of a precipice.

'We remained in the province of Quito for nearly eight months—from the beginning of January till August—and spent the time in visiting the principal volcanoes. We examined in succession Pichincha, Cotopaxi, Antisana, and Ilinica, devoting about a fortnight or three weeks to each, and returning between whiles to the capital; finally, on the 9th of June, 1802, we started for the ascent of Chimborazo.'

The results of the observations made by Humboldt during this sojourn upon the elevated table-land of Quito were early given to the world in his great work, where they appear according to their several classifications, whether botanical, geological, meteorological, hypsometric, or astronomical: it was not until a much later date that he published separate topographical descriptions or monographs of the ascents of some of the volcanoes. He gives as a reason for this delay, that he thought it necessary to wait until he could show 'the relative worthlessness' of his labours, by distinguishing between those geological observations, which had been made on principles since proved obsolete, and those of a character which could not be affected by time.

There is, in fact, little of purely scientific interest, and still less of scientific value, to be gained by the arduous and dangerous ascent of high mountains which rise far above the line of perpetual snow, especially when it is impossible to make a prolonged stay at the summit. The barometer, indeed, affords a ready means for ascertaining the height of mountains, but the results obtained are subject to error from the rise and fall of atmospheric currents, and the irregular decrease of temperature. The structure of the earth's surface is hidden from the scrutiny of the geologist by the covering of perpetual snow; and organic life is wholly absent, except perhaps in the soaring condor or the few insects carried up by currents of air.

Still the arduous and perilous ascent of lofty mountains has excited universal interest in all ages of the world. The inaccessible ever exerts a secret fascination. Man is most powerfully attracted by that which is least attainable, and almost defies investigation.

For all such undertakings Humboldt was pre-eminently qualified by his indomitable courage and unconquerable endurance. Though failing from an attack of giddiness and insensibility to reach the summit of Pichincha on his first ascent on April 14, he determined to repeat the attempt on May 26, since, as he remarked, ' it seemed a disgrace to leave the high plain of Quito without personally examining the crater of Pichincha.'

. . . . ' The attendants and the large instruments were, as on the previous occasion, left below, and I commenced the ascent accompanied only by a very intelligent Creole, M. Urquinaona, and an Indian, Felipe Aldas. We sat down, disheartened, at the foot of the peak. The crater we were in search of lay, no doubt, behind the wall of rock to the west of us, but how were we to reach it and ascend this precipice ? The towering masses seemed too steep, and indeed, in some places, were almost perpendicular.

' In ascending the Peak of Teneriffe, I had greatly facilitated the climb up the cone of ashes, by following the edge of a projecting ridge of rock, and holding on by my hands, though in so doing I was a good deal lacerated ; I therefore resolved to attempt the ascent here, by similar means, following the slope of pumice which lay against the edge of the southern precipice. We made two fatiguing ascents, reaching, in one case, to a height of about 300 feet, and in another to more than 700 feet. The covering of snow seemed to bear us safely, and our hopes of reaching the edge of the crater were all the stronger from the probability that Bouguer and La Condamine had taken the same route, when ascending the snow-covered slopes of the cone of ashes, some sixty years previously. The snow was so hard that our chief peril seemed to be lest by a fall we should roll down the sloping surface with accelerating speed, and come in contact with some of the sharp rocks that projected out of the pumice. Suddenly, with a loud cry of

terror, the Indian, Aldas, who was close in front of me, fell
through the frozen crust of snow. He sank up to the waist,
and as he declared that he could find no support for his feet, we
feared he had sunk into an open cleft. Happily, the danger
was less imminent. Falling with his legs apart, he had in
sinking pressed together by his weight a considerable mass of
snow, upon which he was supported as on a saddle. He rode
as it were upon this mass; and as we perceived that he did not
continue to sink, we were able to labour with all the more
calmness for his extrication. This we effected by throwing
him upon his back, and pulling him out by the shoulders.
We were somewhat disconcerted by the accident. The Indian,
under the influence of superstitious fear, on account of the
near vicinity of the crater, protested against all further progress
upon the treacherous snow.

‘ We retraced our steps, to deliberate upon our future plans.
The most easterly of the pinnacles surrounding the crater
appeared, upon nearer inspection, to be steepest at the base,
and to become much less abrupt towards the summit, the
upper parts of the rock being broken up so as to form rugged
steps. I requested M. Urquinaona to remain below in the
Sienega, resting upon a block of stone, while I attempted the
ascent of the lower and steeper portion, and only to follow
when he saw me reappear in course of time upon the turret-
shaped rock bare of snow. The good-natured Indian was easily
persuaded once more to accompany me. . . . . On reaching
the pinnacle we continued the laborious ascent, by means of
the narrow shelves and isolated projections of rock, in ignorance
of our way, but with ever-brightening hopes, and soon found
ourselves surrounded by a cloud of vapour, which gradually
thickened, but remained inodorous. The ledges of rock became gradually wider, and the ascent less steep. To our great
joy, the snow lay only in patches, from ten to twelve feet
in length, and scarcely eight inches deep. After our recent adventure, we feared nothing so much as the half-frozen snow.
The mist hid everything from us, beyond the ground on which
we trod: every more distant object was concealed. We wandered
in a cloud. A stifling smell of sulphur announced to us the
close proximity of the crater; but we little suspected that we

were actually, so to speak, standing over it. We proceeded slowly along a small bank of snow, in a north-westerly direction; Aldas the Indian in front, and I a little behind him somewhat to the left. Not a word was spoken, we fell into the silence common to all mountain-climbers, when, taught by long experience, they become aware that the path they are treading is dangerous.

'Great was my excitement when suddenly, as I was looking at a block of stone immediately in front of us, which seemed to hang suspended in a cleft, I saw between it and the extreme edge of the sheet of snow upon which we were walking, a light at a great depth beneath us, like that of a flickering flame. I pulled the Indian violently back by his poncho (the name of a garment, of lama's wool), and forced him to throw himself with me flat down on a shelf of rock, to the left. The ledge was free from snow, and had a horizontal surface, scarcely twelve feet long and between seven and eight feet wide.

'We thus lay stretched upon a platform of rock that overhung the crater like a balcony, and we gazed in fearful proximity into the appalling depths of the terrific gulf. A portion of the perpendicular abyss was filled with eddying wreaths of steam. Assured as to the safety of our position, we commenced investigating our whereabouts. We discovered that the platform of rock, upon which we had thrown ourselves, was separated from the snow-covered mass along which we had come by a cleft scarcely two feet in width. The frozen snow, which formed a sort of bridge, did not extend the whole length of the ravine, and on this bridge we had proceeded several steps while walking in the direction of the fissure. The light we had seen through a portion of the cleft, between the snow-ledge and the block of stone wedged in the fissure, was no deception; we again saw it, on our third ascent, at the same spot, and through the same aperture. It occurred at a part of the crater where the dark abyss was frequently illuminated by small flames, probably of sulphurous gas. . . . The point we here attained was, according to the barometric measurements I subsequently took, no less than 14,940 feet above the sea.'[1]

[1] It was not till the close of a sojourn of seven years in the neighbourhood that La Condamine and Bouguer, in the year 1742, made the ascent

' While the Indian descended to the Sienega to fetch my companion, M. Urquinaona, I sat alone at the edge of the crater, and remarked that my boots, which had been completely saturated with snow-water during our first attempt at the ascent, were becoming quickly dried in the stream of warm air that ascended out of the crater. The thermometer, which in the Sienega stood at 41°, rose as I held it over the abyss to 66°. . . .

' After a tedious delay, M. Urquinaona at length made his appearance, when almost immediately we became enveloped in a dense mist—a cloud of steam—produced probably by the mixing of streams of air of unequal temperature. It was now within an hour of sunset, and with feelings of pleasure at having attained our object, we hurriedly left the volcano, and descended into the valley of Sienega—a district filled with pumice. We crossed, fortunately, before night came on, the steep ridge which separates the Sienega from the valley of Yuyucha. Through this valley we proceeded in total darkness (for not a star was to be seen), encountering numerous falls in consequence of the roughness of the path, and arrived at Quito at half-past twelve o'clock at night. Our arduous expedition had occupied us eighteen hours, fourteen of which we had spent on foot.'

Notwithstanding the dangers and fatigue to which Humboldt had thus been exposed, we find him on May 28, after an interval of only twenty-four hours, again upon the frail balcony of rock overhanging the flaming crater, intent upon making a series of observations and experiments. ' My third ascent was to me the most interesting, from the proof I then received of the continuous or renewed activity of the volcano, for about half-past one in the afternoon we experienced several smart shocks of earthquake while standing on the shelf of rock. I counted fifteen shocks in thirty-six minutes.'

The news that the volcano had shown signs of fresh activity gave rise in Quito to the report that the heretic foreigners had thrown gunpowder into the crater, by means of which the recent shocks had been produced.

As previously mentioned, Humboldt also made the ascent of

of this celebrated mountain : they were unprovided with instruments, and remained only a quarter of an hour at the crater.

the several peaks of Antisana, Cotopaxi, Tunguragua and Ilinica: but of all his mountain expeditions none has excited more attention than the ascent of Chimborazo.

On June 9, the travellers left Quito for Chimborazo ; and on June 23, 1802—the Eve of the Festival of St. John, within a day of the anniversary of his visit to the crater of the Peak of Teneriffe, three years before, June 22, 1799—Humboldt climbed almost to the summit of the giant mountain, at that time regarded as the highest in the world, and attained the height never before reached by any human being of 18,096 feet.

We give the following extracts from the detailed account of this expedition.[1]

Upon reaching an elevation of 15,600 feet, ' The path,' relates Humboldt, ' became every moment narrower and steeper. The natives, with one exception, refused to accompany us farther, and were deaf to entreaties and threatenings, maintaining they suffered more than we did from the rarity of the air. We were left alone—Bonpland, our estimable friend Carlos Montufar, a younger son of the Marques de Selvalegre, a half-caste Indian from the neighbouring village of San Juan, and myself.

' By dint of great exertion and considerable patience, we reached a greater height than we had dared to hope for, seeing we had been almost constantly enveloped in mist. In many places the ridge was not wider than from eight to ten inches ! To our left was a precipice covered with snow, the surface of which shone like glass from the effects of frost. This thin sheet of ice was at an inclination of about 30°. On the right was a fearful abyss, from 800 to 1,000 feet deep, from the sides of which projected huge masses of naked rock. We leant over rather more to this side than the other, for it seemed less to be dreaded than the precipice on our left, where the smooth sides afforded no opportunity of checking a fall by catching hold of projecting pieces of rock, and where the thin crust of ice

[1] A. von Humboldt, ' Ueber zwei Versuche, den Chimborazo zu besteigen,' in Schumacher's ' Astronomisches Jahrbuch,' 1837, pp. 176-206. Berghaus's ' Annalen,' 3rd series, vol. iii. pp. 199-216. A. von Humboldt's ' Kleinere Schriften,' vol. i. p. 133.

furnished no security against being precipitated into the loose snow beneath.

'The sloping surface of snow extended to such a distance that light pieces of dolerite (the only substance at hand), when rolled down the incline, were lost sight of before reaching any resting-place. . . .

'The rock became more friable, and the ascent increasingly difficult and dangerous. At certain places where it was very steep, we were obliged to use both hands and feet, and the edges of the rock were so sharp that we were painfully cut, especially on our hands. In addition to this, I had for some weeks been suffering from a wound in my foot, caused by the repeated attacks of the Niguas[1] (*Pulex penetrans*), which had been greatly aggravated by the fine pumice dust to which I had been exposed while taking measurements in the Llano de Tapia. The loose position of the stones upon the narrow ridge necessitated extreme caution, since many masses that appeared to be firmly attached proved to be only imbedded in sand.

'We advanced all the more slowly, as every place that seemed insecure had first to be tested. Fortunately, the attempt to reach the summit of Chimborazo had been reserved for our last enterprise among the mountains of South America, so that we had gained some experience, and knew how far we could rely on our own powers. It is a peculiar characteristic of all excursions on the Andes, that beyond the line of perpetual snow Europeans are always left without guides just at the point where, from their complete ignorance of the locality, help is most needed. In everything Europeans are left to take the lead.

'We could no longer see the summit, even by glimpses, and were therefore doubly anxious to ascertain how much of the ascent had still to be accomplished. We opened the tube barometer at a spot where the ridge was wide enough to allow two persons to stand side by side in safety. We were only at an elevation of 17,300 feet, therefore scarcely 200 feet higher than we had attained three months previously upon the Antisana.

---

[1] The Sand-flea [Chigoe], an insect which, by burrowing beneath the skin and depositing its eggs, produces swelling and inflammation.

'After an hour's cautious climbing, the ridge of rock became less steep, but the mist unfortunately remained as thick as ever. One after another we all began to feel indisposed, and experienced a feeling of nausea accompanied by giddiness, which was far more distressing than the difficulty of breathing. . . . Blood exuded from the lips and gums, and the eyes became bloodshot. There was nothing particularly alarming to us in these symptoms, with which we had grown familiar by experience. Once when upon the Pichincha, though bleeding did not occur, I was seized with such violent pain in the stomach and overpowering giddiness, that I sank upon the ground in a state of insensibility,[1] in which condition I was found by my companions, from whom I had withdrawn for the sake of making some experiments in electricity. The elevation then was not so great, being less than 13,800 feet. On the Antisana, however, at a height of 17,022 feet, our young travelling companion, Don Carlos Montufar, had suffered severely from bleeding of the lips. All these phenomena vary greatly in different individuals according to age, constitution, tenderness of the skin, and previous exertion of muscular power; yet in the same individual they constitute a kind of gauge for the amount of rarefaction of the atmosphere and for the absolute height that has been attained.

'The stratum of mist which had hidden every distant object from our view began, notwithstanding the perfect calm, suddenly to dissipate—an effect probably due to the action of electricity. We recognised once more the dome-shaped summit of Chimborazo, now in close proximity. It was a grand and solemn spectacle, and the hope of attaining the object of all our efforts animated us with renewed strength. The ridge of rock, only here and there covered with a thin sprinkling of snow, became somewhat wider; and we were hurrying forward with assured footsteps, when our further progress was suddenly stopped by a ravine, some 400 feet deep and sixty feet wide, which presented an insurmountable barrier to our undertaking. We could see clearly that the ridge on which we stood continued in the same direction on the other side of the ravine; but I was

---

[1] Schumacher's 'Astronomisches Jahrbuch,' 1837, p. 192.

doubtful whether, after all, it really led to the summit. There was no means of getting round the cleft. On Antisana, after a night of severe frost, Bonpland had been able to travel a considerable distance upon the frozen surface of snow; but here the softness of the snowy mass prohibited such an attempt, and the nature of the declivity rendered it equally impossible to scale the sides.

' It was now one o'clock in the day. We fixed up the barometer with great care, and found it stood at thirteen inches $11\frac{2}{10}$ lines. The temperature of the air was only three degrees below the freezing point; but from our long residence in the tropics even this amount of cold seemed quite benumbing. Our boots were wet through with snow-water, for the sand, which here and there lay on the mountain ridge, was mixed with the remains of former snow-drifts. According to the barometric formula given by Laplace, we had now reached an elevation of 18,096 Paris feet[1] [19,286 English].

' We remained but a short time in this dreary waste, for we were soon again enveloped in mist; which hung about us motionless. We saw nothing more of the summit of Chimborazo, nor of the neighbouring Snow Mountains, far less of the elevated plain of Quito. We were isolated as in a balloon; a few rock lichens were to be observed above the line of perpetual snow, at a height of 16,920 feet; the last green moss we noticed was growing about 2,600 feet lower. A butterfly was captured by M. Bonpland, at a height of 15,000 feet, and a fly was observed 1,600 feet higher; both had been carried up into the higher regions of the atmosphere by the currents of air originating in the warmer plains beneath. We did not, however, see any condors, which are so numerous upon the Antisana and Pichincha, where, in those vast solitudes, from being unaccustomed to the sight of man, they are wholly devoid of fear.

' As the weather became increasingly threatening, we hurried down along the ridge of rock, and from the insecurity of our footing found that greater caution even was necessary than

[1] ' If La Condamine's estimation of the height of Chimborazo be correct, we were only 1,224 feet in a direct line short of the summit, or a distance equal to three times the height of St. Peter's at Rome.'

during the ascent. We delayed no longer than sufficed for
collecting fragments of rock as specimens of the mountain
structure. We foresaw that in Europe we should frequently
be asked for " *a fragment from Chimborazo.*"

'When we were at a height of about 17,400 feet, we encountered a violent hailstorm, which gave place to snow twenty
minutes before passing the limit of perpetual snow, and the
flakes were so thick that the ridge was soon covered several
inches deep. The danger would indeed have been great had
the snow overtaken us, at a height of 18,000 feet. At a
few minutes past two we reached the spot where we had left the
mules.'[1]

When the measurements of the height of the Himalayas, which
created so much interest a quarter of a century afterwards, were
undertaken by some English travellers, Humboldt wrote 'in a
humorous strain' to Berghaus[2] in November, 1828 :—'I have
all my life imagined that of all mortals I was the one who
had risen highest in the world—I mean on the slopes of
Chimborazo ! . . . . and have felt some pride in this elevation! It was therefore with a certain feeling of—envy that I
saw the announcement of the results obtained by Webb and
his companions with regard to the mountains of India. I have
consoled myself over the achievements on the Himalayas—
by supposing that it was through my labours in America that
the English received the first impulse to direct more attention
to the snowy mountains than had been given for the last century
and a half.'

We will complete Humboldt's description of the journey
by the following extracts from a letter to Delambre, dated
Lima, November 25, 1802 :[3]—

'Your letter has been two years in trying to find me among
the Cordilleras of the Andes. I received it the day after
making my second expedition to the crater of Pichincha. This

[1] Chimborazo has since been ascended by Boussingault and Hall, on
December 16, 1831, who reached the height of 19,692 feet; by Jules
Bourrier, in the years 1849 and 1850; and by Jules Rémy and Brenckley,
to the height of 21,457 feet, on November 3, 1856. The height of the
mountain is, according to Humboldt, 21,460 feet.

[2] 'Briefwechsel A. von Humboldt's mit Heinrich Berghaus,' vol. i. p. 208.

[3] 'Annales du Mus. d'Hist. natur.' An XI. (1803) vol. ii. p. 170.

reminds me that it was at the summit of Guaguapichincha, which I frequently visited, and which I regard as classic ground, that La Condamine and Bouguer received their first letter from the Academy; therefore, it seems to me that Pichincha, *si magna licet comparare parvis*, is a mountain of good omen to men of science. . . .

'Long before receiving the letter you addressed to me in your official capacity of Secretary of the Institute, I had written three letters in succession to the Section for Physics and Mathematics, two of which were from Santa Fé de Bogota, forwarding specimens of bark of seven varieties of the Cinchona, together with some carefully dried skeletons and coloured drawings representing the plants and the anatomy of the flowers, which differ considerably in the length of the stamens.

'Dr. Mutis, from whom I have received every kindness, and for whose sake I undertook a wearisome voyage up the Magdalena of six weeks' duration, has made me a present of a portfolio of 100 magnificent drawings, representing some new genera and species from his " Flora of Bogota." It has appeared to me that this collection, equally remarkable for its botanical value as for the beauty of the colouring, could not be in better hands than in those of Jussieu, Lamark, and Desfontaines, and I have therefore offered them to the National Institute as a slender token of my attachment. . . . I despatched a third letter to the Institute, from Quito, with a collection of geological specimens from Pichincha, Cotopaxi, and Chimborazo. It is very distressing to have to remain so long uncertain as to the safe arrival of these treasures ; for instance, we have heard nothing of the rare seeds we sent to the Jardin des Plantes three years ago ! . . .

. . . . 'At the close of a journey occupying eight months we arrived at Quito, only to learn that Captain Baudin had taken the eastern route, and sailed by the Cape of Good Hope. Accustomed to disappointments, we consoled ourselves by the thought that we had been actuated by a noble purpose in all the sacrifices we had made ; and in reviewing our herbariums, our barometric and trigonometric observations, our drawings, and our experiments upon the atmosphere of the Cordilleras, we have no reason to regret visiting countries which, to a great extent, have never before been explored by men of science. We have been made to feel that man ought not to count upon anything but that which he can procure by his own energy. . . .

'I spent a very pleasant time at Quito. The president of the audience, Baron de Corondeles, loaded us with kindness : indeed, for the last three years, I have never had cause on any occasion to complain of the agents of the Spanish Government; they have uniformly treated me with a deference and delicacy of attention which calls for my perpetual gratitude. How times have changed!'

It is scarcely necessary to mention that eagerly as Humboldt was devoted to the study of volcanoes, he was hardly less interested in work of other kinds, and an important part of his labours included the determination of the latitude and longitude of places. The letters of recommendation with which he was furnished by the Court and Government of Spain, supported as they were by his energy in the pursuit of his scientific undertakings, and by the kindness and amiability he displayed in social life, gained for him at Lima, as in other places, the friendship of the most distinguished men of the locality, many of whom felt incited to share with him the fatigues of his mountain ascents.[1] Among these friends, none evinced so marked an attachment to Humboldt as Carlos Montufar, a younger son of the Marques de Selvalegre, an estimable youth, who accompanied him to Europe, and on his return met with the melancholy fate of being shot by order of General Morillo, during the insurrectionary war.

The following particulars of Humboldt are related by the well-known traveller, Professor Moritz Wagner, in his treatise 'On some Hypsometric Labours among the Andes of Ecuador.'[2]

'Of those who were personally acquainted with Alexander von Humboldt during his residence at Quito, there survived, in 1859, only two very old ladies, members of the wealthy and

[1] 'Of the Europeans who accompanied me on my second ascent of the Pichincha, Don Pedro Urquinaona, Don Vincente Aguirre, and the Marques de Maenza, then a mere youth, the latter was still alive in 1853, residing in Europe as a grandee of the highest rank, with the hereditary title of Count of Puñonrostro.'—Humboldt, *Kleinere Schriften*, vol. i. p. 55.

[2] 'Zeitschrift für allgemeine Erdkunde' (Berlin, 1864), new series, vol. xvi. p. 235.

much respected family of Aguirre y Montufar, from whom Humboldt had received great hospitality in 1802. They had both a lively recollection of that time, and could distinctly recall the distinguished man of science, then comparatively young, of whose sojourn in Quito they gave me many interesting particulars. Señora Rosa Montufar—in 1802 a noted beauty of Quito, but much changed when I saw her in 1859, a sister of Carlos Montufar, who accompanied Humboldt in his ascent of Chimborazo—related to me, among other things, the following interesting details, which I inscribed in my journal : —" The baron was always amiable and polite. At table he never remained longer than was necessary to satisfy the claims of hunger and pay courteous attention to the ladies. He seemed always glad to be out of doors again, examining every stone and collecting plants. At night, long after we were all asleep, he would be gazing at the stars. To us young ladies, this mode of life was even more incomprehensible than to my father the marquis."

' The house occupied by Humboldt and Bonpland in Quito, near the grand square, was only slightly injured by the earthquake of the 22nd of March, 1859, which laid so many buildings in ruins. The family of Aguirre have still in their possession a half-length portrait, life-size, of their distinguished guest, painted by a native artist, which is preserved in their country house of Chillo, half a day's journey from Quito, whence Humboldt used to make excursions in the pursuit of geology and botany. The young German baron, at that time (in 1802) thirty-three years of age, is represented in a court uniform of dark blue with yellow facings, a white waistcoat, and white breeches of the fashion of the last century. His right hand rests upon a book, entitled " Aphorism. ex Phys. Chim. Plant." His thoughtful brow is covered by long dark brown hair. The features in the youthful face are strongly marked, especially the nose, mouth, and chin. The peculiar expression of the eyes is the point of resemblance most readily traceable in this picture to Humboldt as I saw him fifty years later, then a venerable old man. The artist has evidently given a faithful representation of the features of Humboldt's countenance. But of the genius of that master mind, as in the glory of his manhood his eye threw a penetrating glance over the magnificent valley of Chillo, where Nature in her grandest forms lay before him—a genius which no doubt shone mightily through every feature of that expressive countenance—the painter has failed to give more than a very inadequate rendering.'

Humboldt left the province of Quito in July, 1802.

CHAPTER VII.

FROM QUITO TO MEXICO.

Acquisition of Manuscripts—The Carib and Inca Languages—Former
    Civilisation—The Road of the Incas—Expedition to the Amazon, and
    Return over the Andes—Caxamarca—First View of the Pacific Ocean
    —Truxillo, Lima, Guayaquil—Guano as Manure—Acapulco—The
    Humboldt Current—Letter to the National Institute.

As all hope of joining the expedition under Baudin was now
finally extinguished, Humboldt came to the determination of
relying, for the future, entirely on his own resources ; he at once
made arrangements for leaving Quito, and undertook an expedition to the River Amazon, on his way to Lima, where he
hoped to observe the transit of Mercury.

The route by which he travelled led him by the ruins of
Lacatunga, Hambato, and Riobamba.

' At Riobamba,' he writes to his brother, ' we spent some
weeks with a brother of our travelling companion, Carlos Montufar, who resides there officially as corregidor—a magistrate
by royal appointment. Here we made by chance a most remarkable discovery. The condition of the province of Quito
prior to its conquest by the Inca Tupayupangi is still involved
in obscurity. We ascertained that the Indian king, Leandro
Zapla, who resides at Likan, and who, for an Indian, is a man of
considerable culture, is in possession of manuscripts written in
the sixteenth century, by one of his ancestors, in the Puru-
guayan tongue. This was, at that time, the universal language
of Quito, though, owing to the introduction of the Inca or
Quichua language, it has since been lost. It is, therefore,
a fortunate circumstance that Zapla is also in possession of a
translation of these papers in the Spanish tongue, the work of
another of his ancestors.

' From this valuable source, we have gathered much interesting information concerning the history of those times, especially with regard to the remarkable eruption of the Nevado del Altar, which must at that time have been the highest mountain in the world, higher even than Chimborazo, and known to the Indians by the name of Capa-urku (Chief of the Mountains). This event occurred in the reign of Uainia Abomatha, the last independent kochokanao (king) of the country, who held his court at Likan. The priests gave the following ill-omened interpretation of this portentous catastrophe :—" The earth," said they, " is changing its form ; a new order of deities are coming. by whom the gods we now serve will be driven away. Let us not withstand the decrees of fate ! " The worship of the sun was in fact introduced by the Peruvians, in place of the old-established religion. The eruption of the volcano lasted seven years, and, according to Zapla's manuscript, there fell so dense and perpetual a shower of ashes, that at the- town of Likan there was no daylight for seven years. Exaggerated as this statement may appear, it would seem not to be wholly without foundation, for Quito has frequently been veiled in darkness by the ashes from Cotopaxi for fifteen and eighteen hours at a time ; moreover, the mass of volcanic material strewed over the plain of Tapia testifies to some enormous eruption, while the two lofty peaks warrant the supposition that the gigantic mountain, asserted at that time to have fallen in, must some time or other have been violently torn asunder.

'The discovery of this manuscript has revived my wish to investigate the early history of the aborigines of these countries, a desire first aroused in me by the traditions I collected at Parime, and by the hieroglyphics I met with in the wilds of the Casiquiare, where there is now no trace of inhabitant. A recent perusal of Clavigero's account of the wanderings of the Mexicans in South America has again directed my thoughts to the subject, which I intend to follow up as soon as I can devote time to the purpose.

'The study of the American languages has also occupied much of my attention, and I have discovered no evidence of the poverty remarked by La Condamine. The Carib language, for instance, combines richness, grace, power, and tenderness. It affords means of expression for abstract ideas—futurity,

eternity, existence, &c.—and is able to express in words every
numerical combination which can be denoted by figures. I am
devoting myself particularly to the Inca language, which is in
ordinary use in this part of the country—Quito, Lima, &c.—
and is so rich in variety and delicacy of expression, that the
young gentlemen, when making themselves agreeable to the
ladies, usually adopt it after they have completely exhausted
the vocabulary of the Castilian tongue.

' These two languages, together with some others of equal
richness, afford sufficient evidence that there once reigned in
America a higher state of civilisation than existed at the time
of the Spanish conquest in 1492; but I am in possession of
proofs of a much more positive nature in regard to this fact.
Not only in Mexico and Peru, but at the court of the King of
Bogota, the priests of those ages possessed sufficient knowledge
of astronomy to draw a meridian line and to observe the actual
moment of the solstice; they changed the lunar into the solar
year by the intercalation of days, and I have in my possession a
stone in the form of a heptagon which was found at Santa Fé,
and was employed by them in the calculation of these inter-
calary days. Nor is this all. At Erivaro, in the interior of the
district of Parime, the natives believe that the moon is inhabited, and know, by the traditions of their ancestors, that its
light is derived from the sun.

' From Riobamba, the route to Cuença led me over the famous
Paramo of Assuay. Before setting out I visited the extensive
sulphur mines at Tiscan. The rebel Indians conceived the idea
of setting fire to these sulphur works, after the earthquake of
1797 ; certainly the most horrible plan ever devised even by a
people driven to despair. They hoped by this means to produce an eruption by which the whole province of Alausi should
be destroyed. On the Paramo of Assuay, at a height of 15,090
feet, the magnificent road of the Incas may still be traced.
This causeway reaches almost to Cuzco, and is constructed
entirely of hewn stone; it is perfectly straight, and resembles
the finest roads of the ancient Romans. In the same neighbourhood are the ruins of the palace of the Inca Tupayupangi,
which was described by La Condamine in the Memoirs of the
Academy of Berlin. I do not know whether he mentions the

so-called summer-house of the Inca. It is a couch cut in the rock, ornamented with arabesque devices. Our English gardens contain nothing more elegant. The good taste of the Inca is everywhere visible; the seat is so placed as to command a most enchanting prospect. In the sandstone rock of a neighbouring wood is to be seen a circular spot of yellow ironstone, which the Peruvians have ornamented with figures, supposing it to represent the sun. Of this I made a drawing.

'We remained only ten days at Cuença, whence we set out for Lima, passing through the province of Jaen, where we spent a month in the vicinity of the Amazon. We arrived at Lima on the 23rd of October, 1802.

'I think of setting out in December for Acapulco, *en route* for Mexico, in the hope of reaching Havana in May 1803. I shall then embark without delay for Spain. You will perceive that I have given up the idea of returning by way of the Philippines. I should have to encounter a sea voyage of prodigious length, to see scarcely more than Manilla and the Cape; and had I made up my mind to visit the East Indies I should have been very inefficiently provided, as the necessary equipments could not have been procured here.'

Additional details may be gathered from the letter to Delambre, already referred to at page 315, dated Lima, November 25, 1802, wherein Humboldt remarks :—

. . . . 'After crossing Assuay and passing through Cuença, where bull-fights were given in our honour, we took the road to Loxa, in order to complete our investigations upon the Cinchona. We spent a month in the province of Jaen de Bracamoros, and visited the Pongo of the Amazon, where the banks of the river are ornamented with the *Andira* and the *Bougainvillea* of Jussieu. It was to me a matter of considerable interest to determine the longitude of Tomependa and Chuchungat, which, from being included in La Condamine's map, gives me a line of connection with the coast. La Condamine was only able to obtain the longitude of the mouth of the Napo, and as chronometers were unknown in those days, the longitudes then taken have great need of revision. My chronometer by Louis Berthoud performs admirably. . . .

'On leaving the Amazon we crossed the Andes, near the

mines of Gualgayoc, yielding sulphuret of silver and copper, at
an elevation of 13,550 feet: their yearly produce is valued at a
million piastres. We descended to Truxillo, by Caxamarca,
where I made drawings of the Peruvian arches in the Palace of
Atahualpa, and thence crossed the desert plains on the coast of
the Pacific Ocean to Lima—a place which for six months in
the year is overhung with dense vapours.' . . .

At the conclusion of this very long letter he writes:—' I am
full of anxiety about my manuscripts, considering how to preserve them in safety and secure their publication. I trust I
shall greet you in Paris some time during the months of September or October of 1803. How I long to be once more in
Paris !'

The latter part of this journey—that is to say, his route from
the Amazon to the coast across the Andes, has been described
by Humboldt in a short treatise, entitled: ' The Plateau of
Caxamarca, the Ancient Capital of the Inca Atahualpa; first
View of the Pacific Ocean from the Crest of the Andes.'[1] In
ancient Caxamarca were enacted the bloodiest scenes of the
Spanish Conquest. Beneath the ruins of the citadel and the
ancient palace of Atahualpa the room is still to be seen, upon
the wall of which is shown the mark indicating the height to
which the Inca engaged to fill his prison with gold, on condition of being set free.

In agreeable contrast with the gloomy associations of this
place was the grand prospect Humboldt first obtained here of
the Pacific Ocean—a moment for which he had so ardently
longed. To one who was so deeply indebted to a fellow-voyager
of Captain Cook, both for the development of his mind and the
formation of his tastes, there was something in this sight peculiarly impressive. ' To George Forster I had early communicated, in general outline, my schemes for travel, when privileged under his auspices to visit England for the first time.
The charming description of Otaheite given by Forster had
awakened, especially in the North of Europe, a universal, I
might almost say a romantic interest in the islands of the
Pacific Ocean. These islands, fortunately for them, were at

---

[1] ' Ansichten der Natur,' vol. ii. p. 315.

that time but seldom visited by Europeans. This rare privilege promised shortly to be mine ; for the object of my present journey was not so much to observe the transit of Mercury as to fulfil an engagement made with Captain Baudin, on my departure from Paris, to join his expedition round the world, upon which he was to start as soon as the French Republic was in a position to furnish the sum of money already destined for that object.'[1]

After visiting the mines of Gualgayoc, Humboldt crossed for the fourth time the chain of the Andes, and passing through Quercotillo and Cascas, reached the shore of the Pacific Ocean at Truxillo. Here he remained a few days in order to obtain the latitude and longitude of the town, and to test the going of his chronometer, and then set out for Lima, travelling along the coast, over a portion of the great plain of Peru, which stretches to the south as far as Pisco and Yca.

The geographical position of Lima had up to that time been very imperfectly ascertained. For its more accurate determination Humboldt compared the longitude given by his chronometer with the results obtained by a series of lunar altitudes. The transit of Mercury was successfully observed at Callao, the port of Lima, on November 9, 1802.

While at Callao, Humboldt's attention was first directed to the valuable properties of guano as manure. Through him its efficacy was first tested in Europe, where its introduction for agricultural purposes may be ascribed to his writings, in which he discusses its formation, and describes its profusion and successful employment upon the sterile coasts of Peru.[2]

On December 5, 1802, Humboldt embarked for Guayaquil, and landed at that port on January 9, 1803. During the voyage he rendered valuable service to navigation by determining the exact position of several places along the coast— the island of Pelado, Points Aguya, Pariña, Mala, &c. His stay at Guayaquil lasted six weeks, during which he made an excursion to the almost impenetrable forests of Babajos, and

[1] 'Ansichten der Natur,' vol. ii. p. 365.
[2] Wilhelm Cohn-Martiniquefelde, 'Alexander von Humboldt und die Landwirthschaft,' in Frühling's ' Neue landwirthschaftliche Zeitung,' 19th annual issue, Part III.

was very nearly being an eye-witness of the frightful eruption of Cotopaxi.

On February 15, 1803, he sailed from Guayaquil for Mexico, and landed in the harbour of Acapulco towards the end of March.

It has often been erroneously maintained—among others by Carl Ritter—that the cold current upon the coast of Peru, known as the Humboldt current, was *discovered* by him during this voyage. The truth is, that Humboldt merely instituted a series of very careful observations, especially with regard to the temperature of the current, and far from having any wish to appropriate to himself or allow others to ascribe to him a merit to which he was not fully entitled, he distinctly repudiated the discovery attributed to him by the remark: ' The existence of this current has been known since the sixteenth century to every sailor-boy accustomed to navigate from Chili to Payta.' [1]

Humboldt had no wish to remain long in the kingdom of Mexico. The motives by which he was induced to curtail the journey as originally planned, and to postpone, at least for the present, his projected tour through a part of Asia and Africa, are specified in the following letter, addressed to the National Institute of France, dated ' Mexico, 2 Messidor IX. (21st June, 1803)' :—[2]

' Our voyage through the Pacific to Acapulco was happily accomplished, notwithstanding a violent tempest which we encountered when more than 300 leagues to the west of the volcanoes of Guatemala—a part of the ocean to which the name of Pacific is scarcely applicable. The damaged state of our instruments occasioned by land transport, in journeys extending over 2,000 leagues, the futility of our efforts to replace them by new ones, the impossibility of meeting with Captain Baudin, for whom we had waited in vain upon the shores of the Pacific, the reluctance we felt to traverse a boundless ocean in a merchant ship which could furnish no facilities for touching at any of those lovely islands so interesting to the natu-

---

[1] ' Briefwechsel mit Berghaus,' vol. ii. p. 284. (See also pp. 160 and 275.)

[2] 'Annales du Museum d'Hist. natur.' An XII. (1804) vol. iii. p. 396.

ralist, but, above all, the rapid advancement of science and the
necessity of gaining acquaintance with the new discoveries
which must unquestionably have taken place during an interval
of four or five years, . . . these are the motives which have
led us to abandon the projected plan of returning by the Philippines and through the Red Sea to Egypt. Although enjoying everywhere the distinguished protection of the King of
Spain, I could not fail, as a private individual travelling at my
own cost, to encounter a thousand difficulties, unknown to those
engaged in expeditions undertaken by Government. The task
to which we shall henceforth devote ourselves will be the reduction of the observations we have made while in the tropics,
and their arrangement for publication. With life still before
us, and inured to danger and privations of all kinds, we yet
linger over the hope of visiting Asia and her adjacent islands
at some future time. With an increased store of knowledge,
and possessed of instruments of greater accuracy, we may perhaps some day undertake another expedition, the plans for
which already allure us as in a seductive dream.'

CHAPTER VIII.

MEXICO AND THE UNITED STATES.—RETURN HOME.

From Acapulco to the Capital—Acquisition of Historical Information
concerning New Spain—The Mines of Moran and Guanaxuato—The
Jorullo—Correspondence with Willdenow—Reminiscences of a Fair
Mexican—Popocatepetl and Iztaccihuatl—The Pyramid of Cholula—
Jalapa, Cofre, Orizaba—Second Visit to Havana—Visit to the United
States—Sojourn with Jefferson at Washington—Return Home—At
Bordeaux—Humboldt an Apparition—Greetings to his Friends.

THE exact determination of the latitude and longitude of the
harbour of Acapulco was of the highest importance to the
geography of America, since this port served as a starting-point
for determining, by means of the chronometer, the position of
all the harbours in use along the north-west coast of America
up to the 60° of latitude.   Setting aside former errors which
placed this harbour full four degrees too much to the west, even
Arrowsmith, in his 'Chart of the West Indies and Spanish
Dominions' (in four sheets), bearing date 1803, is in error as to
its position to the amount of half a degree of longitude, and
seven minutes of latitude.[1]   The correction of this error is due
to Humboldt.

Before visiting the interior, he remained some time at the
coast for the purpose of completing his collections and instituting a series of observations.   On his way to the capital he
traversed the scorching valleys of Mescala and Papagayo, where
the thermometer stood at 104° in the shade, and thence as-

---

[1] In a skeleton map, entitled 'Carte des Fausses Positions,' forming No.
10 of the 'Atlas géogr. de la Nouv. Espagne,' the large errors hitherto
existing in the maps of New Spain are very strikingly shown in the erroneous positions of the three most important places in Mexico, the capital and
the harbours of Acapulco and Vera Cruz.

cended to the high plains of Chilpanzingo, Tehuilotepec, and Tasco—plains which, at an elevation of between 4,000 and 5,000 feet, enjoyed a temperate climate favourable to the cultivation of most of the European cereals and the growth of the oak, the cypress, and the fir. At Tasco, he inspected the silver mines, which are some of the oldest and probably the richest in the country, and thence continued his journey through Cuernaraca and the foggy district of Cuchilague to the capital, where he arrived towards the end of April.

The city of Mexico numbered at that time more than 150,000 inhabitants, and was selected by Humboldt as his head-quarters, as from its central position it afforded facilities for making excursions of varying extent in all directions. The time occupied by his sojourn in the kingdom of Mexico was little short of a year; that is to say, from March 23, 1803, till March 7, 1804.

'I endeavoured to employ the time spent in Mexico not merely in scientific investigation, but in acquiring an accurate knowledge of the political condition of this extensive and remarkable country. The civilisation of New Spain presented a striking contrast to the limited amount of culture, both physical and moral, visible in those countries I had recently visited. I carefully compared all that I had seen upon the banks of the Orinoco, and the Rio Negro, in the province of Caracas, in New Granada, on the slopes of the Andes around Quito, as well as on the coasts of Peru, with the condition of things I found in the kingdom of Mexico. The result of this comparison was to incite me to investigate the causes, as yet but partially developed, which have proved so favourable for the increase of population and of national industry in this country.

'The circumstances in which I was placed were highly advantageous for the prosecution of this object, since in the collection of materials, in which no published book could be of any avail, various manuscripts were placed at my disposal, and I was allowed free access to the public archives.'[1]

---

[1] A. von Humboldt, 'Ueber den politischen Zustand des Königreichs Neuspanien' (Tübingen, 1809). Preface.

Thus originated his celebrated work entitled ' Essai politique sur le Royaume de la Nouvelle-Espagne.'

The courteous and friendly reception which had been accorded to Humboldt at Lima and other capital cities in the Spanish colonies [1] awaited him also at Mexico, where he was graciously received by the viceroy Iturrigary.

After a sojourn of some months in the capital, Humboldt visited the famous mines of Moran and Real del Monte. The whole of this district is of the highest geological interest, affording specimens of basalts, amygdaloid and calcareous rocks of the secondary formation, besides the remarkable columnar porphyry of Actopan.

In the month of July, Humboldt started on a second expedition, to visit the provinces in the north of the kingdom. His route lay first to Huehuetoca, where, at a cost of six million piastres [1,291,770*l.*], a canal has been cut through the mountain Sinoq, for the purpose of draining the valley of Mexico into the river Moctezuma ; thence he visited Queretaro, which in our days has become invested with a melancholy interest, as the scene of the tragic fate of the unfortunate Emperor Maximilian, and after passing through Salamanca, and across the fertile plain of Irapuato, he reached Guanaxuato, a town of 50,000 inhabitants, situated in a narrow ravine, and possessing in its vicinity mines even more prolific than were formerly those of Potosi.

Humboldt spent two months at Guanaxuato occupied with observations of various kinds, and researches as to the geological structure of the district, and on the completion of his labours, he continued his journey to Comagillas, noted for its hot mineral springs, of which he made a chemical analysis. He thence travelled through the valley of the Santiago, which in primeval times had probably been enclosed by basaltic mountains of volcanic origin, to Valladolid, the capital of the ancient kingdom of Michoacan, situated in one of the most beautiful and fertile districts of the New World. Thence he descended amid the ceaseless rain of the autumnal season to Patzcuaro and over the plains of the Jorullo, to the shores of the Pacific. On September 19, he ascended the peak of Jorullo, a volcano

[1] See pp. 293, 299, 305.

rising to the height of more than 1,500 feet, formed in the course of a night, during the year 1759.

Humboldt and his companions descended to a depth of 250 feet into the crater, whence smoke was issuing from more than 2,000 small orifices. With considerable danger, on account of the brittle nature of the blocks of lava, they reached almost to the bottom of thê crater, and there made an analysis of the air, which was heavily laden with carbonic acid. The inhabitants of that locality maintained that the heat of the crater had formerly been much greater, and Humboldt gave it as his opinion that the whole of the surrounding district is undermined by volcanic action.

Humboldt returned to Mexico, across the Plateau of Toluca, and ascended the volcano of that name on September 29. At the capital he again made a sojourn of several months, for the purpose of arranging his botanical and geological collections, and of reducing his barometric and trigonometric observations, while, at the same time, he laboured for the completion of his tables of statistics, and carefully prepared materials for a map of the country.

During his first visit to the capital, Humboldt wrote to Willdenow the following letter, dated April 29, 1803 :—

' I received your welcome letter of the 1st October, 1802, a few days after my arrival in this great and imposing capital of New Spain. My joy in reading it was all the greater from its being the first and only letter I have received from you since I left Europe, although I am convinced you must have written to me frequently. Even from my brother I have received only some five or six letters at most, during the four years that have elapsed since I left Corunna. It seems as if an unfriendly star had presided over our letters, if not over the ships that carried them. But I will not complain, since I have now the prospect of so soon embracing you all again.

' We have already despatched to Europe some ten or twelve consignments of newly gathered seeds; one parcel went to the Botanic Gardens at Madrid, among which, as I learn from the " Annales de Historia Natural," Cavanilles has already discovered some new species; a second parcel was enclosed to the Jardin des Plantes at Paris ; and a third went by way of

Trinidad to Sir Joseph Banks in London. Do not suppose, however, that my stores are thereby exhausted, or that I have altogether forgotten Berlin. I have still by me an excellent collection, the result of my botanical explorations in the districts of Quito, Loxa and Jaen, on the banks of the Amazon, on the slopes of the Andes in Peru, and during my travels from Acapulco to Chilpensingo and Mexico. I shall not venture to trust this invaluable treasure to the care of the postal authorities here, for they are incredibly careless in their arrangements; but as I am about to start for Havana, on my return to Europe, I shall convey this collection to you in person. I have taken the greatest precautions to have the specimens carefully dried. . . .

'My friends in America, moreover, have been kind enough to say that they will always be glad to forward parcels of fresh seeds, as you may require them. During my travels I have met with several assiduous botanists; among whom I may mention Tofalla at Guayaquil, Oliveda at Loxa, Mutis at Santa Fe, and his pupil Caldas, at Popayan.

'I am delighted to find that the botanical specimens which I sent you through Mr. Fraser reached their destination at length in safety.' (See p. 281.)

After giving a summary of his movements since leaving Quito, Humboldt continues:—'I was very anxious to have returned to Europe before the end of the year, but the news that the yellow fever is raging at Vera Cruz and Havana, and the fear of encountering a bad passage across the Atlantic, so late in the year as October, have decided me to postpone my journey. I should not like my travels to end with a tragedy. By adopting what seems to me to be the safest course, my arrival in Europe will be delayed till April or May in 1804.'

In two other letters written by Humboldt during his residence at Mexico,—the one addressed to Cavanilles on April 22, and the other to Delambre on July 29,—he recapitulates the events of the last eighteen months, and describes in detail the incidents of his journey, the difficulties he encountered, and the scientific results of the expedition. At the close of his letter to Delambre, he again announces the abandonment of his original plan of returning by the Philippines, but adds :—

'I have only given it up temporarily, for I have many projects in view with regard to the East Indies, but I am anxious
first to publish the results of this expedition. I hope to be
with you early next year ; the reduction of our observations
will occupy us for two or three years at least. In speaking
only of two or three years, pray do not laugh at my inconstancy,
this *maladie centrifuge* of which Madame . . . . used to ac-
cuse me and my brother. It is a duty incumbent upon everyone to seek that position in life in which he thinks he can be
of most service to his generation, and I believe that to fulfil
my destiny, I ought to perish at the edge of a crater, or be
engulphed by the treacherous deep. This at least is my present
opinion, after experiencing hardships and privations of all
kinds for five years ; but I can easily believe that with advancing age, and the renewed enjoyments of European life, I may
yet live to change my views. *Nemo adeo ferus est, ut non
mitescere possit.*'

In Mexico, as well as at Lima, we are indebted to a lady for a
sketch of Humboldt as he appeared in social life amid the
highest circles in the capital.

Madame Calderon de la Barca, wife of the Spanish ambassador at Mexico, in her journal for the years 1839 and 1840,
mentions among her acquaintance a lady formerly well known
and greatly esteemed in Mexico, under the name of 'the fair
Rodriguez,' who had been regarded by Alexander von Humboldt as the handsomest woman he had met with during his
travels. In conversation with this lady Madame de la Barca
gathered the following particulars, which she thus narrates :—

'We talked of Humboldt, and while speaking of herself
quite as an indifferent person, she recounted to me the details
of his first introduction to her and of his admiration of her
beauty : she was very young at the time, though married, and
the mother of two children, and happened to be in the room,
seated at the window, sewing, when the baron paid a visit to
her mother. Her presence was unobserved until, on his expressing a wish while conversing eagerly on cochineal to visit a
certain plantation, she remarked from her place at the window :
"Oh ! we can easily drive Herr von Humboldt there,"—when
he looked up and stood transfixed before her, exclaiming at

length, " Valgame Dios! who is that young lady ? " From that
time forth he was always at her side, more captivated, it is
said, by the graces of her mind than by the beauty of her
person. She was regarded by him as an American Madame de
Stael. One is led to suspect from these little incidents that
the grave man of science was fascinated by a witchery, from
which all his mines and mountains, his geography and geology,
his fossil shells and Alpine limestone, were alike powerless to
protect him.'

In concluding these reminiscences of the Mexican beauty,
Madame Calderon de la Barca adds, with evident satisfaction :
' It is quite refreshing to see that even the great Humboldt was
not wholly removed from susceptibilities of this nature ! '

In January, 1804, our travellers left the capital to explore
the eastern slope of the Mexican Cordilleras. They measured,
by geometrical observations, the height of Popocatepetl, and
Iztaccihuatl, known as the volcanoes of Puebla : the crater of
the former is inaccessible, though a tradition exists that Diego
Ordaz was lowered down into it by means of ropes, for the
purpose of collecting sulphur, apparently an unnecessary
proceeding, since sulphur abounds on the plain.

The summit of Popocatepetl, which has been ascended to a
height of 16,779 feet by the zealous mineralogist Herr Son-
nenschmidt, was discovered by Humboldt to be considerably
higher than the Peak of Orizaba, hitherto regarded as the
highest mountain of the plateau of Anahuac. He also took
measurements of the great Pyramid of Cholula, a mysterious
erection of unburnt bricks, the work of the Totteken, from the
summit of which there is a magnificent prospect of snowclad
peaks, and the fertile plains of Tlascala.

Continuing their descent to the coast, the travellers passed
through Perote, on their way to Jalapa, which from its
position, 4,423 feet above the sea, enjoys a mild and genial

---

[1] As a matter of curiosity, it may be mentioned that during his sojourn
in Mexico Humboldt wrote in May, 1803, to assert his claim to the invention of an instrument for measuring carbonic acid, in the ' Allgemeine
Literaturzeitung,' 1800, No. 93, unjustly appropriated by Voigt, instrument
maker to the Duke of Saxony. (' Intelligenzblatt zur Allgemeinen Literaturzeitung,' 1803, p. 1487.)

climate. Barometric measurements of height were taken by Humboldt at frequent intervals along this ill-constructed road, cut through almost impenetrable forests of oak and fir, and the results of these labours were afterwards employed as a basis for laying down the line for a new high road. On February 7, 1804, the travellers made the ascent of Cofre, a mountain in the neighbourhood of Perote, the height of which exceeds that of the Peak of Teneriffe by 1,063 feet. Trigonometric measurements were also taken by Humboldt of the volcano of Orizaba.

Thus employed in scientific investigations of various kinds, the travellers prosecuted their journey, until they at length reached the goal of their wanderings, and entered the town of Vera Cruz, the harbour of which is the centre of European commerce, in the West Indies. The town is situated in an arid plain, with neither river nor running stream in its vicinity. From this port they embarked on board a royal frigate on March 7, 1804, for Havana, to take possession of the collections left for safe keeping in the year 1800, and to complete the data they had gathered in Mexico, for the treatise upon which Humboldt was engaged, entitled, 'Essai politique sur l'Ile de Cuba.'

After a sojourn of nearly two months at Havana, Humboldt set sail for the United States, on April 29, accompanied by Bonpland and Carlos Montufar. In the Channel of Bahamas, they encountered a severe storm, which lasted seven days ; but Humboldt and his companions were mercifully preserved through every danger, and after a voyage of twenty days they arrived safely in the harbour of Philadelphia.

A welcome, as courteous as that which had been accorded to Humboldt by the viceroy at Mexico, awaited him from Jefferson, the President of the United States, from whom he received the following invitation to visit Washington :—

'Washington : May 28, 1804.

'Sir,—I received last night your favour of the 24th, and offer you my congratulations on your arrival here in good health after a tour in the course of which you have been exposed to so many hardships and hazards. The countries you have visited are of those least known and most interesting, and a lively desire will be felt generally to receive the information you will be able to give. No one will feel it more strongly than myself, because no one perhaps views this New World with more partial hopes of its exhibiting an ameliorated state of the human condition. In the new position in which the seat of our government is fixed, we have nothing curious to attract the observation of a traveller, and can only substitute in its place the welcome with which we should receive your visit, should you find it convenient to add so much to your journey. Accept, I pray you, my respectful salutations and assurances of great respect and consideration, &c.

'JEFFERSON.

'A M. le Baron de Humboldt.'

Short as was Humboldt's stay in the United States, his visit was productive of important results in the opportunity it afforded him of becoming acquainted with that wonderful political organisation which was then exciting universal interest. He studied the system in its minor details, as well as in the broad principles of its administrative economy, and instituted a comparison between the condition of the United States and that of the Spanish colonies, through which he had just been travelling. He made acquaintance with the most influential men in the country, and obtained from intercourse with them some insight into their views as to the future policy of the re-public. Engaged with considerations of this nature, Humboldt found ample occupation during the three weeks he spent with Jefferson at Monticello; during his visit the president com-municated to him an extraordinary project of his imaginative but somewhat fantastic genius, a project for the future division of the continent of America into three great republics, in which were to be incorporated the Spanish possessions in Mexico, and the States of South America.[1]

Throughout the remainder of his life, Humboldt always retained a pleasing remembrance of this sojourn in the United States, and frequently alluded to it in grateful terms to the various Americans who visited him : the shameful institution of

[1] Silliman, 'A Visit to Alexander von Humboldt.'

slavery was the only cloud which cast a shadow upon these bright recollections, and saddened his noble spirit.

On July 9, 1804, Humboldt bade farewell to the continent of America, and set sail from the mouth of the Delaware, landing, after a prosperous voyage, at Bordeaux on August 3.

The news of his return spread great and universal joy throughout Europe, for notwithstanding the numerous letters received from him, reports of his death had frequently appeared in the newspapers.[1] Even so late as July 17, 1804, Körner wrote to Schiller: 'Pray let me know if you hear anything certain of Alexander von Humboldt. I shall be exceedingly glad if the report of his death proves to be groundless.'[2]

The welcome announcement of the safe arrival of the distinguished traveller was first received at the French capital through his letter to the National Institute, by whom the intelligence was at once communicated to his sister-in-law, the wife of William von Humboldt, at that time in Paris.

The pleasurable emotions experienced by Humboldt upon his safe return find expression in the following letters:—

To Freiesleben.

'Off Bordeaux: August 1, 1804.
'In haste.

' My dearest Karl,—After an absence of five years I am once more upon European soil. We made the entrance of the Garonne two hours ago. We have been most highly favoured in our voyage, accomplishing the passage from Philadelphia

---

[1] In the summer of 1803 it was reported in Paris that Humboldt had perished among the savages in North America. ('Allg. geogr. Ephemeriden,' vol. xii. p. 239.) The 'Hamburger Correspondent' of June 12, 1804, contained the announcement:—'We regret to learn that the celebrated traveller, Herr von Humboldt, has been attacked with yellow fever, and has died at Acapulco.' (Ibid. vol. xiv. p. 510.)

[2] 'Briefwechsel zwischen Körner und Schiller,' vol. iv. p. 366. Humboldt himself wrote in later years when consoling the father of the unfortunate African traveller, Eduard Vogel, for the sad fate of his son:— . . . . 'Consolation can only come from above ! I fulfil a tender duty in reminding you of this, and in assuring you that I am not without hope of his safety. I have myself read the announcement of my own death in the South Seas, which had been promulgated in Paris ; and one evening, when visiting the

in twenty-seven days. I left Mexico in February,[1] and sailed
by way of Havana to North America, where I was loaded with
marks of honour by Jefferson, the President of the States. My
expedition in both hemispheres, extending over a distance of
40,000 miles, has been favoured by fortune to an almost unprecedented degree. I was never once ill, and I am now in better
health, stronger in body, more industrious, and gayer in spirits
than ever. I return laden with thirty cases of treasures of all
kinds, botanical, geological and astronomical, and it will take
me years to bring out my great work. My sectional drawings
of the Andes, based upon no less than 1,500 of my own measurements, a botanical atlas, and a geological pasigraphy, consisting
of new symbols for expressing the various formations, will all
interest you exceedingly. I confess it was with a heavy heart
that I bade farewell to the bright glories of a tropical clime, yet
the thought of being within reach of you, of embracing you
once more, and of gathering the gold out of the quartz with you
by my side, possessed for me a still higher attraction. As soon
as I can get out of quarantine, I shall leave for Paris, that I may
begin my work immediately. I am particularly anxious to
commence the reduction of my astronomical observations. *When*
I shall see you, my dear Karl, and *how soon*, I cannot say. My
friends are all dispersed in Spain, Italy, and elsewhere. I quite
dread the first winter, everything will be so strange, and I shall
be some time in settling down again. But I shall be consoled
through everything by the thought of my safety. Pray remember me to your parents, to Fritzchen, Fischer, and Werner,
for whom my veneration increases year by year, for I saw much
to substantiate his system during my travels in the southern
hemisphere. How shall I ever find time to write to them all!
Remember me to Böhme, and to all our old friends.

' Ever yours,

' HUMBOLDT.

Duc de Crillon, as my name, according to Paris custom, was announced at
the door, a scream was heard, and a lady fainted away. This lady was
Madame La Pérouse, in whom grief at the loss of her husband had been
painfully reawakened by the announcement of my name, as of an apparition
from the world of shadows.'

[1] More correctly, on March 7.

' I used often to speak of you with Del Rio, who is living at Mexico, and is now married. I have with me a piece of native platinum, two ounces in weight, and about this size (a sketch). The platina sand occurs in hyacinth, basalt, and clinkstone.

' Write to me at Paris, under care of M. Chaptal, Ministre de l'Intérieur, and tell me of any geological books I ought to read, and of any new ideas that Werner may have lately promulgated. I do not know the address of our excellent friend Buch, but please send him from me an affectionate greeting.'

To Kunth.

' On board " La Favorite," off Bordeaux:
' In quarantine, August 3, 1804.

' My dear Friend,—Mercifully preserved from all the dangers incident to foreign travel, I have the happiness of finding myself, after an absence of six years, once more on European soil, and I seize an early opportunity of informing you of my return, and of giving you the renewed assurance of my affectionate regard. I am too well acquainted with the kind feelings you generously entertain towards me, not to flatter myself with the thought that these lines will be joyfully welcomed by you, and will be received also with pleasure by the small circle of friends who after my long absence may still be left to me at Berlin. Fortune has never once deserted me during the great expedition successfully brought to a close with my companions Messrs. Bonpland and Montufar. After passing two months very pleasantly in the United States, visiting Philadelphia, Baltimore, and Washington, where we enjoyed the most marked courtesy from Mr. Jefferson and the highest officers of the Republic, we embarked for Europe, and a voyage of twenty-nine days brought us from the mouth of the Delaware to the Garonne. We anchored on the 1st of August, and are now in quarantine, which I have reason to hope, from the respectful manner in which we are treated, will not be greatly prolonged, especially as the yellow fever had not yet reached North America when we set sail. My collections fill thirty-five cases, which I am anxious to despatch to Paris, that I may bring them before the attention of scientific men, and compare them with previous collections. I have a great longing to see my brother, who is, I suppose, at Rome, and therefore I shall most likely pass the winter there. It is five years since I heard from you: surely, my good friend, you have not altogether forgotten me? That is impossible. Write to me as soon as you can, and direct your letter to Paris, under care of M. de Luchesini. Tell me of your health, your means, your mode of life—you know how keenly I am interested in everything connected with yourself. Nearly a year ago I drew upon you to the amount of 10,000 piastres, in favour of Mr. Murphy, of Cadiz. Please let me know if this money has been paid, as in that case Mr. Murphy owes me 6,000 piastres. I should be much obliged if you would kindly send me, by return of post, a brief abstract of the state of my affairs and of my present income, made out, if possible, in French, on a separate piece of paper (without comment), and bearing your signature, for I may probably be in want of such a statement in transacting some affairs in which I am at present engaged. The state of my finances absorbs a good deal of my attention just now. I am not only free from debt, but, if Mr. Murphy has been paid, I have 6,000 piastres in his hands at Cadiz. I am sure you will not misinterpret the expression "without comment," for you well know that any comment or counsel from you is always most highly appreciated by me; but the statement that I venture to ask for must be of a character that I can show to anyone.

'I am much more robust and stouter than I used to be, and more active than ever. Nevertheless, you and I, my dear friend, are beginning to grow old. Write me a long letter. You know the strength of my attachment, and how vast the debt of gratitude I owe you, for you must be aware that the little celebrity I have attained is to a great extent the result of your labours, and I believe you are not so completely insensitive to renown as to be wholly indifferent to this thought. I embrace you with my heartiest affection.

'A. HUMBOLDT.

'I shall leave for Paris as soon as possible, where please address me, under care of M. Luchesini. Tell me about Minette,[1] the

---

[1] The sister of his friend Haften, see p. 144.

Haftens and the Captain, to whom I send my kindest remembrances.'

Humboldt also addressed a letter to his sovereign, Frederick William III., dated September 3, 1804, informing him that, after an absence of five years, during which he had travelled 40,000 miles in South America, he had, a few weeks since, safely arrived in Paris, bringing with him several interesting contributions to the various scientific museums at Berlin.[1]

[1] From the royal private archives.

CHAPTER IX.

HOME LIFE.

New Circumstances, but old Friends—Co-operation with Pictet—Plan for
the Publication of his Works—English Translation—Labours with
Gay-Lussac and Biot—Letters to Friends at Berlin—Visit to Italy—
Visit to Germany—Berlin—Honours and Occupation—Letters to old
Friends—The Fall of Prussia—Humboldt as a Mediator—Consolation
in the Study of Nature.

WHAT changes had transpired in France during Humboldt's
absence! The hopes of freedom had been cruelly blasted.
The Revolution had assumed a new phase, and was proceeding
in quite a new direction ; a sultry atmosphere everywhere gave
painful warning of the outbreak of another political convulsion.
On his departure for America, Humboldt left France a republic;
he returned to find an imperial throne occupied by an ambitious
conqueror; and, on his arrival at Paris, on August 18, he beheld
the people intoxicated with glory and conquest, still tumul-
tuous with the celebration of Napoleon's birthday, which had
been kept publicly, for the first time, on the 15th of the
month.

Meanwhile the interests of science, especially in the branches
of mathematics, physics, and natural science, far from suffering
from the events of war, had been very largely promoted. Rich
endowments were granted to scientific institutions of every
kind, and men of science were not only cheered by tokens of
honourable distinction, but received liberal support in all their
undertakings. Paris had undoubtedly become the high school
for the exact and natural sciences.

There seems some necessity for recalling these circumstances
to mind, since it is only by so doing that a just estimate can be

formed of the importance that was attached to Humboldt's return—an event which, in the midst of scenes of such excitement, could yet arouse the keenest interest. On his return, he was not only met by his old friends with expressions of undiminished affection, but he found himself surrounded by many new acquaintances, among whom Gay-Lussac,[1] already known to fame through his researches in chemistry, and François Arago, whom he met with somewhat later, deserve especial mention. The admiring wonder excited by the display of his treasures, and the interest aroused by the narrative of his journey, were enhanced by the conviction of the personal heroism which had enabled him successfully to encounter so many dangers, and undergo such long-continued exertion. With unexampled success had this important undertaking been carried out by the ability, intelligence, learning, and perseverance, of a private individual, actuated by no motive of personal advantage, but solely by the pure love of science. It was not so much the fulness and depth of knowledge evinced, nor the charm and truth of the narrative, nor yet the generosity and unreservedness of the speaker; but, far more, the unpretending love of truth, the spirit of benevolence, and the wonderful power of organisation displayed in the conception of new theories which, when expressed in a grand flow of eloquence, captivated all hearts and enchained all minds. Humboldt was fêted wherever he went, not only in

[1] The commencement of the friendship between these distinguished men is thus related by Arago, in his Eloge on Gay-Lussac ('Franz Arago's Sämmtliche Werke, herausgegeben von Hankel,' vol. iii. p. 14):—' Among the persons assembled in the drawing-room of the country-house at Arcueil, Humboldt one day remarked a young man of tall figure and of a modest but self-possessed demeanour. In answer to his inquiries, he was informed that it was Gay-Lussac, the physicist, who, in the hope of solving some important problems in physical science, had made some of the most daring balloon ascents at that time on record. "It is he, then," replied Humboldt, " who wrote the severe criticism upon my work on Eudiometry' (see p. 235). Soon repressing the feelings of animosity which such a reminiscence could hardly fail to arouse, Humboldt approached Gay-Lussac, and, after a few complimentary words upon his aëronautic expeditions, offered him his hand, and begged to be allowed the honour of his friendship. It was the unreserved expression of, "Let us be friends, Cinna!" The friendship thus commenced continued unbroken, and was productive of the happiest results.'

the public assemblies of scientific institutions, but in the drawing-rooms of private circles.

In all sincerity, therefore, could his sister-in-law, Frau von Humboldt, express herself in the following terms, in writing to Kunth, on September 10, 1804:—' The crowning pleasure of my visit here has been the happy return of our dear Alexander, and the gratification of witnessing the reception that has been accorded him. It has rarely fallen to the lot of any private individual to create so much excitement by his presence or infuse an interest so universal.'

There was but *one* exception to this wide-spread homage; the Emperor Napoleon, alone, received Humboldt with incivility and a feeling of animosity he scarcely strove to conceal. 'You are interested in botany? My wife also studies it,' were the almost insulting words with which the emperor greeted him upon his presentation at court.[1] In writing to the author, Humboldt on one occasion remarked: ' The Emperor Napoleon behaved with icy coldness to Bonpland, and seemed full of hatred to myself.'

Of Humboldt's manner and personal appearance at this time, we learn the following particulars from the letter of his sister-in-law above cited:—' Alexander has not in the least aged during his six years' absence. His face is decidedly fatter, and his vivacity, both in speech and manner, is if possible increased. He seems much pleased at finding me here, and you may well believe what a gratification it is to me to see him once again. He talks of spending the winter with us in Rome; which I think will be very desirable on account of his health, for after so long a sojourn in tropical regions it is well that he should pass at least the first winter in a moderate climate. We shall not, however, take the journey in company, for he will not be able to leave before January, while I am anxious to return as soon as possible, so as not to be away longer than is necessary.'

Humboldt found abundant occupation in the arrangement of his extensive collections, and in drawing out the plans for

---

[1] Burckhardt, ' Goethe's Unterhaltungen mit Kanzler Friedr. von Müller,' p. 101.

the publication of his works. In Paris he met with every facility for the prosecution of this undertaking, and had the offer of zealous co-operation from numerous scientific friends. He had, however, in the first instance applied to Pictet at Geneva, with whom he had entered into correspondence about the arrangement of his works, and their translation into English : for indeed at one time he appears to have been more attracted to Geneva as a residence than to Paris. ' It is quite possible,' he once wrote to Pictet, 'that I shall some day come and settle myself near you. Such a peaceful retirement forms one of my brightest visions, for I love to indulge the hope that after having explored the tropics, and contemplated the wonders of the universe, I may eventually seek repose upon the borders of your lake.' . . . ' May fortune ever preserve the enviable peace of your happy shores, where I hope one day to find a home—a peace so necessary to the active exercise of genius, and the development of the social virtues.'[1]

Geneva was, in fact, at that time the centre of much useful effort in the cause of science. It is true that Horace Bénédict de Saussure (1740–1769) had passed away; but there were still many distinguished men actively engaged in various branches of scientific labour : the two brothers De Luc, the physicist Marc-Aug. Pictet (1752–1825), Pierre Prévost (1721–1839), Jean Trembley (1749–1811), and the botanist Jean Senebier (1742–1809), who was afterwards succeeded by the elder Decandolle.[2]

Humboldt further wrote to Pictet on February 3, 1805 :[3]— ' I hasten to give you the list of the works we have in progress, which are all so far complete, that even in the event of my death they could be published in a more or less perfect state. For the convenience of the public, and for the sake of reducing the labour of editorship, I propose publishing the results of my travels in eleven distinct works.'

Of these eleven works, the titles were to be as follows :—

1. Plantes équinoxiales.

[1] ' Le Globe, journ. géogr. de la Soc. de Geogr. de Genève,' 1868, vol. vii. pp. 152, 157.
[2] Peschel, ' Geschichte der Geographie,' p. 501.
[3] ' Le Globe,' 1868, vol. vii. p. 158.

2. Nova genera et species plantarum æquinoctialium.
3. Essai sur la Géographie des Plantes.
4. Relation abrégée de l'Expédition.
5. Observations astronomiques et Mesures géodésiques.
6. Observations magnétiques.
7. Pasigraphie géologique.
8. Atlas géologique.
9. Cartes fondées sur des Observations astronomiques.
10. Voyage aux Tropiques.
11. Statistique du Mexique.

All these works were to be published under the joint names of Humboldt and Bonpland; but Nos. 1 and 2 were to appear as 'edited by Bonpland,' and Nos. 3 and 11 'edited by Alexander von Humboldt.' The letter then proceeds:—

' I am anxious that the narrative of the expedition should be written in a manner to interest people of taste. It will naturally contain, besides statistical information, descriptions of the physical aspect of the country and its antiquities, of the manners and intellectual culture of the inhabitants, with data upon commerce and finance, together with a narrative of the personal adventures of the travellers. Considering the remarkable energy of my disposition, I expect to see the whole out of my hands in the course of a couple of years, or at most in two years and a half, for I am now impatient to discharge my cargo, that I may embark on something new. I hope while I am at Rome to be able to draw out a general prospectus of these works, which will be sold separately, though published in a uniform edition; I shall have the prospectus printed in French, German, English, Dutch, Spanish, and Danish, for I am informed that preparations are being made for publishing an edition of my works in these six languages.

' But before this prospectus can appear, do you not think that a bill of fare, like the enclosed, might serve to excite an interest among English booksellers? Though we might engage that all the works should ultimately appear, we must be careful to make a separate contract for each work, as Nos. 3, 4, 8, 10, and 11 ought to be sold at a higher price than the others. I think that No. 3 will be a very important volume, all the more from its being the first to appear. The entire work ought to bring in several thousand pounds, so that there will be something for everybody.

'HUMBOLDT.'

To these plans, which, with characteristic enthusiasm, he had laid out on possibly too extensive a scale, he adds the postscript:—

'Meanwhile, to amuse the public, we must publish something of a comprehensive character. In selecting between Nos. 3 and 4, my philosophy leads me to prefer a comprehensive view of Nature to a narrative of personal adventure. Moreover, No. 3 indicates what I have accomplished; it shows that my attention was directed to phenomena of every description, and above all it appeals to the imagination. The world likes to *see*, and I there exhibit a microcosm in a leaf. I think, therefore, that the charlatanry of literature will thus be combined with utility.'

He then proceeds to furnish Pictet with instructions in reference to the English translation, and the attitude he should assume towards the booksellers and literary men in England.

'You might also drop a hint, that I have some idea of bringing out an edition in America, where I do not hesitate to say that among the anti-federal party the success of my expedition has produced quite an enthusiasm, apparent even in the newspapers that find their way here. The sale in the United States will be very large, and should it be requisite to find subscribers, (a method which always seems to me wanting in delicacy), I am sure that my friends Messrs. Jefferson, Madison, Galatin, Whister, Berton, and others, will be able to procure me a large number. An English edition, therefore, ought at least to consist of 4,000 copies.'

It would thus appear that Humboldt's keen insight into matters of business did not save him from forming somewhat too high an estimate of the sale of his works.

At a subsequent conference, held at Paris on March 7, 1805, to which Pictet had been summoned from Geneva, it was decided that the conduct of the translation into English of certain portions of the work should be entrusted to him, and that he should supply any notes that might be necessary, and superintend the publication in London. The proceeds were to be shared equally by Pictet, Bonpland, and Humboldt. The issue of the work was to commence with the 'Essai sur la Géographie physique des Plantes,' and for this pamphlet, which was to consist of seven or eight sheets, the sum of 200*l*. was to be asked. Many of these arrangements, especially those relating to the classification of the work, afterwards underwent considerable modification.

Closely as Humboldt's attention was engaged by the publication of his travels, he yet found time to spend months together in the laboratory of the École Polytechnique; where, in conjunction with Gay-Lussac, he carried on a series of investigations on the chemical constitution of the atmosphere, and on various methods of testing the purity of the air—a subject which had earnestly engaged his attention before he set out upon his travels. He communicated the most important results of these labours, in a paper he read before the National Institute, on the 1st Pluviose, XIII., entitled 'Mémoire sur les Moyens eudiométriques et la Constitution chimique de l'Atmosphère.'[1] Berthollet drew up a report of this work, for the Section for Physics and Mathematics, and pronounced it worthy of admittance into the 'Recueil des Savants étrangers.'[2]

Humboldt also entered into scientific investigations with Biot, and communicated the results of their joint labours in a Memoir 'Upon the Variations in the Earth's Magnetism in different Latitudes,' which he read before the Physical and Mathematical Section of the National Institute, on December 7, 1804.[3] In a letter from Rome to Vaughan in Philadelphia, dated June 10, 1805, he mentions: 'I have read before the Institute nine Memoirs, all of which are being printed.'[4]

Thus burdened with press of work, Humboldt could yet write as follows to Willdenow, who had asked him for some ferns; the letter is dated, February 1, 1805 :—

. . . . 'Busy as I am with the whirl of my own occupations,

---

[1] 'Journ. de Phys.' vol. lx. pp. 129-158; Gilbert's 'Annalen,' vol. xx. pp. 38-93.

[2] 'Annal. de Chim.' vol. liii. p. 239; Gilbert's 'Annalen,' vol. xx. p. 99.

[3] 'Journ. de Phys.' vol. lix. pp. 429-450; Gilbert's 'Annalen,' vol. xx. pp. 257-299.

[4] De la Roquette, 'Humboldt, Correspondance, etc.' vol. i. p. 183.

I will yet find time to attend to your request. This very evening I am going to Dupetit-Thouard, who may in truth be said to be a very *wooden* [1] sort of man. It is a thousand pities that your good genius did not lead you this year to Paris, instead of Trieste. You might then have had at your disposal the extensive herbarium formed by Bonpland and myself, besides the collections of Jussieu and La Marque, and from these you could have selected whatever would have been of use to you. Vahl adopted this plan, for nothing can be got out of the people here, merely by correspondence.

'With this letter I am sending you a small box of seeds we collected in South America and Mexico. Many of them have germinated at Malmaison, and I hope they will succeed equally well at Berlin.

'I am now bringing through the press the following works: 1. "Tableau physique des Régions équinoxiales;" 2. The first part of the "Plantæ æquinoctiales," with magnificent plates; 3. "Observations de Zoologie et d'Anatomie comparée;" 4. "Observations astronomiques et Mesures exécutées dans un voyage aux Tropiques." They will all appear simultaneously in German.'

On February 16, Humboldt writes to Friedländer :—

. . . . 'Since my return to Europe I have been so incessantly occupied, that I have had no time for enjoyment ; I have indeed begun more than I can well carry out. Three of my works are now being published, of course both in French and German. I say of course, because I have heard with astonishment that it has been reported in Germany that I am having my works translated into German. Such a report can have arisen in no favourable quarter. I must confess that I believe I write the Spanish language with greater correctness than any other, but I am yet sufficiently proud of my country to write in *German*, however inelegantly I may do so.'

To Karsten he wrote, on March 10, 1805, dated Paris, from the École Polytechnique :—

. . . . 'I have destined to you all the minerals I have

---

[1] [Dupetit-Thouard had raised himself to fame by his painstaking investigations on the nature of wood, but his ingenious theories on the growth of trees failed to secure the notice they deserved, from the obstinacy with which he clung to notorious errors.]

collected. The sorting and packing have occupied a good deal of time, but I trust you will on the whole be pleased with the specimens.

'I have entrusted seven large cases to the care of M. Luchesini. You, who know the difficulty and expense of land transport among the Cordilleras, and are aware that, owing to the war, my boxes have been widely dispersed—you who know that I share my collections with my valued friend and fellow-traveller Bonpland, that several of my cases have been dragged about for two years unopened, and that though for five years I enjoyed considerable celebrity, I was furnished with no pecuniary assistance—you who are aware that we brought back with us 60,000 specimens of plants, of which 6,300 are new species, and have experienced how difficult it is, at the same time to observe, draw, and collect, especially when, in a fit of ill-humour, treasures that have been laboriously carried about for months are thrown away in order to lighten the packages—you, my dear friend, will think it no marvel that the collection I send you is so limited.

'Insignificant, however, as this collection may be as to the number of the specimens, I believe it will be found of great importance as regards the interests of science. With each specimen I am able to give the height above the sea at which it was found, together with a description of the stratification and the position in which it lay. No former collection of minerals has ever contained specimens from Chimborazo, Cotopaxi, or Pichincha; and if, while you are unpacking the case, you meet with much that seems to you uninteresting at first sight, it will perhaps cease to be so when you come to read the inscription. The cases also contain some gold medals, some old Mexican statues, and a picture made in feathers. I have tried to write on the labels as much of what was interesting as possible. May I request you to share the duplicates with Herr Klaproth, and to convey to that distinguished man the assurance of my highest esteem. Perhaps you would also be kind enough to allow these specimens to remain separate for a while before incorporating them with your collection of European minerals. This would be of essential service to me in bringing out my works, as I have not retained a single specimen for my own use. I leave here to-morrow with Gay-Lussac, to institute some chemical experiments at Mont Cenis; thence we proceed to Rome.' . . .

Round the margin of this letter Humboldt has added:—' My health is better than ever. I am working with more studied application than formerly, so that I hope that my present productions will be free from the immaturity characterising many of my earlier works. I am making preparations for an expedition to northern Asia, which will doubtless be of great importance for the theory of magnetism, and for the investigation of the chemical composition of the atmosphere, for which the long polar nights will be peculiarly favourable. I cannot hope, however, to carry out this plan for two or three years to come. The emperor has settled a pension of 3,000 francs on my travelling companion, Bonpland, as a remuneration for the journey. To obtain this for him has been the main object of my long sojourn here. I trust you have, ere this, safely received the large piece of platinum I gave in charge to Count Hake.' . . .

On March 12, 1805, Humboldt left Paris to join his brother at Rome. He was accompanied by Gay-Lussac, who, through Berthollet, had obtained leave of absence. The two friends were provided with the best meteorological instruments, with which they conducted a series of experiments at Lyons, Chambéry, St. Jean de Maurienne, St. Michel, Lanslebourg, on the summit of Mont Cenis, &c., and after a short stay at Genoa, arrived at Rome on June 5.

William von Humboldt had, since the close of the year 1802, been resident at Rome, as Prussian Minister to the Papal Court. The position was one for which both by taste and inclination he was pre-eminently adapted, for he enjoyed almost equal distinction as a diplomatist, a man of letters, and a discriminating patron of the arts. Whether at Rome, Ariccia, or Albano, his house was ever the centre of the most distinguished society; his acquaintance was sought out and prized alike by princes and statesmen, by poets and literati. His house was especially the resort of artists, to whom Frau von Humboldt always proved a zealous friend and patroness, particularly to those who could claim her as a fellow-countrywoman. Among these were Gmelin, Grass, Tieck, the two Rippenhausen, Carsten, Schick, Thorwaldsen, Rauch, and Schinkel, men who by their labours in painting and sculpture were inaugurating a new era in the history of art.

To Alexander Humboldt the society assembling at the house of his brother afforded the keenest enjoyment, and most agreeable stimulus. While enchanting a circle of eager listeners by his descriptions of the grandeur of nature in the New World, the new life aroused within him by the contemplation of the impressive monuments of Roman antiquity inspired him with fresh impetus and new images for instructive comparison. While contributing to his brother's philological studies the valuable data he had gathered among the dialects of America, he found among the treasures of the libraries and museums at Rome both manuscripts and ancient memorials, which afforded valuable elucidation of the antiquities he had met with in America. In these researches he met with a willing guide and assistant in the celebrated archæologist Zoëga. Humboldt was also indebted to this visit to Rome for much valuable assistance in other branches of his undertaking; and many of the maps and illustrations in his geographical and pictorial atlases bear the name of the artists with whom he made acquaintance at the villa of William von Humboldt.[1]

Even Nature herself seemed intent on favouring Humboldt during this visit to Italy: Vesuvius gave warning of an approaching eruption, and afforded an inducement so irresistible for visiting Naples, that, on July 15, Humboldt and Gay-Lussac set out thither, accompanied by Leopold von Buch, who had just then arrived in Rome. It may well be imagined the interest this expedition must have had for Humboldt, in being able to discuss with such companions his experiences in America, and compare the phenomena he had witnessed there with the volcanic disturbances of Vesuvius.

All the time that could be spared during their stay at Naples

---

[1] 'I have had a great many drawings executed here; I have met with artists who have been able to make pictures from even my slightest sketches, and they have drawn for me the Rio Vinagre, the bridge of Icononco, and the Cayambe. . . . Dergia is in possession of a valuable collection of Mexican manuscripts, from which I intend to publish several plates, and am having them engraved here.'—*Humboldt to Bonpland,* 'Rome, June 10, 1805.' (De la Roquette, ' Humboldt, Correspondance, etc.' vol. i. p. 177.)

from the contemplation of the burning mountain, was spent in visiting the various museums, where they were treated with extreme courtesy by the curators, and honoured with marks of especial attention from the Duca della Torre and Colonel Poli. In striking contrast to this behaviour was the reception they met with from Dr. Thompson, who, when applied to by the travellers, accompanied by a Neapolitan gentleman of distinguished learning, for permission to see through his museum, acceded to their request with the exceedingly discourteous remark :—'You must please divide your party, gentlemen ; I can keep my eyes upon two, but four are too many to watch.'[1]

On their return to Rome the travellers made but a short sojourn in the Eternal City, and on September 17 started on their homeward route to Germany. They took their journey by way of the mountains, for the sake of instituting a chemical analysis of the celebrated springs at the baths of Nocera, and on September 22 they entered Florence. They visited the splendid galleries of the city of the Arno, under the guidance of Fabbroni, a distinguished connoisseur, who gave them the astounding intelligence in answer to their inquiry as to the amount of variation in the magnetic needle at Florence, that the fine instruments in the Grand Duke's Museum of Natural Science had never been brought into use, for fear of injuring the polish of the metal.

Passing hastily through Bologna and Milan, where they were at some pains to find out Volta, the three friends crossed the pass of St. Gothard on October 14 and 15. A short summary of the remainder of the journey is contained in the following letter, addressed probably to Spener or Sander, a bookseller in Berlin :—

'Heilbronn: October 28, 1805.

'Dear Sir,—I have been prevented by the war from carrying out my original intention of visiting Vienna and Freiberg. I should have very much liked to have compared my manuscript with the Vienna Codex, since I have had engraved portions of the Mexican Codex in the Vatican, and of the one in the

[1] Franz Arago's ' Sämmtliche Werke, deutsche Uebersetzung,' vol. iii. p. 20.

Borgia collection; but as I am accompanied by my friend M. Gay-Lussac, I have been obliged to avoid the Austrian dominions. The sciences are no longer a Palladium in these frightful Mahratta wars in which Europe is now perpetually engaged. We found, however, some compensation in a visit to my old friend Volta at Como. But the pass of St. Gothard! With what storms of rain, snow, and hail were we greeted by the Alps! On the road from Lugano to Lucerne we really encountered a good deal of annoyance, and we hear that in the beginning of October the whole of Swabia lay deep in snow. Yet this is called, apparently in jest, the temperate zone! We go hence to Heidelberg and Cassel; and as I do not intend to stay long at Göttingen, supposing the Russians allow us to go there at all, I hope soon to find myself at Berlin, where I shall devote myself exclusively to my American works. The second part of the " Plantæ Æquinoctiales " has appeared.' . . .

On November 4 they reached Göttingen, where they spent some days among their friends, both professors and former fellow-students; and on November 16 they arrived at Berlin.

During his stay in Paris, Humboldt had received the most gratifying proofs of the lively interest which his friends at Berlin had taken in his return to Europe, and of the high appreciation they entertained of his valuable services to science. From all quarters came letters of greeting and congratulation, numbering sometimes as many as forty in a week, to only a few of which had he been able to reply. On the proposition of the Royal Academy of Sciences, he had been nominated an honorary member, by a royal order in council bearing date August 4, 1800[1]—an honour he was not likely to appreciate too highly (see p. 212). He was distinguished with especial marks of favour by the king. 'The king showers favours upon me,' he writes to Pictet;[2] 'his attentions are almost oppressive, for they take up too much of my time. He has granted me a

---

[1] Humboldt was introduced to the Academy as an *ordinary* member upon his arrival at Berlin. By a royal order in council of November 19, 1805, he was, ' as Member of the Academy of Sciences,' awarded a pension out of the funds of the Institution.—Humboldt's letter to the Academy, conveying his thanks for his election, is dated ' Paris, September 4, 1804,' and was written on receiving the notification through Kunth.

[2] ' Le Globe,' &c. p. 179.

pension of 2,500 thalers, equal to 10,000 francs, with no condition of service.' The appointment of Royal Chamberlain was
also conferred upon Humboldt; but in writing to Pictet he
requests :[1] 'Pray do not mention' (in the introduction to the
English edition of his American Travels) 'that on my return
to my country I was made a—a Chamberlain! Say something,
however, at the close complimentary to the king, for he has
really treated me with great distinction.'[2]

Though Humboldt might view distinctions of this kind with
tolerable indifference, he could not but be filled with abhorrence
and grief at the condition of things prevalent in official life
and society of every grade in Berlin, where it was evident
no improvement had taken place since his departure for
America. The inducements therefore that could have led
Humboldt to quit Paris for a visit to Berlin, at a time when
he had just commenced the publication of his travels in
America, must have been the yearning to meet with the early
friends of his youth, and the hope of inducing Willdenow to
undertake the editorship of a portion of the botanical department of his work, in which case he hoped to procure for him
by personal influence the necessary leave of absence. Once
at Berlin he met with much to detain him, not only in arranging with Oltmanns for the reduction and publication of
his astronomical observations, to be completed in three years,
but in preparing many of the maps for his ' Atlas géographique
et physique de la Nouvelle-Espagne,'[3] executed under his super-
vision by Friessen, a talented young architect.

The industry he displayed during this sojourn at Berlin
eclipsed anything he had before evinced. In addition to the
labour inseparably connected with the publication of his great

[1] ' Le Globe,' &c. p. 189.

[2] Humboldt relates on this occasion an amusing conversation which took
place between a chamberlain and the celebrated traveller, Reinhold Forster,
who, as is well known, was exceedingly bluff in manner. — The Great
King himself was accustomed to say : ' Chambellan, ce n'est qu'un titre
chimérique,' and that ' Chamberlain, in plain German, meant nothing more
than court lackey.' (Preuss, ' Urkundenbuch zur Lebensgeschichte Fried-
richs des Grossen,' vol. iv. p. 302.)

[3] See Humboldt, ' Essai polit. s. l. r. de la Nouv.-Espagne,' I., LXII. and
LXIV. ed. 1811, in 4to. ; ' Deutsche Turnzeitung,' 1859, No. 2.

work (see p. 349), and the pressure of a voluminous corre-spondence which consumed a considerable amount of time, he read the following papers before the Academy :—'On the Laws regulating the Decrease of Temperature in the Higher Re-gions of the Atmosphere and the Limit of Perpetual Snow;' 'On Steppes and Deserts;' 'On the Cataracts of the Orinoco;' 'Thoughts on the Physiognomy of Plants.' Several smaller treatises, of which he made presentations to the editors, ap-peared in various periodicals. Among these may be cited:— 'On the Aboriginal Inhabitants of America;' 'Experiments upon Electrical Fish;' 'Upon the Various Species of Cinchona;' 'Capture of Electric Eels by means of Horses;' 'Observations on the Influence of the Aurora Borealis upon the Magnetic Needle, instituted at Berlin on the 20th of December, 1806;' 'On the Scientific and Mathematical Instruments devised by Nathan Mendelssohn,'[1] &c.

Labours of a purely experimental character also occupied the attention of Humboldt; at this time he commenced a series of observations upon the earth's magnetism, at first, in conjunction with Gay-Lussac, which were carried on in a small building constructed for the purpose entirely without iron in the garden of Herr George, a wealthy brandy distiller, a site now occupied by Renz's Circus. Varnhagen remarks in his journal, on July 4, 1857:—'Humboldt was relating to me yesterday that when he was carrying on magnetic observations in an out-building in George's garden, he was at one time so intently occupied in his labours that he spent seven days and nights in succession with only snatches of sleep, as every half-hour he had to take the readings of the instrument; subse-quently he was relieved in this duty by his assistants. This was in 1807, just fifty years ago: I have often seen this building when visiting Johannes von Müller, at the time he too was conducting observations in a garden-house. Humboldt further related, that when George, as an old man, used to show strangers over his garden, he never failed to boast of "his scientific friends." "Here I used to have the famous Müller, here Hum-

---

[1] Nathan Mendelssohn, the youngest son of Moses Mendelssohn, was an admirable mechanician, and one of the most zealous founders of the Polytechnic Society of Berlin; he died in January 1852.

boldt, and here Fichte, who, however, was only a philosopher." [1]

In addition to these various labours he was also engaged in
preparing for the press his ' Aspects of Nature,' which made its
appearance shortly after his return to Paris. The work consists in part of fragments from the lectures delivered before the
Academy, and is in truth a master-piece, from the æsthetic
treatment of the subjects brought under notice. Humboldt
himself called it his 'favourite work,' [2] and described it as
' a book written from a purely German point of view.' [3] In
alluding to a review that appeared in Paris, Humboldt wrote
on June 24 ,1808, to Malte-Brun : [4]—' Good Heavens! How
you have loaded my book with praises. One journal accuses it
of being too full of German metaphysics. A singular charge,
which seems to savour of the *Couvent de Munich*,' which in
the ' Briefwechsel mit Berghaus,' i. 121, he terms ' the *spelunca
maxima* of the German Ultramontanism.'

In view of an industry so versatile and comprehensive, we
need not wonder that Karsten, in writing to Von Moll on
January 7, 1807, should state : [5]—' Humboldt is so universally
sought after, that he is seriously thinking of withdrawing from

---

[1] The building used by Humboldt was not the one in Mendelssohn's garden, where the manor-house now stands, as stated by Dove in his 'Éloge of
Alexander von Humboldt,' p. 22 ; the one referred to by Dove was not erected
till 1827.—The following note to Henriette Herz, in Hebrew characters,
belongs to the period referred to in the text :—

' I am quite grieved, my dear friend, at having unintentionally deceived
you. I forgot when I accepted your invitation that Schleiermacher's visit
will take place at the equinox, therefore just at the time when, on account of my nocturnal magnetic observations, I am not my own master.
I am somewhat exhausted by my watchings for the past seven nights,
and dare not undertake even this short excursion, as I cannot leave my
magnetic instruments for that length of time. My reason is thus fighting
against my inclination. Were I only to consult the latter, you and your
worthy friend, from whose society when I last met him here I derived so
much enjoyment, would prove a magnet too powerful for me to resist.
Pray do not scold me, nor be angry with your sincere friend,

' Berlin : September 23, 1806.   HUMBOLDT.'

[2] ' Briefe an Varnhagen,' p. 244.

[3] ' Briefe von A. von Humboldt an Bunsen,' p. 115.

[4] De la Roquette, ' Humboldt, Correspondance, etc.' vol. ii. p. 36.

[5] Von Moll, ' Mittheilungen aus Briefen, &c.' p. 358.

social life, in order that he may devote his time exclusively to the interests of science. With unprecedented industry, and a perseverance which never falters, he applies himself incessantly to the completion of his works.'

From Humboldt's letters to several of his friends at this time we may gather something of his own state of feeling :—

To Dr. Beer, at Glogau.

'Berlin: April 22, 1806.

. . . . 'Although it is fifteen years since I last wrote, my affection for you has remained unchanged. . . . I rejoice with you in your happiness. I am living, my dear Beer, isolated and as a stranger, in a country which has become to me like a foreign land. Even before I left Europe the ground was burning under my feet. . . . I have renewed with Frau Herz all the old sympathy and friendship. She is, indeed, an admirable woman and loves you as much as I do, and that is a good deal. Nathan Mendelssohn is also here, and has grown up to be an interesting man.[1] Only come and see me, and you will find me as simple as formerly, only somewhat less sprightly.

'Your affectionate friend,
'HUMBOLDT.'

To Frau Caroline von Wolzogen.

'Berlin: May 14, 1806.

'An attack of rheumatic fever with severe toothache—an illness from which I have frequently suffered since my arrival in this uninhabited desert—prevented me from sending you a few lines with William's magnificent poem, to thank you, my dear friend, for the kind note I received from you. Whatever you may say in joke (for malicious it is not in your nature to be) on the score of my universality, you will at least give me credit for enough of German feeling to think daily with affectionate regard of you and of Goethe, and of our deceased friend, and to be ever conscious that it was something great over which to glory that I once occupied a position not wholly insignificant near such a trio.

[1] See p. 356.

'Though vast mountain ranges and an immeasurable ocean, and other aspects of Nature if possible more impressive and sublime, have intervened between those days and the present —though a thousand wondrous forms have since passed before my mind, yet " the new has always been interwoven with the old," the unfamiliar has been assimilated with the associations of bygone days, and I have been constrained to admit, while ranging the forests of the Amazon, or scaling the heights of the Andes, that there is but One Spirit animating the whole of Nature from pole to pole—but One Life infused into stones, plants, and animals, and even into man himself. In all my wanderings I was impressed with the conviction of the powerful influence that had been exerted upon me by the society I enjoyed at Jena, of how, through association with Goethe, my views of Nature had been elevated, and I had, as it were, become endowed with new perceptive faculties.

'The kind way in which you also speak of my little work on the " Physiognomy of Plants " has made me very happy. My whole being seems fuller of life and energy, but I dare not always give expression to my feelings. Just now I brood much over myself, for I lead a sad and isolated life. There is nobody with whom I can sympathise, and this is a very melancholy and depressing sensation. Will not you and Goethe be coming to Lauchstädt? for I might hope to see you there, dear friend. Remember me to dear Schiller, kiss the little ones for me, and convey to Goethe the assurance of my filial regard.

'Ever yours,
'HUMBOLDT.'

To Von Zach.

'Berlin: September 19, 1806.

. . . . 'I shall be occupied for two years in arranging the materials for the publication of the results of this journey. . . . . I only hope the astronomical portion may please you and prove worthy of you, for to you it owes its existence. But for you the starry heavens in their tropical splendour would have had no charm for me; to you am I indebted for the calm and peaceful happiness afforded by the contemplation of Nature in one of her grandest aspects—certainly one of the purest of all enjoyments. These feelings of deep gratitude have inspired me with the wish to dedicate to you in conjunction with M. Delambre the astronomical portion of my work, and I earnestly solicit your permission in the name of Herr Oltmanns and myself. . . . It is an inexpressible happiness to me to have met here with Herr Oltmanns ; he is a wonderful young man, entirely self-educated, and full of talent, modesty, and astonishing perseverance. He seems to live solely for astronomy, and for a fortnight together he scarcely leaves his work ; he possesses great facility in the higher branches of calculation, and is exceedingly well read. Men who love science for its own sake are rare. . . . Although I give but little time to sleep, and am by no means idle, I cannot get on as fast with the publishing of my great work as the world seems to expect. I am anxious to prepare something of undoubted value, and I am therefore indifferent at finding myself somewhat unjustly censured by the unfriendly portion of the public.' . . .

To Wattenbach.

'Berlin : April 10, 1807.

' Dearest Wattenbach,—In days gone by, when you were destined for the Church, and I was studying book-keeping in preparation for a business career, we used to be very intimate. Since then you have fulfilled my part in life though I have not entered upon yours ; yet notwithstanding this change of plan, and my long absence, I am sure you have not forgotten me. I at least can recall with pleasure the happy intercourse we had together as youths, and the pleasant home circle, enlivened by the amiable cheerfulness of your noble disposition. With the confidence inspired by these bright reminiscences, I venture to recommend to your notice the bearer of these lines, Herr Moritz Robert . . . . a young man for whom I have great regard, brother of a lady of remarkable intellectual endowments.[1]

' Poor Dohna has to act host to a number of emperors and kings who have invited themselves into his domain. I saw Gill at Barcelona, very interesting, but rather melancholy, and in shattered health. I should be very glad to hear something of Dashwood, for I have lost sight of him for the last fifteen

---

[1] Rahel.

years. I met his brother in the West Indies. Maclean is in Danzig still; wealthy, industrious, and as noble-hearted as ever. I live in hopes of meeting you once more.

'Ever yours,

'HUMBOLDT.'

As the insertion of these letters has somewhat disturbed the chronological order of events, it will be well to give a glance at the occurrences which had transpired. Gay-Lussac had returned to Paris in the spring of 1806,[1] and Bonpland had paid a short visit to Berlin; the printing of Humboldt's great work was meanwhile progressing satisfactorily, both in Paris and Stuttgart. Then burst forth the frightful political catastrophe, which scattered even the circles of the learned. The disastrous battle of Jena had been fought. Prussia was annihilated; the king put to flight; Napoleon, as conqueror, was seated at Berlin. — — —

Humboldt found himself, on all sides, called upon to act as mediator between the Prussians and their inexorable foes; it was an office for which he was peculiarly fitted by his knowledge of languages, his courtly manners, his personal acquaintance with the parties with whom he had to treat, his influential position, and his sincere patriotism. Yet the circumstances of the time were too strong, even for him.

The following letter to Frederick Augustus Wolf [2] gives some account of his fruitless efforts to save the University of Halle from impending dissolution:—

'Berlin: November 18, 1806.

'I hasten, my dear friend, to reply to your welcome letter of the 14th, which reached me safely. I hope you will receive my answer, which is to be committed to the post unsealed, soon enough to relieve me from the reproach of having remained inactive at a crisis so important for the happiness of mankind and the cause of intellectual culture. No, my dear friend, at the time your letter reached me I had been engaged upon

[1] At least it is so stated by Arago in his éloge of Gay-Lussac; but Humboldt, in his Autobiography in Brockhaus' 'Conversations-Lexikon,' mentions his return as having taken place in the winter of 1805-6.

[2] Körte, 'Leben und Studien Friedrich August Wolf's,' vol. i. p. 359.

these affairs for a week past, and, during one forenoon alone, had been occupied upon them from six to eight hours. The most influential French statesmen, such as Maret, the Secretary of State, and Daru, the Comptroller of the Household, both certainly very superior men, are so incessantly occupied with business, that one may wait about for hours, or even half a day, without being able to speak to them for a single moment. I have, however, made repeated intercession with them both, as well as with Governor-General Clarke, but hitherto without success, notwithstanding the good-will they have to the cause, and the interest they take in the University. To attempt to influence the emperor is not to be thought of, for he is one of those monarchs with whom everything must originate, and from whom everything must emanate. I have, besides, never had a personal interview with the emperor (except, for a few moments, when presented to him at Paris),[1] and of course no private individual can be admitted to his presence except by his express orders. From what I hear, the emperor seems to be always irritated afresh whenever any mention is made of Halle. M. Daru made another attempt, the day before yesterday, on presenting to him an application from the University, but without success. The ill-will of the emperor seems to be grounded upon his belief in certain facts which are unknown to me. I do not attempt to express how deeply I feel all this, and how much it grieves me. I expect time will produce much alleviation, and, in anticipation of this change of sentiment, I shall continue my indefatigable exertions. I cannot believe that the various scientific institutions of which Halle can boast will all be dissolved. Halle is near Leipzig, on the further side of the Elbe. I fear the emperor has designs connected with political events which the peace only will reveal.

'I am grieved to write you such a letter.

'I need not repeat how completely I share in my brother's feeling of attachment, veneration, and love to yourself.

'A. HUMBOLDT.

'Postscript.—I have just been fortunate enough to find an old Hamburg newspaper containing the protest made by the

[1] See p. 344.

University against the warlike temper of the students. I immediately sent it on to M. Daru, with another very earnest appeal.'

It was under the influence of circumstances of this depressing character that in May 1807 the 'Aspects of Nature' was dedicated to those *oppressed spirits* 'who, glad to escape from the stormy waves of life, accompany me with sympathising interest through the dense glades of primeval forests, across immeasurable plains, and over the rugged heights of the Andes. To such the following lines of the chorus speak with sympathetic voice :—

> Auf den Bergen ist Freiheit! Der Hauch der Grüfte
> Steigt nicht hinauf in die reinen Lüfte;
> Die Welt ist vollkommen überall,
> Wo der Mensch nicht hinkommt mit seiner Qual.' [1]

The same tone of sadness pervades also the following passages : [2]—

'Let those who, wearied with the tumult of discordant nations, are striving to obtain the enjoyment of intellectual calm, turn their attention to the silent life of vegetation, and therein contemplate the mysterious processes of Nature, or, yielding to those impulses implanted in man from the earliest ages, let them gaze with wondering awe upon those celestial orbs, which are ever pursuing, in undisturbed harmony, their ancient and unchanging course.' . . .

' Thus, when the charms of intellectual life wither, when, under the relentless hand of time, the productions of creative art are gradually perishing, the earth with indestructible vitality is ever putting forth new life. With wondrous prodigality, Nature continues to unfold her countless buds unconcerned, though man, with whom she ever stands at variance, should, in his presumption, trample on her ripening fruit.'

In the hope that by means of negotiation with the Emperor Napoleon the burdens imposed by the shameful peace of Tilsit might be somewhat diminished, the Government, in the spring

---

[1] 'Ansichten der Natur,' 3rd. ed., Preface, p. 10.
[2] Ibid. vol. i. pp. 38, 286.

of 1808, decided to send to Paris Prince William of Prussia, a
younger brother of the king, who was distinguished alike for
personal bravery and graceful manners; greatly to his surprise,
Humboldt was selected by the king to accompany the prince
upon this difficult political mission, on the ground of his great
experience and intimate acquaintance with persons of influence. The prince, in whose suite as adjutant was General von
Hedemann, the future son-in-law of William von Humboldt,
remained in Paris till the autumn of 1809, when he returned
to Berlin. Humboldt, however, in his character as one of
the eight foreign members of the Paris Academy of Sciences,
sought and obtained the permission of the king to remain in
France, foreseeing the impossibility in the actual state of
Germany of bringing out his great work in that country,
unsupported as he was by Government.

The poem addressed to him by his brother from Albano, in
September 1808, as a grateful acknowledgment of the dedication prefixed to the 'Aspects of Nature,' reached him almost as
a parting greeting on his quitting Berlin. In this poem to
'Alexander von Humboldt,'[1] William von Humboldt follows
in a strain of lofty thought the course of the world's history,
then passes to his brother's personal achievements, and, in conclusion, refers to his projects of further travel in the following
lines :—

> Glücklich bist Du gekehrt zur Heimaterde
> Vom fernen Land und Orenoco's Wogen.
> O ! wenn—die Liebe spricht es zitternd aus—
> Dich andern Welttheils Küste reizt, so werde
> Dir gleiche Huld gewährt, und gleich gewogen
> Führe das Schicksal Dich zum Vaterherde,
> Die Stirn von neuerrungnem Kranz umzogen !

[1] Wilhelm von Humboldt's 'Gesammelte Werke,' vol. i. p. 361.

TRAVELS IN ASIA.

'Je me ferai Russe, comme je me suis fait Espagnol.'

A. von Humboldt to Von Rennenkampff.
Paris: January 7, 1812.

CHAPTER I.

PROJECTED SCHEMES.

Proposals from Russia in 1811—Negotiations with Von Rennenkampff—Preparatory Studies—Projected Arrangements with Prussia, 1818—Munificence of Frederick William III.—Acquisition of Asiatic Languages.

A JOURNEY through Asia and Northern India to the Himalayas and Thibet had ever formed, as we have repeatedly observed, one of Humboldt's most cherished projects. With this object in view, he had commenced the study of the Persian language, under the guidance of Sylvestre de Sacy and André de Nerciat. But zealously as he carried on his preparations for the journey, carefully as he weighed and organised his plans,[1] and near as he often believed the fulfilment of them to be, some hindrance always presented itself to the accomplishment of his wishes, in the events of war, financial difficulties, or the delay in the publication of his works.

It was not till the close of the year 1811, when Russia was preparing an expedition to Kashgar and Thibet, that these

---

[1] 'Herr von Humboldt is still busy in Paris with preparations for his scientific expedition to Tartary and Thibet. He has recently completed an excellent treatise on the Mexican Calendar, and, by the comparisons he has instituted between it and the calendars of Peru, Japan, China, Mongolia, Thibet, and Hindostan, new light has been thrown upon the earliest history of the world and its primeval inhabitants.'—*Allg. geogr. Ephemeriden,* 1811, vol. xxxvi. p. 376.

long-cherished hopes received any promise of fulfilment. Count Romanzow, Chancellor of the Empire, was personally acquainted with Humboldt, for whose zeal and scientific knowledge he had a high appreciation, and through his influence Von Rennenkampff, at that time in the service of Russia, but subsequently High Chamberlain of Oldenburg, was commissioned to invite Humboldt to take part in this mission to Central Asia. The proposal was received by Humboldt with the liveliest satisfaction : we extract the following passages from his reply, dated ' Paris, The Observatory, Rue St. Jacques, January 7th, 1812 :—

. . . . ' Besides the publication of my works upon America, I am now engaged in studies preparatory to an expedition into Asia. I had conceived a project to this effect before my return to Europe, and I fully intend to carry it out, but I cannot leave Paris until I have completed my work, of which more than two-thirds is now finished. . . .

' The object of chief interest to me in a journey to Asia would be the high mountain range extending from the sources of the Indus to those of the Ganges. I should of course wish to see Thibet, but a visit to that country would constitute an object of but secondary interest. Most probably I should go by way of the Cape of Good Hope, for I have long contemplated a work upon the declination of equatorial stars. I should like to remain a year at Benares ; and should I not be able thence to visit Bokhara or Thibet, I should at least be within reach of the coasts of Malacca, the islands of Ceylon, Java, or the Philippines. I should prefer reaching India by this route, as once landed I should be sure of an interesting journey, replete with discoveries of every kind.

' The political situation of Europe, at the time of my departure, will determine the course of my route, whether by way of Constantinople, Bussorah, or Bombay. As India and the mountains of Central Asia, situated between 35° and 38° of latitude, form the principal objects of my expedition, I am quite indifferent as to the line of route.

' Such are some of the plans at present engaging my attention. I am extremely flattered by the interest shown me by the Court of St. Petersburg. . . . Count Romanzow, Minister of

Trade, did me the honour to make me some proposals during his stay in Paris. . . . I shall be delighted to accept any propositions that may be made to me by Government through an official channel, *provided I am furnished with geographical details concerning the districts to be explored.* It will cost me a great deal to give up all hope of visiting the banks of the Ganges, the home of the banana and the palm. I am now forty-two years of age, and I should be glad to undertake an expedition that would occupy about seven or eight years, but I should be loth to renounce visiting the tropical regions of Asia, unless I had the prospect of a very extensive and comprehensive journey. I should be less attracted by the Caucasus than by Lake Baikal and the volcanoes in the peninsula of Kamtschatka. Will it not be possible to get as far as Samarcand, Cabul, and Cashmere ? Must I lose all chance of ascertaining the height of Mustag or the plateau of Shamo? Is there a person in the Russian Empire who has visited Lhassa or Thibet by any other than the ordinary route of Teheran, Casween, and Herat, or else *viâ* Calcutta ? Is Russia at war with all the nations along her southern frontier, and can nothing be done except amid the tumult of arms ? An expedition of this character, on account of the extent of country traversed, would doubtless yield a vast accession of data in various branches of science ; in geography with regard to the superposition of rocks, the identity of formations, and the distribution of vegetation ; in meteorology, the theory of magnetism (inclination of the needle, declination, intensity of force, hourly variations), and in experiments with the pendulum. The study of the various races of men, their language—one of the most enduring monuments of ancient civilisation—the possibility of opening up a road to commerce from the south, together with a thousand other kindred subjects, will engage our earnest attention. In order to grasp at once the field of operation, I should like to traverse first the whole of Asia, under latitudes from 58° to 60°, passing through Iekaterinbourg, Tobolsk, Ieniseisk, and Iakoutsk, to the volcanoes in Kamtschatka and the shores of the Pacific. As these districts are all inclined towards the north, the most recent formations will be exposed to view ; in returning, it would be desirable to select the route under the

48° of latitude, by way of Lake Baikal, in order to carry out the
investigations to be made south of this parallel, which would
occupy four or five years.

'Expeditions of this nature would not be very costly, for
though the instruments required must necessarily be of the
most perfect construction, they are but of small dimensions.
It would be well that the scientific men selected for the expedition should be Russians, for they are more hardy and possess
much greater powers of endurance, and will be less impatient
to return. I do not know a single word of the Russian language, but I shall make myself a Russian, as formerly I
made myself a Spaniard. Whatever I undertake, I execute with
enthusiasm.' . . .

After pointing out the advantages that would accrue from
such an expedition into the interior of the continent, not only
to science, but also to the Russian Empire, Humboldt continues :—

'You will naturally infer from the hopes I thus indulge
that I have made up my mind to accept any proposals that I
may have the honour of receiving, provided the plans for the
expedition be conceived on a scale sufficiently extensive to
be worthy of the monarch by whom half the old world is
governed. The apprehension that exists of a war breaking out
in the North may perhaps delay the execution of these grand
projects; I cannot but hope that peace may be preserved, but
should I be disappointed, it may at least be expected that at
the close of a war, Government will be all the more disposed
to entertain those projects which, calling for only a small outlay
of the public funds, tend to increase the internal prosperity of
the country. It will be impossible for me to visit St. Petersburg before the winter of 1814; but this delay will not prove
detrimental to the expedition. More than a year will be
required for the preparation of various scientific instruments;
and it will be desirable to have them constructed by the best
opticians in Europe—Fortin, Breguet, and Lenoir, of Paris;
Troughton, Mudge, and Ramsden & Son, of London; and
Reichenbach, of Munich. It will take some time to make the
necessary arrangements with artists and men of science, and to
acquire all the information needed in regard to the southern

frontiers of the Empire, and the possibility of penetrating beyond them. . . . I have written to you with the same frankness which I made use of when addressing the Court of Aranjuez, in 1799. . . .

' I am too well acquainted with your delicate consideration to deem it needful to request that the contents of this letter should be communicated only to those persons directly interested in the execution of a plan likely to be of service to the advancement of science. I should think it no humiliation to offer my services to a Prince, under whose government the arts and sciences have flourished throughout the length and breadth of his vast dominions, did not my position entirely prohibit such a step. I would refuse nothing which claimed for its object a glorious or useful end ; I would unhesitatingly undertake a journey from Tobolsk to Cape Comorin, even were I assured that out of nine persons one only would accomplish it in safety; but simple in my tastes, and preserving a moral independence, in which I am sustained by a strong will, I prefer to follow my scientific career, relying merely upon my own resources. I should do violence to my character if, instead of replying to the questions you proposed to me, I were to make any propositions. . . .

'ALEXANDER VON HUMBOLDT.'[1]

It cannot now be ascertained how far these negotiations with the plenipotentiary commissioner of the Russian Government proceeded ; it is certain, however, that, through the war which soon after broke out between France and Russia, Humboldt was obliged to relinquish all hope of comparing the geological conditions of the Himalayas and Kuen-lün with those of the Andes. Neither did the years of peace which succeeded, bring the project any nearer to a successful issue. New problems, on the contrary, were being presented to his consider-

[1] When the permission of Humboldt was sought for the publication of this letter, he replied : ' I do not disavow any of the motives by which I was actuated in writing to Baron von Rennenkampff; seventeen years later, in 1829, I undertook, by command of the Emperor Nicholas, the expedition described in my " Asie centrale." This letter may be printed either before or after my death. It is the expression of a strong will!

' Berlin : October 18, 1853. A. VON HUMBOLDT.'

ation, to the solution of which his active mind was earnestly directed.

Upon his return from America, in the year 1804, he was unable to institute comparisons between the limits of perpetual snow on the Cordilleras, and the snow line on the Himalaya, Hindu-Khu, Ararat, or the Caucasus. It was not till 1812, that Moorcroft visited the elevated plain of Daba, in Thibet, while the important series of geodetic and hypsometric observations undertaken by Webb, Hodgson, the Brothers Gerard and William Lloyd, were not carried out till some years later, between 1819–1821. In Asia, as in every other quarter of the globe, adventurous explorers had preceded any expedition of a scientific character, and many discussions had been raised as to the accuracy of the measurements made by such travellers, of the height of the mountains in India, and especially as to the extreme height of the limit of perpetual snow upon the northern slopes of the Himalayas—a height which had been received with incredulous surprise.[1] These circumstances gave rise to a Memoir, 'Sur les Montagnes de l'Inde,' written by Humboldt in the year 1816, which aroused considerable attention, especially in England.

At the congress of Aix-la-Chapelle, in 1818, where, by desire of the king, Humboldt was in close attendance upon him, the long-cherished project of an expedition into Asia was again revived; and from being supported by the favour of his sovereign and the long-established friendship of Prince Hardenberg, Chancellor of State, there seemed at length to be abundant warrant for the hope of its final accomplishment.

To meet the expenses of the journey the king granted a yearly stipend of 12,000 thalers. For the purchase of astronomical and other scientific instruments, as well as of maps and books, requisite for such a journey, the sum of 12,000 thalers was in May of the following year remitted to Humboldt, in Paris, with the accompanying injunction, that, upon the completion of the journey, the instruments should be deposited in the Royal collection; 'it is not, however, at all the intention of his majesty to render this condition in the least of an oppressive character. As in the course of such an expedition, instruments are liable to be broken, or by frequent use

[1] 'Annal. de Chim. et de Phys.' vol. iii. p. 103.

to become no longer available, or from other causes to receive injury, it would not be just to require the redelivery of these instruments intact. The instruments which the king has graciously accorded to Baron von Humboldt may, therefore, be used by him without restriction, and, upon his return, be delivered to the Royal collection in whatever condition they may be.'

His preparations were completed, and Humboldt had selected his travelling companions and scientific fellow labourers,[1] when he was again compelled to relinquish his journey; this time probably through the illiberal policy of the East India Company.

In contemplation of this second scheme for penetrating into the interior of Asia, which seemed to promise fulfilment, Humboldt devoted himself to the study of the Asiatic languages, and the diligent search for all documents yielding information as to the climate of Asia and its mountain structure. As the immediate fruit of this work, appeared his second treatise, entitled, ' Mémoire sur les Montagnes de l'Inde,' with considerable additions, and bearing the distinctive title, ' Sur la Limite inférieure des Neiges perpétuelles dans les Montagnes de l'Himalaya et les Régions équatoriales.'[2] By frequent intercourse with the great linguists Abbé Grégoire, Abel Rémusat, Letronne, Hase, Freytag, Klaproth,[3] Villoisin, Champollion, the Persian

[1] Among these was Dr. Baeyer, then Captain and now Lieutenant-General of Dragoons, and President of the Royal Geodetic Institution of Prussia. To him, as an excellent geodetic observer, were to be committed all matters relating to the topographical surveys of the countries they visited, such as observations for latitude and longitude, determinations of the heights of mountains, &c.; under the direction of Professor Weiss he had also qualified himself as assistant-observer in mineralogy and geology.—Dorow ('Erlebtes,' vol. iii. p. 65 fg.) relates : ' Goerres expressed himself in the most enthusiastic manner concerning Alexander von Humboldt's projected journey to Persia and Thibet, and wished to induce Dorow to exert himself to join the expedition. . . . Goethe and Reinhardt, on the contrary, doubted much whether Humboldt would accept Dorow as one of the party, seeing he was scarcely likely to prove a very energetic assistant.'

[2] 'Annales de Chim. et de Phys.' vol. xiv. (Paris, 1820).

[3] Klaproth, the most renowned Sinologist of that time and the greatest authority on the geography of Asia and the various nations peopling that country, and author of the 'Tableau historique de l'Asie,' was then engaged upon a work in which Humboldt took a lively interest, ' Carte de l'Asie

traveller Andrea de Nerciat, and the celebrated Sylvestre de
Sacy, the most distinguished Oriental scholar of this century,
he was gradually acquiring an extensive acquaintance with the
Asiatic tongues, when, early in the year 1827, his studies were
interrupted by a change of residence from Paris to Berlin.

centrale, dressée d'après les cartes levées par ordre de l'empereur Khian-
loung, par les missionnaires de Pékin et d'après un grand nombre de notions
extraites et traduites de livres chinois.'   (See Humboldt's 'Briefwechsel
mit Berghaus,' vol. ii. p. 1.)

CHAPTER II.

TRAVELS IN ASIATIC RUSSIA.

Proposal and Negotiations—Journey from Berlin to St. Petersburg—Visits to St. Petersburg and Moscow—Line of Route—Distance travelled—Personal Adventures—Diamonds in the Ural Mountains—Return and Reception of Honours.

COUNT CANCRIN, the Minister of Finance in the Russian cabinet, addressed a letter to Humboldt on August 15, 1827, consulting him as to the employment of platinum—of which a considerable quantity had recently been discovered in the Ural Mountains—for purposes of coinage, and requesting information as to the proportionate value of that metal with respect to gold and silver. In the communication it was casually mentioned ' that the Ural mountains would well repay a visit from a man of scientific eminence.' Humboldt, in his reply, dated November 19, 1827, treated the subject from a truly scientific point of view; he discountenanced the project of instituting a platinum coinage, on account of the impossibility of its maintaining an unalterable value with respect to gold and silver, and concluded his letter with the courteous assurance, that he hoped some day to have the pleasure of paying his respects in person to the Minister at St. Petersburg. ' The Ural Mountains, Ararat now soon to be included within the Russian dominions ; yea, even the Lake of Baikal, present images before my mind, which awake my keenest interest.' These expressions were hints that fell not unheeded. On December 17, 1827, Cancrin wrote to express the wish of the Emperor Nicholas, that Humboldt should undertake a scientific expedition to the east of Russia, the cost of which was to be borne by the emperor. This flattering proposal was eagerly accepted by Humboldt, with the proviso, that he was not to start before the completion of the works upon which he was engaged, a task which was likely to occupy him till the spring of 1829. With regard to the offer of means, he should have no hesitation in availing himself of the emperor's munificence if his sojourn in Russia could be of service to the interests of science. On this subject he expresses himself more fully in a letter, dated January 10, 1829:[1] 'I have expended the whole of my inheritance (100,000 thalers); and as I have sacrificed my fortune in the cause of science, I make the confession without fear of blame. The king, with whom I stand in a purely personal relationship, grants me a yearly pension of 5,000 thalers, with a munificence greater than I, as a scientific man, consulted only occasionally in the affairs of administration, can in anywise deserve. So far, as I am but a bad manager in household affairs, and am always too ready to help young students, I have been spending yearly somewhat over my income. It is therefore to be hoped that should I be granted a safe return, and take up my residence again, either in Berlin or Paris, I may not find that the expedition to the waters of the Irtysch has placed me in a worse position or encumbered me with heavier financial difficulties.' The cost of the journey from Berlin to St. Petersburg and back, estimated by him at about 3,000 thalers, he was anxious to defray from his own resources; but the minister empowered him to obtain for this purpose the sum of 1,200 ducats at Berlin. On his arrival at St. Petersburg, 20,000 roubles were placed at his disposal for his personal expenses during the expedition.[2]

[1] 'Im Ural und Altai. Briefwechsel zwischen Alexander von Humboldt und Graf Georg von Cancrin, aus den Jahren 1827–1832' (Leipzig, 1869). This pamphlet contains the correspondence between Humboldt and Count Cancrin, prior to and during the course of the Asiatic expedition; out of forty-three letters twenty-eight are from Humboldt. The particulars of the journey are derived mainly from the letters written by Humboldt to his brother during the expedition, and from various unpublished letters to Cancrin. The former were mostly written to cheer and console his brother, grieving over the loss of his wife, and are characterised by the tenderest expressions of feeling. ('To-morrow, my dearest brother, is your birthday; I shall celebrate it in Asiatic Ural amid the copper-mines of Gummeschewskoi. I am deeply moved as I write these lines. How gladly would I form one of your family circle to-morrow!)

[2] Of this sum, 7,050 roubles were in excess of his requirements; and,

In addition to these financial arrangements, the minister made the most careful and systematic preparations for the comfort of the travellers. In an autograph memorandum furnished to Humboldt under date of January 1829, there occurs among other entries the following: ' I will arrange to send with you a mining official [1] speaking either French or German, as well as a courier, one, if possible, acquainted with German, or by his majesty's gracious permission a rifleman, for the purpose of transacting the details of the journey, such as the ordering of post-horses, arranging for hotel accommodation, &c. These attendants will be in receipt of a daily allowance. . . . The expenses of post-horses, gratuities, and carriage repairs, will be defrayed by the officer accompanying you from the Mining Department, for which purpose a sum of money will be placed at his disposal, for which he will be held responsible. The cost and trouble of such arrangements will in this way be spared you. The route to be taken in the journey, and the objects to be attained, will be left entirely to your selection. The aim of Government is solely the advancement of science, and, as far as it is practicable, the increase of the Imperial revenue by the extended working of the mines in Russia. All governors, vice-governors, mining officials, shall receive instructions to further in every way the object of your journey. Suitable quarters will be assigned to you when visiting the mines, and

on the refusal of the Government to accept its return, Humboldt devoted it to a scientific journey, conducted by two young mineralogists, Von Helmersen and Hofmann, who had been commissioned by Government to join his party at Miask.

[1] The mining official selected for this office was Von Menschenin, a manager of smelting works, and subsequently overseer of mines. He was always alluded to by Humboldt in favourable terms, although Helmersen, while admitting him to be a very well-informed man, spoke of him as having frequently forgotten his subordinate position. This gave rise to many uncomfortable scenes, which were always passed over by Humboldt with friendly patience. His consideration for his ill-humoured and often not very courteous travelling companion was carried so far that he would never sit down to dinner without waiting his arrival, however tardy that might be. Soon after his arrival at Berlin, Humboldt addressed several friendly letters to Herr von Menschenin, one of which was accompanied by the gift of a small but beautifully executed universal instrument. As he received no answer even to this letter, Humboldt inquired from Herr von Helmersen, ' What is the reason of this anger? '

labourers, and all necessary help afforded you at the cost of the crown in any experiments you may wish to make. As soon as your route is more accurately planned, directions for the journey shall be prepared, pointing out all that is of interest in the districts you may visit, in addition to which any further information you may require shall be communicated to you.'

Humboldt's request that his two friends, Professors Ehrenberg and Rose, might be allowed to accompany him was acceded to in the most gracious manner. Furnished with letters of recommendation from the King and Crown Prince to the Imperial Family, in which he was styled, 'Actual Privy Counsellor with the honorary title of Excellency,' Humboldt left Berlin on April 12, 1829. The cortége consisted of two travelling carriages, loaded with numerous instruments and apparatus of every kind. In the low plains of the Vistula and the Nogath, the transit was rendered both difficult and dangerous by the floods and the floating ice, but the travellers reached Königsberg in safety on April 15, where they had the pleasure of meeting with Bessel. The journey beyond Königsberg was still more difficult. 'The roads are really very tolerable, but since leaving Dorpat we have been surrounded by all the horrors of winter, and see nothing but snow and ice as far as the eye can reach; we are considerably delayed by the rivers, which are either blocked up with ice, as was the case with the Dwina and Narowa, or can only be approached by roads so much washed away that the front wheels become buried in the mud, and travellers have to provide themselves with planks, that, with extra horses and the manual assistance of the peasantry, the carriages may be helped over the deepest holes. These are some of the ordinary occurrences of returning spring. . . . Owing to these difficulties, the journey to St. Petersburg will probably cost 900 thalers.' Up to April 29, the carriages had been ferried over rivers seventeen times. At length, the travellers reached St. Petersburg on May 1, and Humboldt took up his quarters in apartments prepared for him at the residence of the Prussian Ambassador, Lieutenant-General von Schöler.

'My success in society is most surprising,' he writes to his brother, from St. Petersburg on May 10. 'I seem to be surrounded by excitement wherever I go, and it would be impossible to be treated with greater distinction or with more generous hospitality. I have dined nearly every day with the Imperial Family in their strictest privacy—covers being laid for four—and I have spent the evenings with the empress, with the most delightful freedom from constraint. The heir-apparent has also entertained me to dinner, " in order that he may remember me in time to come." The young prince has been told to ask for my portrait, from the original to be painted by Sacszollo. Czreitszef, the Minister of War, has given me a collection of maps published at the Government press. . . . The carriages are very beautiful, and cost each 1,200 thalers.' A third carriage was provided for the courier and the cook.[1]

On May 20, the travellers left St. Petersburg. During a fortnight's stay at Moscow, Humboldt had the pleasure of again meeting with Loder and his old friend Fischer, ' who is styled his Excellency, drives about in a carriage and four, and is in receipt of a pension of 7,000 francs.' The consideration showed them by Government was beyond all praise, and the honours and distinctions they received became, at last, almost oppressive. ' We are continually the objects of attention with the police, government officials, Cossacks, and guards of honour. Unfortunately, we are scarcely a moment alone; we cannot take a step without being led by the arm like an invalid. I should like to see Leopold von Buch in such a position.' Even the receptions given in their honour by the highest officials, and the entertainments they received from men of learning, became ' bien fatigants.'

The route selected by Humboldt lay through Vladimir and Mourom to Nijnii-Novgorod, and thence down the Volga to Bulgara and Kazan. After a short excursion to the Tartar ruins of Bulgari, the travellers proceeded through Perm to Iekaterinbourg, on the eastern slopes of the Ural, that exten-

---

[1] With praiseworthy zeal the Russian Government equipped three other scientific expeditions in the same year, 1829: one to Ararat, under Parrot the younger; another to Mount Elbrus and the Caucasus, under Kupffer; while a third, under Hansteen, Due, and Erman, was despatched for the purpose of determining the magnetic lines from St. Petersburg to Kamtschatka.

sive mountain chain composed of several parallel ridges, rising
to a height of about 4,600 feet, which, stretching in the same
direction as the Andes, due north and south, extend from the
tertiary formations near the Sea of Aral to the primary rocks
of the Arctic Ocean. June 16 and 17 were occupied in visiting
the Imperial laboratories for the cutting of the topaz, the beryl
and the amethyst, the gold washings of Schabrowski, the
quarries of Rhodonit, and the iron furnaces of Nijnii-Issetsk;
and during the following week, Humboldt and his companions
visited Beresow on the lake of Schartasch, Polewskoi, and
Gummeschewskoi.

An excursion of greater length was then taken northward,
through Pischminsk and Neviansk to Nijnii-Tagilsk, the property of Prince Demidoff, where, on June 27, so late as nine
o'clock in the evening, the travellers paid a visit of inspection
to the mines. Humboldt found that the platinum fields of this
district, as well as those of Sucho-Wissim and Rublowskoi,
yielded an amount of ore equal to that procured from Choco in
South America. The party then proceeded northward, by way
of Kuschwa, Laja, Blagodad, and Nijnii-Turinsk to Bogos-
lowsk, for the sake of visiting the extensive gold mines ; whence
they made the return journey to Iekaterinbourg, in the midst
of a succession of thunderstorms, through Werchoturje, Alopa-
giwsk, Mursinsk, where the topaz and the beryl mines were
visited late at night, and lastly Schaitausk.

The travellers left Iekaterinbourg on July 18, and continued
their journey, by way of Tioumen, to Tobolsk, on the banks of
the Irtysch. There Humboldt decided to deviate from the
route he had laid down ; and, instead of going by way of Omsk
and Semipalatna, he went from Tobolsk through Tara and the
steppes of Barabinsk, notorious for their dreaded swarms of
stinging insects, to Barnaul, on the river Obi. He proceeded
thence to the picturesque Lake of Kalywan, and the rich silver
mines of the Snake Mountains, and those of Riddersk and
Zyrianowskoi, on the south-western slopes of the Altai Mountains.

After passing the small fortress of Ust-Kamennoigorsk, the
travellers arrived on August 19 at Buchtarminsk, on the
borders of Chinese Tartary, where Humboldt obtained per-

mission to cross the frontier, and visited the Mongolian station of Baty or Khoni-Mailakhu, to the north of the lake of Dzaisang, one of the most central towns in Asia.

From Ust-Kamennoigorsk, the party journeyed, through the central horde of Kirghissia, to Semipolatinsk and Omsk, and thence to the southern part of the Ural, through the Cossack lines of Ischim and Tobol. In the rich gold fields of Miask, which occupy but a limited extent of ground, there had been found three large nuggets of gold only a few inches below the surface, two of which weighed fifteen pounds and one twenty-four pounds. Various excursions were made in the neighbour-hood of Miask, to the lake of Ilmen, to Slatoust and to Soimonowsk. In the southern part of the Ural chain they visited the beautiful quarries of green jasper near Orsk, where the mountain chain is broken through from east to west by the passage of the river Jaik, remarkable for an abundance of fish. Here they turned westward to Guberlinsk, and reached Orenburg on September 21; thence they pursued a southern course to the Caspian Sea, visiting on their way the celebrated mines of rock salt at Iletzkoi, in the steppe of the small horde of Kirghissia, and passing by way of Urlask, the chief settle-ment of the Ural Cossacks. They next visited Saratow on the left bank of the Volga, where a colony of Germans had been es-tablished by Government, and made an excursion to the large salt lake of Elton in the plains of the Kalmucks; thence passing through the beautiful Moravian settlement of Sarepta, they arrived at Astrachan.

In visiting the Caspian Sea the chief objects of the expedition were the chemical analysis of the waters, the institution of baro-metric observations for comparison with those made at Orenburg, Sarepta, and Kasan, together with a collection of specimens of the various kinds of fish to be found in this inland sea, as a con-tribution to the great work on Fish, then engaging the attention of Cuvier and Valenciennes.

On October 21, the travellers left Astrachan on their home-ward journey, crossed the isthmus which separates the Don from the Volga at Tischinskaya, and, after traversing the terri-tory of the Don Cossacks, reached Moscow by way of Voronesch and Toula, on November 3, and St. Petersburg on the 13th of the same month.

According to the accounts rendered by Menschenin, the party had travelled 14,500 versts, or 9,614 miles, during their twenty-five weeks' absence from St. Petersburg—that is to say, from May 20 to November 13, 1829 ; of this distance 690 versts were by water, exclusive of a voyage on the Caspian Sea of 100 versts. The number of post stations passed through was 658, and 12,244 post horses had been employed on the journey. The passage of rivers had been effected fifty-three times—the Volga had been crossed ten times, the Kama twice, the Irtysch eight times, the Obi twice. Humboldt computed that during the nine months he had been absent from Berlin— from April 12 to December 28—he had travelled 11,505 miles.

As we have only been able to give a slight sketch of the expedition, we shall be obliged to confine our descriptions of the personal adventures of the travellers to a few of the most characteristic incidents. We are indebted for the following sketch to General Helmersen, a Member of the Academy of St. Petersburg :—

' Humboldt, at this time in his sixtieth year, walked with a tolerably erect gait, his head being a little inclined forward. During the journey and in his travelling carriage, we never saw him in any other costume than in a dark brown or black dress-coat, a white neckerchief, and a round hat. He usually wore a long over-coat of a similarly sombre hue. His gait was slow and measured, cautious but secure. During the excursions he never rode ; when a carriage could no longer proceed, he would alight and go on foot, climbing high mountains without any show of fatigue, and clambering over fields of rough stones. It was evident from his mode of walking that he had been accustomed to bad ground.

' His meals were always characterised by his well-known moderation ; and even after hours of great fatigue, he would persist in declining the abundant supply of food which the Russians, with well-meaning hospitality, were accustomed to press upon their guests. His manner of rejecting these kindly meant attentions was invariably marked, whatever the rank of his host, by the gentle courtesy ever distinguishing the true aristocrat.

' Humboldt's fame had penetrated even among all classes of

the population in the Ural districts, but the kind of authority with which he had been delegated was interpreted in various ways. In general, it sufficed that he was entitled, Actual Privy Counsellor, and, that the Emperor Nicholas had issued orders that he was to be received with the honours befitting a senator and a general. The commandants of the small fortresses in the military district of Orenburg, through which Humboldt travelled, presented themselves in full uniform, and laid before him, according to military custom, a report upon the condition of the troops under their charge. When the travelling cortége, consisting of three equipages, arrived at any garrison town, there was always waiting an expectant crowd of Cossacks, soldiers of the line, Kirghissians, Baschkirians, Tartars and Russians, besides women and children of all ages.

'On one occasion, while the horses were being changed at the fortress of Tamalyzkaja, there suddenly came forward from a crowd of this description a Baschkirian, who, making up to Humboldt's carriage, near which I was stationed, with raised voice and lively gesticulations, addressed a speech to the great traveller in Turkish jargon, not one word of which could be understood by any of our party. Upon Humboldt inquiring from me, in the most courteous manner, "What does this *gentleman* want?" I summoned an interpreter, and, through him, learnt that the supplicating Baschkirian had, on the previous night, been robbed of his horses by the neighbouring Kirghissians; and, on hearing there was a great personage coming *who knew everything*, he had presented himself with the urgent request that this distinguished person would tell him who the robbers were and how he was to recover his lost treasures. As the police rushed up at this moment to seize the unsummoned intruder, and prevent any mischief, Herr von Humboldt, who laughed heartily at the occurrence, begged that mercy might be shown to this simple son of the desert.'

As a smile from Humboldt sufficed to preserve this poor Baschkirian from severe punishment, on another occasion, a word from him was the means of alleviating the condition of a young political prisoner, Witkiewicz, a Pole who, though self-taught, had attained a remarkable degree of intellectual culture.[1]

[1] See 'Briefwechsel Humboldt's mit Berghaus,' vol. ii. p. 279.

With these exceptions, Humboldt, often as he was compelled to witness occurrences distressing to his feelings of humanity, never allowed himself while in Russia to interfere in any way with the affairs of Government or the arrangements of social life, carefully avoiding even an expression of opinion upon such subjects. In his 'Essai politique sur l'Isle de Cuba,' he expresses himself as follows :—'It is the duty of a traveller who has been called to witness scenes involving the suffering and degradation of human nature, to convey the complaints of the unfortunate victims to those who have it in their power to relieve their distress; ' and in the preface to the same work he remarks : 'To this portion of my book, namely, the improvement of the condition of the slave, I attach much higher importance than to the arduous labours of determining by astronomical observation the latitude and longitude of places, of registering the variations in the force of the magnetic current, or of collecting statistical information.' Yet Humboldt was compelled to write in a very different strain to the Russian Minister, in a letter he addressed to him from Iekaterinbourg, dated July, 5 (17), 1829 :[1]—'It will be taken as a matter of course that we (Humboldt and Rose) confine ourselves to the investigation of inanimate nature, and avoid everything connected with Government or the condition of the lower classes of the people: statements made by foreigners unacquainted with the language are usually regarded as incorrect, and when referring to subjects so complicated as the relationship between the rights that have been acquired by the higher orders of society and the duties incumbent upon them towards the lower classes, such statements irritate without producing any beneficial result.'

This quotation may suffice to show that the position he occupied while travelling in Russia differed very materially from the relationship in which he stood to the Spanish Government while travelling in America. In this change of circumstances lay, no doubt, the reason why Humboldt, in the midst of so much honour and public recognition, and with all the conveniences accompanying an imperial commission to travel, would

---

[1] 'Im Ural und Altai,' pp. 74, 79.

yet never accede to the pressing solicitations of the emperor
to undertake another expedition through his vast dominions.[1]

While at Miask, in the month of September, the two mining
students previously mentioned—Ernst Hofmann, subsequently
Lieut.-General, and Von Helmersen, now Lieut.-General and
Director of the Imperial Institute of Mines in Russia—were
appointed as guides to Humboldt and his party during their
visit to the Ural district. They remained in Humboldt's service
nearly a month. From Miask, Humboldt wrote to the minister
on September 15:—'Yesterday I completed my sixtieth year,
and spent my birthday on the Asiatic slopes of the Ural Mountains. I have therefore passed through the most important
period of my life, and reached a turning-point, which makes
me regret that so much still remains unaccomplished before
the strength of manhood declines with advancing age. Thirty
years ago I was among the forests of the Orinoco and on the
heights of the Cordilleras. To *you* am I indebted for rendering this year one of the most important of my restless life,
by enabling me to collect so vast an assemblage of facts, in
the course of this extensive journey, a journey during which we
have travelled more than 9,000 versts since leaving St. Petersburg.' In the same letter occurs the passage which, from the
truth of its prophetic tenor, has become so justly celebrated:—
'The Ural Mountains are a true El Dorado, and I am confident
from the analogy they present to the geological conformation of
Brazil—an opinion I have maintained for two years—that even
under your administration diamonds will be discovered in the
gold and platinum washings of the Ural Mountains. I gave
this assurance to the empress before I left Russia; and, though
neither I nor my friends were able to make the discovery,
the object of our journey will be accomplished if the attention

[1] On May 22, 1843, Humboldt wrote as follows to Schumacher :—'It has
cost me a great effort to dedicate the three volumes of my 'Asie centrale'
to the Emperor of Russia; it was an unavoidable step, since the expedition
was accomplished at his expense. My relationships with the monarch have
undergone considerable alteration since 1829, on account of my political
missions to Paris. The dedication, about which I consulted Arago, is appropriate and dignified on my part. The emperor has sent me his portrait;
I should have been better pleased to have received no answer or only a
formal acknowledgment.'

of others be aroused to this subject.' The truth of this pre-
diction was verified, as is well known, a few days after it was
written, in the gold and platinum washings belonging to Count
Polier.

The distinguished traveller was received on his return to
Moscow with every mark of honour, and entertained at State
festivals by the highest officers in the Government. A caustic
though in many points an accurate description of one of these
entertainments is given by Alexander Herzen,[1] at that time a
student in Moscow :—

' On his return from the Ural Mountains, a State reception
was given to Humboldt by the Scientific Society connected with
the University ; an institution comprising amongst its members
various senators, governors, generals, and other officials, who
had never been in any way concerned with scientific pursuits, and
were quite unacquainted with science in its higher branches.
The fame of Humboldt had reached them as of one who was
Privy Counsellor to the King of Prussia, and invested by the
emperor for the time being with all the honours due to one
ennobled by the Star of St. Anne ; as such they resolved to
prostrate themselves in the dust before the man who had
climbed the summit of Chimborazo, and had lived in the
palace of Sanssouci.

' The affair was treated very seriously. The governor-
general and the military and civil dignitaries appeared in gala
uniforms, decked with the ribbons of their orders ; the professors strode along trailing their swords in military fashion
with their three-cornered hats under their arm. Humboldt,
who had anticipated nothing of this sort, presented himself in
a simple blue coat, and was taken completely by surprise.
From the stairs to the saloon in which the " men of science "
were assembled a row of seats had been arranged ; here stood
the rector, there a dean, on the right a professor just starting
in life, on the left a veteran at the close of his career speaking
with the measured accents of age. From everyone he received
a few words of welcome, from one in German, from another in

---

[1] Julius Eckardt, ' Jungrussisc
turgeschichtliche Aufsätze ' (Leipzig, 1871), p. 149.

Latin, from a third in French, standing all the while in the corridor, where it was impossible to remain for an instant without being chilled through for months. Humboldt patiently listened to all these harangues with uncovered head, and replied to each of them. I do not believe that the great investigator ever experienced as many discomforts from the painted savages among whom he had travelled, as were prepared for him by the ceremonious festivities of his reception at Moscow.[1]

'When Humboldt had at last entered the saloon and taken his seat, the proceedings were still conducted in the same ceremonious manner. Pissarew, the curator of the University, considered it necessary to read out a kind of order of the day in the Russian language before a silent audience, on the merits of his Excellency the great traveller. Then Sergei Glinka, with his hoarse soldier's voice, recited a poem beginning: "Humboldt, Prométhée de nos jours."

'Meanwhile Humboldt's intentions had been to discuss his observations on the deviation of the magnetic needle, and to compare the meteorological data he had obtained in the Ural Mountains with the results registered at Moscow. Instead, he was expected to look with devout admiration on a plait made from the venerated hair of Peter the Great, shown to him by the rector. It was with great difficulty that his com-

---

[1] Even during his first visit to Moscow Humboldt complained to his brother, in a letter dated May 14 (26), that 'This perpetual ceremonial (sad necessity of my position and of the noble hospitality of the country) becomes very fatiguing.' At Kazan it was still more annoying; fête followed after fête. At one o'clock in the morning he wrote to his brother:—'We are to start to-morrow morning at five o'clock, and the professors and authorities threaten to come at half-past four to take leave. We are not left alone a single instant.' At Iekaterinbourg, on June 21, he was obliged to dance a quadrille! At Miask, on the sixtieth anniversary of his birthday, he was honoured by the presentation of a sword from the mining officials. In the plains near Orenbourg a Kirghissian festival was held, in which the entertainment consisted of races, wrestling matches, and, 'unfortunately, also of vocal music performed by the Tartar sultanas.' The ceremonial presentations, the fêtes, the distinctions of all kinds repeated everywhere, 'd'après des ordres émanés d'en haut,' drew from him the remark—': This excess of politeness deprives me of all opportunity of being alone and at my ease.'

panions Ehrenberg and Rose found opportunity to relate their discoveries.'

At St. Petersburg Humboldt was again the object of a demonstration of favour almost oppressive, especially in the reception given him at court.

' It is really a shame, my beloved friend,' writes Humboldt to his brother on November 20, 1829, ' that though I have been here a week, I have not found before to-day the quiet necessary to give you this sign of my affection and well-being. . . . My journey across nearly every part of European Russia has so enlarged my social relationships, and increased so considerably the chimerical notion that I can be of use to every one, that I am almost overwhelmed by the amount of drudgery imposed upon me by this position. My health is excellent; while we were at Moscow the emperor, with characteristic delicacy, conferred the Order of St. Anne of the second class upon Rose and Ehrenberg, while the Order of St. Anne of the first class, with the imperial crown (equivalent to the decoration in diamonds which is no longer given) was sent to me on the day of my arrival here, accompanied by a very flattering letter. His Majesty expressed his regret that he was still prevented by indisposition from seeing me, that he might " profit by my discoveries." I hope he may recover before my departure, which I am anxious now to arrange for the 1st of December. The empress has done me the honour to grant me a most gracious reception ; yesterday I dined with the heir-apparent, and I am again summoned by the empress this morning ; I am, in fact, treated with a kindness that seems daily to augment.' . . .

A few days before leaving St. Petersburg, he writes :[1]—' In the morning I and my two friends were nearly an hour and a half with the empress, and in the evening I had an audience with the emperor from half-past eight to eleven o'clock. I have been loaded with favours.' . . .

He wrote further to his brother on December 9 :—

' I can only write you a few lines. I hope to leave here by the 12th or 14th ; although I am obliged to await another audience from the emperor, who did me the honour to receive me for a couple of hours on Sunday—a distinction all the

[1] To Cancrin, ' Im Ural und Altai,' p. 118.

more conspicuous from the emperor not having yet received all his ministers. He is, however, rapidly progressing towards recovery. He has loaded me with tokens of affection and esteem. "Your sojourn in Russia has been the cause of immense progress to my country; you spread a life-giving influence wherever you go." I have been presented with a sable cloak worth 5,000 roubles, and a vase as beautiful as any in the palace, standing, with pedestal included, seven feet high, worth from 35,000 to 40,000 roubles.[1] I shall not be able to write this morning to Madame Kunth.[2] Do me the favour to write to her, and acquaint her that I have to-day written a letter to the king, mentioning her misfortunes and the worth of her husband, and begging that her case may receive special attention. I wrote by the first opportunity after receiving the news of her loss in a letter from young Kunth of the 23rd of November. What happiness to be with you so soon! With affectionate regards to all your family,

'ALEXANDER HUMBOLDT.'

Early on December 15, Humboldt left St. Petersburg in company with his two travelling companions, Ehrenberg and Rose, at a temperature of 13° below zero, and yet was able to write to Cancrin from Königsberg in a most cheerful strain on December 24 :—'We have spent an interesting day at Dorpat with Evers, Struve, Ledebuhr, and Engelhardt. It unfortunately stands recorded in the theory of probabilities that one cannot travel 18,000 versts without being at least once upset. This doctrine of probabilities has maintained its rights in the spirit of a Nemesis. We were upset with so much force upon a mill-bridge at the foot of a slight hill, near Engelhardtshof, two posts before Riga, by the whirling round of the carriage upon the smooth surface of the ice, that one side of the carriage was completely smashed. One of the horses fell a distance of eight feet into the water. As may be supposed, the parapet gave way and we lay in a very picturesque position about four inches from the edge of the bridge. Nobody was hurt, and though Ehrenberg and I were in a carriage enclosed by glass,

---

[1] This magnificent malachite vase is in the royal palace at Berlin.
[2] Kunth died on November 22, 1829.

we both escaped the slightest injury, for which we have to
thank Providence.   As two men of science and an accomplished rifleman were of the party, there was no lack of con-
tradictory theories as to the cause of the accident.   This much,
however, is certain, that the carriage swung round, and that,' as
he good-naturedly adds in exculpation of the man in charge
of the conveyance, ' the postilion was quite free from blame.'

On December 28, Humboldt and his companions arrived at
Berlin.   The results of the journey were published in the
following separate works:—' Fragmens asiatiques,' 1831, ' Asie
centrale,' 1843, both edited by Humboldt, and ' Journey to the
Caspian Sea and the Mountains of the Ural and the Altai,'
1837–42, edited by Gustav Rose.   Ehrenberg published in a
separate form his descriptions of the Siberian tiger and the
northern panther, while in his ' Micro-Geology,' 1854, he discusses the forms of organic nature existing in Russia, Siberia,
and Central Asia, amid the mountains of the Ural and those of
Altai, drawing special attention to the life visible only through
the microscope—a kind of organism which had hitherto been
unsuspected.   His botanical collections have never yet been
published.

Amid the universal applause that greeted Humboldt upon
his return to Berlin, he was yet the victim of an irritating
attack made upon him in England.   Unfortunate speculators
in the Mexican mines challenged the correctness of some of
his views, and attributed to him motives which ought never to
have been imputed to a man of science, least of all to such a
one as Alexander von Humboldt.   To this he refers in the
following passage, in a letter to Cancrin, dated April 3, 1830 :—

' It is really quite inhuman thus to attack a man who has
never given any evidence of selfishness, at the moment of his
return from a distant journey in the interests of science !   Is it
my fault that the accounts I published fifteen years ago,
of the richness of the Mexican mines (the correctness of which
has never been called in question by anyone living in Mexico)
should have led John Bull in the most foolish manner to trust
millions into the hands of ignorant speculators ?   I have always
declared from the first that I would have nothing to do with
speculative schemes ; I declined the appointment of Consulting

Director in Europe, with full liberty of action, by which I should have gained 20,000*l.* sterling, and returned a large gold snuff-box which was presented me as a token of gratitude from those who had made a fortune by speculating in the mines, and ascribed their success to my writings; in short, I have always manifested a wish to keep aloof from such affairs. All this is well known in England, and therefore I have offered no reply beyond correcting some figures of an exaggerated character, which appeared in German newspapers, and which might readily have been ascribed to me, although I never publish anything, even in newspapers, without my signature. Immediately on my return to Berlin, I placed myself in communication with the former Director-General of Mines in Brazil, Baron von Eschwege, now resident at Lisbon, in order to lay before him some enquiries as to the yield from the gold and platinum sands at Brazil. I hope to give your Excellency, in my next work upon the Ural Mountains, an interesting comparison between them and the mines of Brazil. So far the returns seem in favour of the Ural districts.'

The scientific results obtained by Humboldt in his Asiatic journey proved of high and lasting importance, though it cannot be denied that his expedition to Asia is greatly inferior in interest to his travels in America—not only as to its extent and duration, the personal adventures of the travellers, but also as to the range of subjects under investigation, and the variety and multiplicity of the results obtained. Yet during the Asiatic journey a surprising number of new facts and ideas were acquired, and an astonishing amount of observations accumulated, the results of which were far from being confined to the benefit of the internal economy of Russia. Additional knowledge of a much more accurate character was acquired concerning the configuration of the interior of Asia, the features of the mountain ranges, the phenomena of climate and the magnetic changes, the distribution of Flora and Fauna, and the grand historical roads, which had for ages been the means of communication with the interior of the country, so that the boundaries of science had been widely extended in the various branches of Physics, Geography, and History. The views which Humboldt had ever entertained upon the connection existing between all telluric phenomena received their fullest illustration and most perfect development during this expedition.

Humboldt, upon his return from Asia, was in his sixty-first year. The earnest longing to visit foreign lands, which he had cherished from childhood, had received its fullest gratification in the widely extended travels which had furnished such important results, and with this journey his unsettled career as a traveller was brought to a close. Henceforth commenced the peaceful years of unwearied labour, when, in the retirement of his study, he occupied himself in giving forth to the world the embodiment of the final results of his scientific labours and his comprehensive views of nature in his celebrated philosophical work ' Cosmos.'

APPENDIX.

---

1. The Humboldt Household. *Page* 34.

An interesting glance into the home-life of the Humboldts is given by Frau Caroline de la Motte Fouqué, née von Briest.[1] She writes to her sister in January 1785, that she has been spending the morning with Frau von Humboldt, and remarks:—

'Everything at the Humboldts' is just as it was. Nothing in the house is changed, neither the people nor anything else. Of course I shall never cease to feel his absence, for his pleasant, cheerful conversation formed always a charming contrast to the quiet and reserved manners of his wife. She, I can assure you, looks exactly as she always did and as she always will do. Her coiffure just the same as it was ten years ago or more, always smooth, neat, and simple. The same pale countenance and delicately formed features, upon which no trace of emotion is ever visible, the same soft voice, the same cold though sincere greeting, and the same unalterable faithfulness towards all with whom she is connected. She still offers a home to her brother-in-law, his daughter, and the aged aunt ; the old dog Belcastel is still permitted to lie snoring on the sofa ; her equanimity is alike undisturbed by contradiction or domestic troubles. One could almost assert that as one left the family on saying adieu, so one would find everything upon returning after a year's absence.

'Of the youths I can assure you that William, with all his learning, is anything but a pedant ; on the contrary, he is always ready to raise a laugh, and is made an idol of by all the house. It is possible that Fräulein U. . . . may have a little weakness for her young cousin, but the aunt is equally his sworn ally. Alexander is somewhat of a *petit esprit malin*. He is, however, remarkably talented, and drew both heads and landscapes even before receiving instruction. The walls of his mother's bedroom are hung round with

---

[1] Caroline Baronin de la Motte Fouqué, geb. von Briest, 'Der Schreibtisch, oder alte und neue Zeit' (Cologne, 1833), pp. 6, 7.

these productions. He has now arrived at an age when he begins to exhibit some gallantry. He wears two long steel watch chains, joins in the dance, and makes himself agreeable to visitors in his mother's boudoir—he begins, in short, to play a part in life. He reminds one strongly of his father.'

2. VISIT TO ENGLAND IN 1790. *Page* 91.

Fragment of Journal of a Tour with Forster. From the Radowitz Collection of Autographs, in the Royal Library at Berlin, No. 6,255.

From a crowd of notes and comments made by Humboldt upon Somersetshire, Gloucestershire, &c., we shall only subjoin a few extracts as proof of the wide range of his observations :—

'Wiltshire.—Many cloth manufactories in Wiltshire, Gloucestershire, and Somersetshire, especially at Trowbridge and Bradford, in the neighbourhood of Bath. In these south-western counties the sheep are shorn twice in the year, while in the counties of Leicester, Lincoln, Warwick, they are usually shorn but once. The high price of wool in the year 1790 has been ascribed to the establishment of the cloth manufactories. The highest price given for Welsh wool was 1s. 3d. per lb.; for South Down, 1s. 2d.; for Norfolk wool, 1s. 1d.; and for the wool of the western districts, 8½d. (Upon the breeding of sheep in England, and the antiquity of the wool trade—woollen cloths having been made at Winchester for the Roman emperors—see Report of the Committee, by D. Anderson.) The dyers all crowd to the Avon, as the waters of that river are peculiarly favourable for their art. There is still great need of accurate chemical investigation into the influence of various kinds of water upon the processes of dyeing and brewing. It is quite as unreasonable to ascribe the superiority of English beer and English dyes to the quality of the water in England, as it is premature to deny the influence of the vapours of the atmosphere in chemical processes rendered complicated by the presence of fermentation, which is of a nature too subtle to admit of the application of any test.

'Chippenham was the residence of Alfred the Great, who in 884 expelled the Danes from London and Rochester. At that period the west of England appears to have been most under cultivation, while at the present day the east seems to be the most highly cultivated. Was not the former condition of things a more natural arrangement than the present, seeing that the entrance to the Bristol Channel is more convenient than that of the Thames—the west of England is richer in exports, the communication with Ireland is more direct, and Milford affords a safer harbour for the fleet than Portsmouth ? Assuredly it was ; and foreign interests have alone elicited an opposite tendency. The Flemish trade peopled the Channel with a fleet of merchantmen, the Hanseatic counting-houses of London brought the trade of the country into notice, English wool began to be thence exported, the ceaseless disputes with France rendered a fleet and a harbour of safety in the Channel necessary ; the bar-barism in which Ireland had lain so long has prevented the existence of a flourishing trade between the two countries. . . .

' Matlock.—Lich. calcar., Lich. saxatil., Lich. tartar., and Verruc. pertusa grow wild nearly over the whole of Northern Europe, yet after reading the patriotic treatises of Damboucney (" Procédés sur les Teintures solides que nos végétaux indigènes communiquent aux laines, etc.," Paris, 1786), and Hoffmann (" De vario Lichenum usu." Erlangæ, 1786), we continue to import from the Canaries, Cape de Verd Islands, the Grecian Archipelago, or the South of France, dyes which might readily be procured from our native lichens. We import litmus from the Dutch and the English, and forget that it has been by them manufactured from Lich. saxatil., which grows with us on every tree, post, or stone. (See Ferber's " Neue Beiträge zur Mineralgeschichte," i. 455, on the manufacture of litmus at Leith, in Scotland, where 200 men are employed in the collection of Lich. saxatil.) Our ignorance is so extreme, that we still allow our-selves to be deceived by Jacobson's statement that litmus is manu-factured from Croton tinctor., or the Tournesol of Grand Gatargues, and not from Lich. tartar. (See Demachy, " Laborant im Grossen," ii. 273, " Technologisches Wörterbuch," ii. 544.) The study of crypto-gamia is not therefore of so little value as is generally supposed. Sound political economy would even make lichens contribute to the wealth of a nation.

'Poole's Hole.—560 yards in length. It lies to the south-west of Buxton, towards the limestone mountain of Axe-Edge, on the banks of the little river Wye. As the cave is narrow, more beautiful stalactites are formed in it than in the Peak Cavern. The two largest have received the names of the Flitch of Bacon and the Queen of Scots' Pillar, the latter called after the unfortunate Mary Stuart, who is supposed to have visited the Cave while a prisoner at Chatsworth. A small stream issues from the cavern. On the road from Buxton to the Cave I found some quantity of Saxifraga granu-lata and S. tridactylites. At the entrance of the Cave I noticed Viola montana, Alchemilla vulgar., and Polypod. vulg.

' In the neighbourhood of Poole's Cavern are several lime-kilns in the open air ; for in England lime is burnt in the open air, and tiles in conical furnaces, exactly the reverse of the custom in Germany.'

3. Humboldt's Passport on leaving Paris in 1798.   Page 238.

'We *Alphonse de Sandos Rollin,*

'Chamberlain to His Majesty the King of Prussia, and Prussian Minister Plenipotentiary to the French Republic,

'Beg and require all whom it may concern to grant a free and safe pass to

'*M. Frédéric Alexandre de Humboldt, Counsellor in the Court of Mines of H. M. the King of Prussia, born at Berlin, aged twenty-eight years, height five feet four inches [five feet eight inches English], light brown hair, gray eyes, large nose, rather large mouth,*[1] *well-formed chin, open forehead, marked with the small-pox. Travelling for the acquisition of knowledge,*

proceeding to *Marseilles and Algiers,*

without permitting him to be hindered or annoyed in any manner, but on the contrary granting him every assistance that he may require.   In faith of which we have given him this passport, available for *eighteen months,* signed with our signature, and sealed with the customary seal of the Royal Arms.

'Paris, 24th Vendémiaire, year VII. (new style), 15th October, 1798.

'De Sandos Rollin.

'Signature of the Bearer :

'Frédéric Alexandre de Humboldt.'

The passport had received no fewer than twenty-four visés before Humboldt left Marseilles—a proof of the perpetual annoyances to which travellers in those days were subjected.   The most important visé—interesting on account of the signature—is the first on the list :—

'The Minister of Foreign Affairs certifies the authenticity of the signature of M. de Sandos Rollin here attached.   Paris, 26th Vendémiaire, 7th year of the French Republic.

'Ch. Maur. Talleyrand.'

4. Warrant of Arrest against Alexander von Humboldt.

Among Humboldt's papers was found the following letter from Baron von Eschwege, the distinguished explorer of Brazil, who subsequently entered the service of Portugal :—

---

[1] On the cover in which the passport was kept, the following remark is written in Humboldt's own hand :—'*Grosses Maul, dicke Nase,* aber menton bien fait.'

Lisbon: March 27, 1848.

' Sir,—While looking over my Brazilian papers the other day, I came across the copy of a document, the contents of which are probably quite unknown to you, and I take the liberty of enclosing the paper to your Excellency, since as it might have been productive of the most disastrous consequences to the journey prosecuted by your Excellency into the interior of America, it may prove an interesting contribution to the account of your travels.

' This document was given to me in Brazil by my revered friend and benefactor, Count de Barca, who died as Minister there, with whom, as Antonio de Aranjo e Azevedo, your Excellency may probably have made acquaintance during his many years' residence in Europe, when Ambassador at the Courts of the Hague, Paris, St. Petersburg, and, if I mistake not, also of Berlin. He told me that as soon as he was aware that an order of this nature had been issued by the Ministry, he wrote immediately to the Prince Regent, urging him at once to recall the order for the arrest of your Excellency, whereby he would incur the odium of the whole of Europe, and suggesting that an order should be issued instructing that your Excellency should receive every assistance — an application which proved successful.

' It thus appears that it was entirely owing to the influence of my revered friend, who took a lively interest in science, and was in possession of all the works published by your Excellency, that you escaped being arrested in the upper regions of the Orinoco, or on the frontier of Brazil, which, if I mistake not, you crossed; and further escaped imprisonment in Ceazò, where you might have been detained at least a year before an order for your release could have been received from Portugal.

' Hoping that your Excellency is in good health, I have the honour to remain, &c.,

' BARON VON ESCHWEGE.'

The document is a warrant for arrest, and is as follows:—

' Notification of the 2nd of June, 1800, from Don Rodrigo de Souza Coutinho to Bernardo Manuel de Vasconcelos, Governor of the Province of Cêará.

' The Prince Regent has commanded that your Excellency be informed that in the " Gazeta de Colonia " of the 1st of April of this year, it is stated that a certain Baron von Humboldt, a native of Berlin, has been travelling in the interior of America, making geographical observations for the correction of certain errors in existing maps, collecting plants, of which he has secured 1,500 new varieties, with the intention of continuing his route through the upper portion of the Province of Maranhão, for the exploration of desert regions, hitherto excluded from scientific research. As in the critical position of the affairs of Government the presence of a foreigner, who, under a pretext of this kind, might possibly conceal plans for the spread of new ideas and dangerous principles among the faithful subjects of this realm at a time when the temper of the nation is in a condition so dangerous and so difficult to deal with, it is expressly commanded by our illustrious Prince—independent of the fact that, by the existing laws of his Royal Highness, no foreigner is allowed to enter his dominions without the express permission of His Royal Highness—that your Excellency shall investigate whether the said Baron von Humboldt or any other foreigner has ever travelled through the interior provinces of this kingdom, or is now travelling there, as it would be in the highest degree injurious to the political interests of the Crown of Portugal were such the case. It is expected, therefore, by His Royal Highness, that your Excellency will zealously and industriously bestow in an affair of so much importance all the ability and penetration which from your experience in the service of the Crown may be expected from you, and, should travellers be in the country, that your Excellency will obviate the evil likely to ensue, and put a stop to any such illegal investigations by arresting and imprisoning, not merely foreigners, but even such Portuguese as render themselves open to suspicion by travelling without a royal permit or the necessary license from the Governor of the Province. Finally, it is desired by His Royal Highness that your Excellency will act in this respect with the most scrupulous circumspection, and communicate immediately to His Royal Highness, through the medium of the Chancellor of State, everything that transpires respecting this affair, in order that the Prince may take whatever further measures may be deemed necessary.'

Upon the original document is inscribed in Humboldt's handwriting :—

'*I desire that this order may somewhere be published after my death.*

'ALEXANDER HUMBOLDT.

'March 1854.'

Varnhagen remarks in his 'Tagebücher,' under date of August 11, 1855 :—' Humboldt has lately been presented with the Grand Order of Brazil, for his services as arbitrator between Brazil and Venezuela, respecting the right of possession of a considerable tract of country. "In former days they wanted to imprison me in Rio Janeiro as a dangerous spy, and send me back to Europe : the warrant that was made out for this purpose is still shown there as a curiosity; now I am chosen as their arbitrator! My decision has naturally been in favour of Brazil, for I was ambitious of possessing the Grand Order, and the Republic of Venezuela has none to bestow!" I interrupted this strain of bitter irony with the exclamation, "How times have changed!" "Yes," he at once replied, "first the warrant for arrest, and then the Grand Order!"'

5. AIMÉ BONPLAND.

Aimé Bonpland stands intimately associated with Alexander von Humboldt, not only as his faithful companion during his travels in America, but also as his able coadjutor in the publication of his works. He was Humboldt's junior by four years, having been born on August 22, 1773, at La Rochelle. Adopting the profession of his father, he devoted himself to the study of medicine, and in 1793 served for a short time in the navy as surgeon on board a frigate of the Republic, which was cruising in the Atlantic during the war with England. He subsequently visited Paris, and attended the lectures of Corvisart, Dessaut, and Bichat. The interest he took in medical science was, however, far surpassed by the keen zest with which he followed the achievements in zoology and geography of Buffon and Daubenton, and the labours of Jussieu in botany; with industrious zeal, he sought to follow in the track of these distinguished investigators. His extensive acquirements and scientific endowments had already met with recognition when he received the appointment of botanist to the expedition under the command of Baudin, fitted out by order of the Directory. He was probably at that time already acquainted with Humboldt.[1]

Bonpland was without means, while Humboldt was in possession of a considerable fortune, which, in a spirit of true disinterestedness, he was willing to expend in the service of science. It was not long, therefore, ere the bond was formed which gave completeness to each of these men. While Humboldt had devoted himself with peculiar energy to the study of inanimate nature, to mathematics, and the observation of terrestrial phenomena, directing his attention mainly to the unity of Nature, the comparison of her various phenomena, and the mutual relationships existing therein, Bonpland was more

---

[1] 'How did you first become acquainted with Bonpland?' enquired Dove on one occasion from Humboldt. In the simplest manner in the world,' he replied. 'You know that when giving up the key of one's apartments on going out, one generally exchanges a few friendly words with the porter's wife. While so doing I often encountered a young man with a botanist's satchel over his arm; this was Bonpland; and in this manner we made acquaintance.' (H. W. Dove, 'Gedächtnissrede auf Alexander von Humboldt,' Berlin, 1869, p. 9.)

at home in the realm of organic life, among Flora and Fauna, and was best adapted to the observation of detail. From the mutual interchange of knowledge and experience sprang that wide range of observation, and completeness of investigation, by which their joint labours were distinguished. In the Spanish Colonies of America every German was in those days thought to be a miner, and every Frenchman a doctor ; and certainly this supposition was substantiated most fully in the case of Humboldt and Bonpland.

Of the mutual esteem and devotion which these travellers so heartily reciprocated, several instances have already been given.[1] It will be suitable to adduce here the following entry from Humboldt's American Journal (vol. v.):—' I have given to M. Elhuyar a sealed letter, to be opened in the event of my death : it contains a bequest to Bonpland of 50,000 francs.'

Soon after his return from America, Bonpland, on the recommendation of Corvisart, surgeon to the Empress Josephine, was appointed by the empress, who was passionately devoted to flowers, superintendent of the gardens at Malmaison and Navarre. Through Humboldt's zealous interposition, he received, in addition to his salary, a yearly pension of 3,000 francs, in acknowledgment of the herbarium presented by him to the Jardin des Plantes.[2]

---

[1] See pp. 278, 279, 287.

[2] The author received on one occasion (1858) from Humboldt the following authentic summary of the plants collected by him and Bonpland :—During the five years spent in their travels, Humboldt and Bonpland collected from 5,800 to 6,000 specimens of plants. Among the 5,500 specimens of phanerogamous plants, 3,000 were found to be new species. The localities of all these plants, as described in the ' Nova Genera et Species Plantarum in Peregrinatione ad Plagam Aequinoctialem Collectarum,' are denoted with the barometric determination of the height above the sea, a detail which has never before been introduced into any botanical work. The number of plants actually described during the journey amounted to 4,528, and the descriptions filled six volumes—three folios and three quartos. These volumes were employed by the celebrated botanist Kunth, Director of the Botanic Gardens at Berlin, in editing at Paris the ' Nova Genera et Species.' As only about a fifth of these descriptions are from the pen of Alexander von Humboldt, the volumes on the death of Professor Kunth were sent by Humboldt, in acknowledgment of the indefatigable industry of his fellow-traveller, to the Museum of Natural History at Paris, where they were preserved as the property of Bonpland. The plants themselves, since the number of duplicates was such as to allow of it, were divided by the travellers on their arrival in Paris into three distinct herbariums ; the most complete collection was retained by Bonpland, who subsequently took it out with him to Buenos Ayres. A second herbarium was presented to the Museum of the Jardin des Plantes, in acknowledgment of which Bonpland received, as stated in the text, a yearly pension of 3,000 francs ; and the third collection was placed at the disposal of Alexander von Humboldt, by whom—in accordance with his invariable custom of reserving nothing for himself, but of presenting everything he collected, whether

Bonpland assisted not a little in laying out the gardens at Malmaison, which at that time excited universal admiration; in the midst of his official employment he enjoyed sufficient independence and leisure to assist in the publication of the results of the American journey, though he did not devote himself to the work with energy and perseverance sufficient to preclude the necessity of calling in the help of Willdenow, and securing the prolonged labours of Kunth.

Associated with Humboldt in his work, in the enjoyment of friendly intercourse with Gay-Lussac, Arago, Thénard, and the most noted scientific men of the day, honoured with the kind consideration of the empress, who entered with quick intelligence into all his pursuits, and unrestrained in the management of one of the most beautiful flower-gardens in the world, Bonpland was now in the enjoyment of the happiest period of his life.

But these days were numbered. Napoleon's divorce of Josephine soon brought over them a cloud; to the watchful guardian of her flowers this noble woman was accustomed to confide the anguish of her grieved and troubled heart. ' Ce n'est pas la perte de la couronne qui m'afflige,' she expressed herself to him on one occasion, ' mais c'est la perte de l'homme que j'ai plus aimé que ma vie, et que je ne cesserai d'aimer jusqu'au tombeau.' The misfortunes of the emperor broke her heart, and she died on May 30, 1814. Bonpland stood beside her on her death-bed.[1]

After this event, Bonpland appeared unable to settle to any other occupation, and seemed as if irresistibly impelled to quit Europe. Even the continuation of his magnificent work, ' Description des Plantes rares cultivées à Malmaison et à Navarre ' (Paris, 1813), as well as the further arrangement of the plants collected in company

in America or Asia, to public or private museums—it was given to Willdenow, his friend and early instructor. By the subsequent purchase of the herbariums belonging to Willdenow and Kunth, the whole of the collections made by Humboldt and Bonpland in America became incorporated in the extensive herbarium of the Royal Botanic Gardens of Berlin. Many of the duplicate specimens from the Humboldt-Bonpland collection are still preserved among the herbariums of Spain and England. The plants collected during the united expedition to America must not be confounded with those collected subsequently by Bonpland during his residence at Buenos Ayres; it is to this more recent collection that he chiefly refers in his letters of later date. In the year 1858 the herbarium belonging to the Royal Botanic Gardens at Schöneberg was transferred to the university buildings at Berlin, where it has been rearranged, and is preserved in admirable condition.

[1] It will not be out of place here to correct the erroneous statement (repeated in the Supplement to No. 197 of the Augsburg ' Allgemeine Zeitung ' for July 16, 1858), that it was Bonpland who advised the emperor in the days of his calamity at Fontainebleau to choose Mexico for an asylum, as a central point whence to

with his distinguished friend, had lost all charm for him.[1] The
desire to supplement the herbariums hitherto contributed from
tropical countries with the collections to be obtained in the temperate zone, a region which had been but imperfectly explored by
the botanical researches of Feuillée, induced him towards the close of
the year 1816 to undertake a second expedition to South America.
He sailed to Buenos Ayres, and took out with him a supply of
fruit trees and various common vegetables.

How completely in the mean time had everything changed in
South America! The invasion of Spain by Napoleon had severed
the ties by which the Spanish Colonies had been united to the
mother country. From Mexico to Buenos Ayres, everywhere floated
the standard of revolt. The struggle, though bloody, was short, and
peace now reigned, except where plots and party strife aroused
dissension. On the banks of the La Plata were established the
first free Republics, and this river was the first to open to unre-
stricted commerce.

At Buenos Ayres Bonpland met with a most flattering reception.
He was at once appointed Professor of Natural History—an office
which he retained only for a limited period; since, owing to the
jealousy, envy, and evil machinations to which, as a foreigner, he
was exposed, the Government was soon prejudiced against him.
In the year 1820 he undertook a journey for purposes of exploration, with the intention of visiting the Pampas districts, the province
of Santa Fé, the desert of Gran Chaco, and of penetrating through
the district of Bolivia to the foot of the Andes. In the course of
his expedition he reached, in sailing up the Paraguay, a former settlement of the Jesuits, lying on the left bank of the river, a few
miles from Itapua.

At this spot he unfortunately entered upon a territory the pos-

observe the events transpiring in either world. A magnificent idea! will be
exclaimed by those who remember the part subsequently played in international
politics by this isthmus, which has since then been rendered so familiar to us.
But this is mere tradition. 'Bonpland,' wrote Humboldt on one occasion to the
writer of this sketch, 'had no conversation with Napoleon either before or after
the battle of Waterloo; he never went to Fontainebleau, where various persons
were proposing impracticable schemes to the emperor. At that time I had
daily intercourse with Bonpland. He may possibly have spoken of Mexico to
some of his acquaintance as a suitable place of refuge, but not to the emperor, of
whom he saw nothing, and to whom he remained a complete stranger.'

[1] Of the two works exclusively edited by him, 'Plantes equinoxiales' and
Monographie des Mélastomes' (two volumes in folio, with 120 plates), the latter
had already given rise to considerable discussion, for, according to Robert Brown,
among the various Melastomæ therein described there was not one to be recognised as genuine. (See Martius, 'Denkrede auf Alexander von Humboldt,' p. 25,
note.)

session of which was a subject of dispute between the Government of Paraguay and the Argentine Republic. He, indeed, at once wrote to Dr. Francia, the Dictator of Paraguay, acquainting him with his arrival, and craving permission, in the interests of science, to collect some information concerning the Maté plant;[1] but Francia, full of suspicion, caused him to be attacked by a troop of horse, who surprised him during the night of December 3, 1821, and, after slaying his defenceless servants, and wounding him on the head with a sabre, loaded him with chains, and carried him off into the interior of Paraguay. Here, in the neighbourhood of Santa Maria, was the friend and travelling companion of Humboldt condemned to inaccessible imprisonment. He was employed by Francia as surgeon to the garrison, and as overseer in matters of trade and agriculture, but his movements were restricted to a limited range.

No sooner was Humboldt made acquainted with the fate of his friend than he exerted every effort to obtain his release. The interest of Cuvier, and the influence of the National Institute, were enlisted on his behalf. In the hope of effecting his deliverance, the well-known traveller Grandsire was despatched to Rio Janeiro, and furnished with letters of recommendation to the French Consul-General there, from the Vicomte de Chateaubriand, Minister for Foreign Affairs ; at the same time Humboldt addressed an intercessory letter to Francia, and forwarded with it several of the works which had been published by him and Bonpland conjointly. But the Grand Seignor of Paraguay, then sixty-two years of age—a tyrant full of distrust and irritability—repulsed every advance on the part of Grandsire, whose mission had no further result than to ascertain that Bonpland was well, that he practised the art of medicine, was engaged in the distillation of brandy from honey, and daily made additions to his extensive herbarium. Through Canning, then Prime Minister, Humboldt attempted further to obtain the influence of the English Government towards effecting the release of his friend, but all his efforts were in vain.

By what means the release of Bonpland was ultimately effected—whether by the pressing solicitations of Mandeville, the French Consul-General at Buenos Ayres, or by the unconcealed threats of Bolivar, the President of Colombia—it is impossible now to say : so much only is known, that on February 2, 1830, he received an intimation of his freedom, and was at the same time informed that ' la Excelencia el Supremo' granted him permission to go wherever he liked. Thus ended Bonpland's unjustifiable captivity of nine years' duration. On receiving his freedom, he still remained in the neighbourhood, and settled down in the small town of Santa Borja, on

[1] [The Paraguay tea-plant.]

the frontier of Brazil, a few miles from the left bank of the Uruguay, where he continued to reside, with frequent visits to the Estancia of Santa Anna, for thirteen years.

The release of Bonpland excited universal joy throughout Europe, and marks of distinction were showered upon him from all quarters in the wish to compensate him for the severe experiences he had encountered. From Berlin, where he had spent several months during the summer of 1806, he received the order of the Red Eagle. In announcing to him the presentation of this honour, Humboldt remarked: ' I am aware of your philosophic principles, but we thought that in your relationships with Brazil (if you have any) it might perhaps be of use to you.' A doctor's diploma, and various honorary degrees from different literary institutions, were conferred upon him in addition. The Imperial Leopold-Charles Academy for Natural Science could find no more euphonious name for their official organ for comprehensive botany than that of ' Bonplandia.'

The interest manifested throughout Europe in everything con-nected with Bonpland was not more remarkable than the number and variety of the reports in circulation concerning him. At one time he was represented as wishing to return to France with his extensive collections, at another time he was said to be devising vast schemes to be carried out in the country of his adoption. After his release a medium of communication between him and Humboldt was opened through the good offices of Herr von Gülich, Prussian Chargé d'Affaires and Consul-General in Chili, and from this correspondence a lively picture may be gathered of Bonpland's mode of life, of the affectionate nature of his disposition, of his in-defatigable industry, of the freshness of his heart and mind, as well as of the hopes and plans still cherished by him in imagination, even when at an advanced period of life.

His correspondence with Humboldt extended over nearly thirty years. He yearly forwarded his certificate that Humboldt might draw for him the pension granted by the French Government. With sympathising interest he followed the illustrious career of his friend. ' Mon illustre ami,' he writes to him from Buenos Ayres on July 1, 1832, ' je te vois tous les jours plus grand, et chaque instant je t'admire davantage.' On July 12 in the same year he asks for a line in Humboldt's own hand, although he could not be seriously apprehensive of his being attacked with cholera, ' parce que tu as une grande force d'âme répressive pour tout ce qui t'est con-traire.' The intelligence of the death of William von Humboldt, communicated to him by Alexander on September 14, 1835, called forth from him the most lively expressions of sympathy, with many allusions to the past, and he made a resolution henceforth to keep

Alexander's birthday as a fête day. Notwithstanding the debilitating effect of two severe attacks of illness, his health became so entirely re-established by his simple and regular mode of life, that, in a letter dated June 14, 1836, he alluded seriously to a project of returning to Paris. It seemed to him that an important duty connected with his visit there would be the addition of Notes to Humboldt's 'Travels in America.' Only five volumes of this work had reached him, but these 'had filled his mind with a crowd of recollections.' While enjoying the mild breezes of a southern clime loaded with the scent of orange blossoms, he compassionates his friend doomed to a dreary existence at Berlin, and recalls the time when they had together rejoiced in the magnificent climates of Ibague at the foot of the Cordilleras, of the Islands of the Hyères, and of the coast of Caleja, between Barcelona and Valencia. After a stay at Paris, he proposed making an expedition to Algiers, which he thought would be well worth visiting.

On returning from Europe in the year 1854, Herr von Gülich conveyed to Bonpland several presents from Humboldt, consisting of books and many likenesses of himself. These were most acceptable gifts, and in contemplating the portraits it was with no small pleasure that, in the altered features of the grey-haired man, Bonpland could still trace the once familiar countenance of his former intimate companion. The idea of returning to Europe was still the cherished wish of his heart. Once more to hold Humboldt in his embrace and revel with him in recollections of the past, was a desire that sometimes seized him with a passionate longing. 'Quelques heures d'entrevue,' he exclaimed (October 2, 1854), 'nous donneraient—il me semble—dix années d'existence!' He derived great pleasure from a perusal of 'Cosmos' and the 'Tableaux de la Nature;' he fancied as he read that he heard the voice of his old fellow-traveller sounding in his ears, and bitterly bemoaned his separation from him:—'Man needs a true friend to whom the feelings of his heart can be unburdened.'

Bonpland continued to follow with sympathetic interest the changes both social and political which were taking place in his native land. On August 8, 1856, he announced to Herr von Gülich that his herbarium was almost ready to be despatched to Paris, and that he should like to be the bearer of it himself, that he might present it to the emperor in person for the enrichment of the Museum. His interest had been keenly excited by the achievements of Louis Napoleon, the events of the Russian war, the revolutions that had taken place in European politics since 1816, and the introduction of railways—a mode of locomotion to which he was still a stranger. But amid all, renewed intercourse with Humboldt stood

ever in the foreground as the chief desire of his heart. Yet doubts
often rose before him as to whether, after so long an absence, he
could renew the relationships from which he had been so long
estranged. 'What compensation could I find in the noise and
bustle of Paris?' he inquired. 'Should I labour in some garret
there for any bookseller who might undertake to print my books?
Should I live there debarred from all other consolation save that of
watching from time to time a rose blooming at my window? I
should have to lose that which is to me of all things most precious,
the society of my beloved plants, which have been my companions
through life.' Even the last letter he addressed to Humboldt,
dated from Corrientes, July 7, 1857, is characterised by the same
remarkable display of mingled feelings—the yearning to behold once
more his distant native land, and the hermit's pleasure with which
he delighted in the solitary home of his choice. 'I intend,' he
writes, 'to be myself the bearer of my collections and manuscripts
to Paris, where I hope to deposit them in the Museum. My visit
to France will be but of short duration; for I shall wish to return
to my home at St. Anna, where I lead a life as happy as it is peaceful. There I trust I shall die, and my grave will be shadowed by the
numerous trees that I have planted. How great will be my delight,
my dear Humboldt, to see you once again, and to renew the many
reminiscences we have in common! On the 28th of next August I
shall have completed my eighty-fourth year, and I am three (four)
years your junior. A man has recently died in this province at the
age of 107. What a prospect is thus presented to two travellers
who have passed their eightieth year!' This passage, as Humboldt
remarked, seems to indicate a strong clinging to life in an old man
of eighty-four. It was his desire to lie buried beneath the shade of
orange trees and palms, among the trees that he had tended and
cherished during life; but in the contemplation of it he again
and again reverted in imagination to the journeys he had taken
with the friend of his youth; he could yet vividly recall the delight
experienced by his friend on first beholding in Spain the glory of
southern vegetation, when, intoxicated by the new sensation, he
desired to pass his life amid scenes of such beauty. A sense of duty
detained Humboldt amid the cheerless surroundings of his native
home, while Bonpland, less conscientious in the indulgence of his
wishes, was content to spend his days in quiet inaction, in the
enjoyment of a life of contemplation. He died on May 11, 1858, at
the age of eighty-five years. The French, usually so zealous for the
honour of their nation, have accorded to the merits of their country-
man no greater meed of praise than is conveyed in the expression

that '*Aimé Bonpland was a fellow-labourer of Alexander von Humboldt.*'

Last Days of Aimé Bonpland. By Dr. Robert Avé-Lallemant.

Humboldt was accustomed to speak in high terms of the industry of Bonpland. Yet this activity and perseverance was always more displayed in the endurance of hardships, in the collection of Flora and Fauna, than in the plodding labour of the desk. Bonpland's sphere of action was pre-eminently amid the wilds of nature, in the forest, on the llanos and the pampas, in the canoe or on the galloping steed. Writing was to him an uncongenial employment, which he carried on with vacillating energy, and never with great accuracy; his works contain much that is incorrect, and several even gross errors. This characteristic of Bonpland was frequently the cause of great perplexity to Humboldt, until at length the irksome task was relinquished. Whether in resigning his occupation he acted with the consent of Humboldt does not appear, but it is certain that they parted in perfect amity and with the most sincere friendship on both sides. Of this evidence is afforded by the following letter,[1] addressed by Humboldt to his faithful fellow-traveller after his settlement at his new home on the banks of the La Plata:—

'Paris : January 28, 1818.

'I avail myself of M. Thounin's departure to give you some news of my welfare, and to assure you again of my inalienable attachment. I have already written to you before this week by M. Charles de Vismes. I am not personally acquainted with M. Thounin, but I have heard a great deal in his favour, and I have been earnestly requested to give him a letter to you. Alas! my dear friend, everyone around me, M. Delille, Lafon, Delpech, have received letters from you telling them of your present position and domestic happiness, while I have not had a single line from you since your departure, except the one short note brought me by M. Alvarez.' . . . He then proceeds to inform him that he (Bonpland) had been elected a corresponding member of the Academy of Sciences, mainly through the influence of Arago, Gay, Thénard, Chaptal, Laplace, and Berthollet, and mentions his intention to enclose him various publications of his own. Finally, he asks him to return the plants which he had—'même contre sa volonté'—carried off with him in some of his packages. He concludes with 'Adieu,

---

[1] De la Roquette, vol. i. p. 206.

my dear old friend; present my affectionate remembrances to
Madame B. With renewed assurance of my devoted attachment,

'AL. DE HUMBOLDT.'

On the liberation of Bonpland, after a captivity of nine years,
Humboldt writes in a joyous strain to Guizot,[1] on November 2,
1832 :—

. . . 'I am glad to have news at length of my unfortunate friend
M. Bonpland. I wish you could obtain for him that decoration
peculiar to France given so often with profuse prodigality !' . . .
And again, on May 25, 1833, he writes :[2] . . . 'It is also a pleasure
to me to express my gratitude for your having so kindly fulfilled
the wish I expressed to you last autumn concerning my unfortunate friend M. Bonpland, in procuring him the decoration of
the Legion of Honour. This has caused me the most lively satisfaction. I had some reason to fear that my fellow-traveller would
experience a fate not uncommon in human affairs : as long as he
was in the clutches of the Dictator Doctor, the Republican tyrant,
he was the object of sympathetic interest—inquiries as to his fate
were reaching me continually, from the banks of the Thames to the
s hores of the Obi. The drama having come to an end, he is but a
man of science who is travelling for the collection of plants. There
was some cause to fear that he would be forgotten ; but such forget-
fulness is impossible in a soul so generous as yours ! Our mutual
friends MM. Benjamin and François Delessert have written to me
more than once to tell me of the noble assistance you have rendered
i n support of the efforts I have been making to secure to M. Bonpland the arrears that are due to him since 1820. I am wrong in
saying the arrears are due to my friend, for I am aware that by law
no payments can be claimed beyond a period of five years. The
Minister of Finance, therefore, in paying up the arrears for the last
five years only, did, no doubt, all that he was empowered to do,
although, in consideration of the peculiar circumstances attending
the case of M. Bonpland, hopes of some more liberal arrangement
had been raised by the Committee of Finance and the Council of
State. I feel constrained to plead the cause of my fellow-traveller
with a minister so generously disposed as yourself to relieve the
misfortunes of men of letters. I have ventured to-day to write to
the king himself, not to urge a right, but to crave a special favour.
May I implore your kind interest in this affair, which is one of great
importance to the ruined fortunes of M. Bonpland ? My letter to
his Majesty will be of no avail unless you supplement the applica-

[1] De la Roquette, vol. ii. p. 95.
[2] Ibid. vol. ii. p. 106.

tion. The pension of 3,000 francs accorded to M. Bonpland was given in consideration of the herbarium, collected during my journey, which I presented to the Jardin des Plantes. I relinquished the possession of it in order that I might be useful to my friend. I am now without a single botanical specimen, or even the smallest memento of Chimborazo! The pension, therefore, is one of a very peculiar character. The treasures which I have surrendered still exist, and imprisonment alone has interrupted the regularity of the payments. Such are the considerations which I have imagined to myself must constitute my rights, yet I urge them only to plead a favour. I venture to think that the Chambers— should the Government stand in need of such a sanction—would not oppose such an act of munificence in favour of a Frenchman whose misfortunes have attained a world-wide celebrity. Pray excuse the length of this letter, which is, I fear, but indifferently expressed. You will not, I am sure, condemn the motives which have inspired it.' . . . In a postscript, Humboldt expresses his regret that Bonpland's collections had not yet arrived.

Visit to Santa-Anna.

For many long years Bonpland had resided at Uruguay, until there remained scarcely anyone in Europe who remembered the eccentric man of science. But the remembrance of him recurred very forcibly to my mind when, in the spring of 1858, I arrived at Rio Pardo, and thence proceeded on horseback to the German colony of Santa-Cruz. On April 8 I reached Santa-Borja, where Bonpland had resided for thirteen years, before proceeding farther up the country in 1853. During his residence there he had enjoyed friendly intercourse with the Governor and inhabitants of the opposite Province of Corrientes, the President of which was at that time Dr. Pujol, an able and intelligent man, and had been the means of instituting a National Museum. As an expression of the thanks of the community, he was presented with a larger tract of land on the right bank of the Uruguay, a few miles south of the spot where the town of Restauracion, in the province of Corrientes, faces Uruguayana, a new and rising town in the empire of Brazil, although severely injured during the invasion of the Dictator Lopez—the son. Upon this plot of pasture land he had erected his *rancho* or cottage, and laid out his new and last *Sanssouci*, to which he gave the name of Santa-Anna. From this residence he kept up a constant communication with Santa-Borja. In the year 1857 he ascended the river Paraguay, in the French war steamer 'Bison,' to Assuncion, the capital of Paraguay, in order that he might revisit the country where for nine years he had endured the 'hospitality' of

Francia. As a Frenchman he was not well received by Francia's successor, General Lopez—the father ; for owing to the treatment Bonpland had received in the country, Paraguay had been placed publicly under a proscription by France, as well as by every other civilised nation.

From the French priest Gay, I learnt that Bonpland had been suffering in health for a long time, and as nothing had been heard of him for several weeks, he was supposed by many to be already dead. I felt it therefore to be an imperative duty to make a personal visit to the old hermit of Santa-Anna.

I started on horseback on April 13 from Santa-Borja, accompanied part of the way by the priest, and on the following day I reached the little town of Itaqui, where I hired a chalana—a river boat—and was conveyed a short way down the stream of the Uruguay.

When I reached the town of Uruguay, I learnt from Kasten, a merchant in the place, and a friend of Bonpland, that he was still alive, but in very bad health. Herr Kasten politely accompanied me across the river to the opposite town of Restauracion, where he assisted me in making arrangements for my ride to Santa-Anna.

On the following morning a peão—groom—a dusky native of the Pampas, remarkable for his taciturnity, presented himself at my door in charge of two tall steeds, one of which I mounted. Without uttering a word, he rode slowly before me until we were out of the town, then pressing his great iron spurs into the sides of his matungo, away we galloped for three German miles in a westerly direction, at first through forests of palm, and afterwards along a kind of high road.

A complete Pampas-plain now spread out before us. One sea of grass seemed to succeed another. Here and there was a miserable mud cabin, always far away from the road. Scarcely did we meet a horseman, scarcely the cart of a strolling pedlar. The solitary riders exchanged a silent greeting as they galloped by. Cattle were seen grazing both far and near, besides numbers of wild horses, which on our approach fled hastily away. A herd of deer sprang out of a swamp, where they had been feeding under the shade of the mimosa, and rushed by with lightning speed, while the ungainly ostrich, trotting like a horse, sped more leisurely across the grassy plain.

The second half of my morning's ride was accomplished through a country in which the track was even less defined. Without a word of explanation, my guide suddenly deviated from the path and took a southerly direction, right across country, over the ocean-like plain of grey-green herbage. After galloping at full speed for a further distance of three German miles, he drew up as we reached a slight elevation, and turning round to me exclaimed, while pointing with his dusky hand in a southerly direction, the first words he had uttered during a ride of six German miles : ' That's where Don Amado lives ! '

In front of a green orchard I noticed two grey cottages, standing at right angles the one to the other, which appeared all the more dirty and wretched the nearer I approached them.

Could it be that in these huts, these miserable sheds, in the midst of this dreary wilderness of Pampas, Bonpland had for so many years led the life of a cynical patriarch ? I could learn nothing concerning the Madame Bonpland to whom Humboldt sent a message of greeting in 1818. In Uruguayana I was told that many years ago the foreign philosopher had united himself to a native woman, one of the class called 'China,' by whom he had several children; wearied probably of the old man, and the solitude in which he lived, she, however, decamped one fine morning, leaving her children behind her at Santa-Anna.

I dismounted amid the violent barking of four great dogs, and upon loudly clapping my hands there came forward a pleasing-looking young girl, whose countenance gave evidence of mixture of race, and who, with some show of timidity, asked me in Spanish what I wanted. I gave her a letter to Bonpland, which she carried into one of the buildings that was meant to pass for a house, but returned shortly to conduct me into the other hut, which served as drawing-room and strangers' apartment.[1] A board supported upon a couple of casks, a bench, two chairs, and two empty bedsteads, constituted the whole of the furniture of this long shed, which being without windows, was very imperfectly lighted by the open door, and by sundry rents in the walls. At the further end lay cow-hides, some worn-out saddle-harness, a heap of onions, and a number of other objects, the form of which I was not able to distinguish in the gloom. The girl, who was a strange mixture of French levity and native simplicity, told me that for some months past Don Amado had been exceedingly ill and weak, but that he managed to get out a little every day, and would come and speak with me.

I had not waited long when the eccentric old man, whose very existence we had begun to doubt, appeared before me. His eighty-five years had not impaired the erectness of his form, but his genial blue eyes beamed from a countenance deeply furrowed, and his voice

[1] A view of the farther side of Bonpland's dwelling, taken from a sketch I made upon the spot, appears on the title-page of the first volume of my ' Reise durch Süd-Brasilien ' (Leipzig, 1859).

betrayed the weakness of infirmity.   His thin figure was clad
in a shirt and trousers of white flannel, while on his stockingless
feet he wore wooden slippers.   He held out his hand to me in
friendly greeting, and as I grasped it I felt the feverish heat that
seemed to be consuming him.   The whole scene in the midst of this
desert region, without one of the attractions of civilized life, impressed me with an indescribable feeling of melancholy.

Some roast meat was set before me upon a pewter plate; knife
and fork there was none—I was obliged to help myself with my
bowie-knife and my fingers.   The old man then became very
talkative, but there was no method in his discourse, and he confounded the date and order of events, and mixed up persons and
things in the most extraordinary way.   The Seine, the Paranà,
and the Orinoco, all flowed in close proximity.   Paris and Assuncion
were associated together; the Cordilleras and the Atlantic Ocean
were placed side by side; and the names of Humboldt and Francia
were mentioned in the same breath.   At last his thoughts became
concentrated upon Humboldt, and the remarks he made upon him
were sufficiently surprising.   It was evident that Bonpland viewed
with envy the immense superiority of his friend.   He thought that
Humboldt had published many things as his own discoveries that
had properly belonged to Bonpland; he believed that Humboldt
had rejoiced to see him start for America the second time, because
he had entered into some special engagements with Kunth, and that
he continued to work with him in the publication of his books
without waiting for Bonpland's return to Europe, and that moreover
he had often advised Bonpland against the return he had projected.
He complained also that his second visit to America had not excited
in Europe the attention it deserved, nor had the collections he had
sent over been estimated at their proper value.

It is unnecessary to defend Humboldt from such accusations.   If
Bonpland failed to secure an independent position, if his name will
be preserved only to posterity as an appendage to the more brilliant
name of Humboldt, the reason lies in his lack of industry and his
unconquerable propensity to postpone every kind of labour.   A
return to France was certainly never seriously contemplated by the
cynic of La Plata, for whom Europe had ceased to possess any
attraction.

After our conversation had been carried on for a considerable
time, I noticed that the aged invalid showed symptoms of fatigue,
and upon my urgent solicitations he retired to take a little repose,
while I employed the interval in visiting the garden, which lay at
some distance from the house.   Oranges and peaches flourished in
perfection; Bonpland's rose garden was in full bloom; fig trees and

APPENDIX. 411

castor-oil bushes grew together in luxuriant interlacement, but weeds were also thickly springing up in every direction. The small plantation formed a marked contrast with the boundless expanse of grass; the grass grew close up to the ruined walls, even within the door of the enclosure. No cattle were seen pasturing in the immense plain; only two ostriches did I notice as they trotted across in the distance. Towards the south-east the horizon was bounded by the forests of the Uruguay.

In the evening Bonpland sent for me into his own dwelling, which differed but little from the apartment already described, with the exception of the bed on which the old man lay stretched. 'It is only about a month ago,' he remarked with a smile, 'that I had a proper bed made for myself; I used before that to sleep anywhere, just wherever I happened to lie down.' Again his thoughts wandered feverishly through the various scenes of his long life. While he was speaking two half-caste boys had entered, who were brothers of the young girl. I wished him good-night, and made up a bed for myself in the other hut.

On my visiting him the next morning I found him very feeble; he had slept badly, and his wasted hands were burning with fever. I begged that he would allow me to nurse him, and assist him in the arrangement of any of his affairs, or conduct him to his friends at Uruguayana; but he declined all offers of assistance. Hopeless as his condition appeared, he would not entertain the thought of death; he seemed to think that as he had been accustomed to put off work all through life, so death might be postponed even at the last. With a cheerful air he invited me to visit him again in the course of 'a few years;' cattle would then be grazing in the fields, the garden in beautiful order, and his dwelling completed and fitted up with all necessary furniture. And as if to begin at once with arrangements for the latter, he commissioned me to ask Herr Kasten to send him a dozen knives and forks. He also gave me a letter to Dr. Pujol, the Governor of Corrientes, requesting me to post it at Restauracion.

I begged him to give me his autograph as a remembrance, and he wrote upon the back of an old letter: 'Aimé Bonpland.' 'That is badly written,' he remarked, and wrote his name a second time, but even less successfully. 'Ah,' he exclaimed, 'I have no longer the power to write;' and it seemed to me as if a tear stole down his cheek. Probably this was the last time that he ever wrote his name.

I had committed the imprudence on the previous evening of acceding to the request of my guide, urged on an apparently reasonable plea, to pay him the sum stipulated for the journey. During the night he decamped with both horses. Bonpland freely offered me the use of his riding horse, only regretting that he could not furnish me with a guide. I saddled my steed, and full of sadness took my leave. The old man pressed my hand between his two wasted hands. As I left the apartment he called after me: ' Come and see me again, and remember me to Humboldt. Bon voyage ! '

On May 11, twenty-three days after my visit, Bonpland died. The words used by Humboldt upon the death of Blumenbach—' The death of Herr Blumenbach, who, like many other men of science, has had the misfortune to survive a literary reputation somewhat too readily acquired '—are still more applicable to Bonpland, who had certainly long outlived his celebrity.

END OF THE FIRST VOLUME.

LONDON: PRINTED BY
SPOTTISWOODE AND CO., NEW-STREET SQUARE
AND PARLIAMENT STREET

Printed in the United States
By Bookmasters